D1714847

General Topology and Homotopy Theory

I. M. James

General Topology and Homotopy Theory

Springer-Verlag
New York Berlin Heidelberg Tokyo

I. M. James
Mathematical Institute
Oxford University
24–29 St. Giles
Oxford OX1 3LB
England

AMS Classifications: Primary: 54-01, 55-01. Secondary: 18-01, 22-01.

Library of Congress Cataloging in Publication Data
James, I. M. (Ioan Mackenzie)
General topology and homotopy theory.
1. Topology. I. Title.
QA611.J33 1984 514 84-5435

With 3 Illustrations

Typeset by Composition House Ltd., Salisbury, England.
Printed and bound by R. R. Donnelley & Sons, Harrisonburg, Virginia.
Printed in the United States of America.

9 8 7 6 5 4 3 2 1

ISBN 0-387-90970-2 Springer-Verlag New York Berlin Heidelberg Tokyo
ISBN 3-540-90970-2 Springer-Verlag Berlin Heidelberg New York Tokyo

Preface

Students of topology rightly complain that much of the basic material in the subject cannot easily be found in the literature, at least not in a convenient form. In this book I have tried to take a fresh look at some of this basic material and to organize it in a coherent fashion. The text is as self-contained as I could reasonably make it and should be quite accessible to anyone who has an elementary knowledge of point-set topology and group theory.

This book is based on a course of 16 graduate lectures given at Oxford and elsewhere from time to time. In a course of that length one cannot discuss too many topics without being unduly superficial. However, this was never intended as a treatise on the subject but rather as a short introductory course which will, I hope, prove useful to specialists and non-specialists alike. The introduction contains a description of the contents. No algebraic or differential topology is involved, although I have borne in mind the needs of students of those branches of the subject. Exercises for the reader are scattered throughout the text, while suggestions for further reading are contained in the lists of references at the end of each chapter. In most cases these lists include the main sources I have drawn on, but this is not the type of book where it is practicable to give a reference for everything.

I should like to thank the many topologists who have helped me complete this work, particularly Albrecht Dold and William Massey, who read and commented on a rough draft, and Wilson Sutherland, who read the proofs. I should also like to thank Sarah Harrington and Val Willoughby for typing the manuscript.

Oxford, May 1984

Contents

Introduction

Nowadays most students of mathematics are acquainted with the language of categories and functors. In this book we shall be dealing not only with the basic category of topological spaces and continuous functions but also with various other categories associated with it. Consequently we begin with a preliminary chapter in which some of the common features of these categories are discussed in general terms. This avoids a certain amount of repetition and enables the essential points to be made more clearly. Illustrations are taken from the category of sets, which underlies all the other categories we shall be considering.

Potential readers of this book will almost certainly be familiar with the basic concepts of point-set topology. However, instead of merely summarizing the results to be used later, or relying on references, it seemed better to develop everything from first principles. The reason is that only a small part of the material in the classic texts is relevant to our purposes and by concentrating on that, with certain additional material which is not usually given, it becomes possible to achieve greater coherence. This comprises Chapter 2.

The third chapter is concerned with the various "comma" categories associated with the basic category of topology, notably the category of spaces over a base, which is of central importance. One of the themes of this book is that many of the concepts of ordinary topology can be extended in a natural way to fibrewise topology, i.e. the topology of spaces over a base. This is relevant to fibre bundle theory, for example.

One of the most striking trends of the past decade or so has been the increase of interest in equivariant topology—topology from an equivariant point of view. Accordingly Chapter 4 is devoted to an outline of the basic theory of topological groups, particularly topological transformation groups. Illustrations are taken from the theory of covering spaces and of fibre bundles in general.

The second half of the book follows on directly from the first and is devoted to various aspects of homotopy theory. One begins with the idea of classifying the points of a space into path-components. This leads on to the idea of classifying continuous functions, from one space to another, by homotopy or continuous deformation. In Chapter 5 we consider not only the homotopy theory of ordinary topology but also the homotopy theory of fibrewise topology and equivariant topology. Again covering spaces provide valuable illustrations of various points.

One of the central aims of homotopy theory is the classification of continuous functions. This is greatly facilitated if one has available either the homotopy lifting property or the homotopy extension property. The homotopy lifting property is enjoyed by fibrations, such as covering projections. The homotopy extension property is enjoyed by cofibrations. Both of these concepts are discussed in Chapter 6, with various applications of the results obtained.

One of the most important results of homotopy theory is the classification theorem for fibre bundles with given structural group. Some kind of restriction is inevitable in this theorem. In the version to be given in Chapter 7 the restriction takes the form of a numerability condition on the bundle structure. We also prove a number of other theorems where a similar condition is needed, including a theorem due to Dold which provides an important bridge between fibrewise homotopy theory and ordinary homotopy theory.

In the final chapter we return to the homotopy extension property. In the cofibration approach a strong condition is placed on the domain side of the continuous functions to be classified. There is, however, an alternative approach, where the main condition is placed on the codomain side, although some restriction on the domain side is necessary as well. In this alternative theory the codomains are called extensors (absolute retracts) or neighbourhood extensors (absolute neighbourhood retracts). As with so many of the other concepts considered in this book, although the basic ideas refer to the category of spaces there are also fibrewise and equivariant versions of extensor theory. Here too there is a bridging theorem which connects the equivariant theory with the other theories.

CHAPTER 1

The Basic Framework

Category theory as such plays no real part in this book. However it is both convenient and illuminating to be able to use the language of that theory to some extent. At the very least it saves a certain amount of repetition. In fact all the categories we shall be dealing with are of the type known as concrete, i.e. they consist of sets with additional structure and functions which respect that structure. However there is little to be gained by restricting attention to such categories at the outset.

Recall that a category $\mathscr{C}$ is a class of objects (denoted by capital letters) together with two functions satisfying certain conditions as follows. One function assigns to each pair (X, Y) of objects of $\mathscr{C}$ a set $\hom(X, Y)$. An element f of this set is called a morphism with domain X and codomain Y, or simply a morphism from X to Y, and may be written either $f: X \to Y$ or $X \xrightarrow{f} Y$. The other function assigns to each triple (X, Y, Z) of objects of $\mathscr{C}$ an operation

$$\hom(Y, Z) \times \hom(X, Y) \to \hom(X, Z)$$

called composition. For morphisms $f: X \to Y$ and $g: Y \to Z$ this operation is written $(g, f) \mapsto g \circ f$, and the morphism $g \circ f: X \to Z$ is called the composite of g with f; in practice the symbol $\circ$ is often omitted. The class $\mathscr{C}$ with these two functions constitutes a category provided the following two conditions are satisfied.

Condition (1.1). *If $f: X \to Y, g: Y \to Z, h: Z \to W$ are morphisms of $\mathscr{C}$ then*

$$(h \circ g) \circ f = h \circ (g \circ f).$$

Condition (1.2). *For each object X of $\mathscr{C}$ there exists a morphism* $\mathrm{id}_X: X \to X$ *such that*

$$\mathrm{id}_X \circ f = f, \qquad g \circ \mathrm{id}_X = g$$

for each morphism f with X as codomain and for each morphism g with X as domain.

Evidently id_X here is unique; it is called the identity morphism of X. In practice the suffix is often omitted.

Due to the associativity condition (1.1) brackets can be omitted in the formation of multiple compositions, such as $h \circ g \circ f$.

A morphism $f: X \to Y$ determines, by precomposition, a function

$$f^*: \hom(Y, Z) \to \hom(X, Z)$$

for each object Z. A morphism $g: Y \to Z$ determines, by postcomposition, a function

$$g_*: \hom(X, Y) \to \hom(X, Z)$$

for each object X.

Definition (1.3). Let $f: X \to Y$ be a morphism. A morphism $g: Y \to X$ is a left inverse of f if $g \circ f = \mathrm{id}_X$. A morphism $h: Y \to X$ is a right inverse of f if $f \circ h = \mathrm{id}_Y$. If f has both a left inverse g and a right inverse h then $g = h$ and f is an equivalence.

When f is an equivalence the (left and right) inverse of f, which is also an equivalence, is denoted by f^{-1}. If there exists an equivalence $X \to Y$ the objects X and Y are said to be equivalent.

A monoid is a category with precisely one object. A groupoid is a category in which all the morphisms are equivalences. For example, let G be a group and let H be a subgroup of G. Then a groupoid is defined in which the objects are the right cosets Hx of H in G and in which the morphisms of Hx into Hy are the elements $g \in G$ such that $Hxg = Hy$; composition of morphisms is given by the binary operation of the group G.

Let $\mathscr{C}$ be a category. For each object X of $\mathscr{C}$ the set $\mathrm{end}(X) = \hom(X, X)$ forms a monoid under the operation of composition while the set $\mathrm{aut}(X) \subset \mathrm{end}(X)$ of equivalences of X with itself forms a group.

The opposite of a category $\mathscr{C}$ is defined to be the category $\mathscr{C}^{\mathrm{op}}$ which has the same objects as $\mathscr{C}$ but, for the morphisms, domain and codomain are interchanged. Explicitly, for each pair of objects (X, Y) the set $\hom(X, Y)$ of morphisms of $\mathscr{C}$ is the same as the set $\hom(Y, X)$ of morphisms of $\mathscr{C}^{\mathrm{op}}$, and the composition operation is modified accordingly.

The basic example of a category is the category $\mathscr{S}$ of sets. Here the objects are sets and the morphisms are functions, with the terms domain, codomain, composition and identity having their normal meanings. In practice the commonest examples of categories are those where the objects are sets with some kind of additional structure while the morphisms are functions which respect that structure. This rather vague description can be made precise as follows.

A concrete category $\mathscr{C}$ is a class of objects with two functions satisfying certain conditions. One function assigns to each object X of $\mathscr{C}$ a set $|X|$, called the underlying set of X. The other assigns to each pair (X, Y) of objects of $\mathscr{C}$ a set $\hom(X, Y)$ of which the elements are functions from the set $|X|$ to the set $|Y|$. The elements of $\hom(X, Y)$ are called the morphisms with domain X and codomain Y. The class $\mathscr{C}$ with these two functions constitutes a concrete category when the following two conditions are satisfied:

Condition (1.4). *For each triple (X, Y, Z) of objects of $\mathscr{C}$ if $f \in \hom(X, Y)$ and $g \in \hom(Y, Z)$ then $g \circ f \in \hom(X, Z)$.*

Condition (1.5). *For each object X of $\mathscr{C}$ the identity function $\mathrm{id}_{|X|}$ belongs to $\hom(X, X)$.*

Evidently a concrete category is a category in the previous, more general sense. As it happens all the categories we shall be using later will be concrete categories, and the phrase "underlying set" will often occur. In practice, however, the special notation $|X|$ for the underlying set of an object X is rather unnecessary and we shall tend to use the same symbol X for both.

In algebra one of the many important concrete categories is the category $\mathscr{G}$ of groups. Here the objects are sets with binary operation, etc. and the morphisms (homomorphisms in the usual terminology) are functions which respect the binary operation, etc. Similarly with the category of rings, the category of vector spaces, and so forth.

In topology, however, the additional structure is of a different type. Each object X consists of a set $|X|$ together with a family of subsets of $|X|$, called the open sets of X. The open sets have to satisfy certain conditions which we need not go into at this stage. Various categories can be defined with this class of objects according to the definition of morphism chosen. One possibility is to say that a function $f: |X| \to |Y|$ is a morphism from X to Y if the direct image fU of each open set U of X is an open set of Y. Another is to say that f is a morphism if the inverse image $f^{-1}V$ of each open set V of Y is an open set of X. Further possibilities are to substitute closed sets for open sets (the closed sets are the complements of the open sets). And of course one can require f to satisfy more than one of these requirements. As we shall see later, some of these ways of constructing a category are more fruitful than others.

A subcategory $\mathscr{C}'$ of a category $\mathscr{C}$ is a category such that each object of $\mathscr{C}'$ is an object of $\mathscr{C}$, and such that each morphism of $\mathscr{C}'$ is a morphism of $\mathscr{C}$. Moreover it is necessary that for such morphisms the domain and codomain, and the operation of composition, are the same in $\mathscr{C}'$ as they are in $\mathscr{C}$. For example the category $\mathscr{S}$ of sets has a subcategory in which only injective functions are morphisms, and a subcategory in which only surjective functions are morphisms.

Two special kinds of subcategory play an important part in our work. In the first, the objects of $\mathscr{C}'$ are a subclass of the objects of $\mathscr{C}$ but, for pairs of such objects, the morphisms in $\mathscr{C}'$ are the same as the morphisms in $\mathscr{C}$. This is known as a full subcategory. For example, the category of abelian groups is a full subcategory of the category of groups. In the second special kind of subcategory the objects of $\mathscr{C}'$ are the same as the objects of $\mathscr{C}$ but the morphisms in $\mathscr{C}'$ are a subset of the morphisms in $\mathscr{C}$. For example one may take the morphisms in $\mathscr{C}'$ to be the equivalences in $\mathscr{C}$. For another example take the category $\mathscr{S}$ of sets and consider the subcategory in which the morphisms are functions such that the inverse images of elements of the codomain are always finite.

Another important method of constructing categories from a given category is to form the Jth power $\mathscr{C}^J$ of $\mathscr{C}$, where J is a (non-empty) indexing set. The objects of $\mathscr{C}^J$ are J-indexed collections of objects of $\mathscr{C}$ and the morphisms of $\mathscr{C}^J$ are J-indexed collections of morphisms of $\mathscr{C}$. Specifically, if $X = \{X_j\}$ and $Y = \{Y_j\}$ are J-indexed collections of objects of $\mathscr{C}$ then a morphism $f: X \to Y$ is a J-indexed collection $\{f_j\}$ of morphisms of $\mathscr{C}$, where $f_j: X_j \to Y_j$. The operation of composition in $\mathscr{C}^J$ is derived from that in $\mathscr{C}$ in the obvious way. In case J is finite, say $J = \{1, \ldots, n\}$ for $n \geq 1$, we may write

$$\mathscr{C}^J = \mathscr{C} \times \cdots \times \mathscr{C} \quad (n \text{ factors}).$$

The product of an indexed collection of categories can be defined similarly.

Yet another way of constructing other categories from a given category $\mathscr{C}$ is to form the category of pairs, usually denoted by $\mathscr{C}(2)$. The objects of $\mathscr{C}(2)$ are the morphisms of $\mathscr{C}$ and the morphisms of $\mathscr{C}(2)$ are the commutative diagrams of $\mathscr{C}$. Specifically, if f and f' are objects of $\mathscr{C}(2)$ (i.e. morphisms of $\mathscr{C}$) as shown below, then a pair (ξ, η) of morphisms of $\mathscr{C}$ is a morphism of $\mathscr{C}(2)$ if the condition $f' \circ \xi = \eta \circ f$ is satisfied.

$$\begin{array}{ccc} X & \xrightarrow{\xi} & X' \\ {\scriptstyle f}\downarrow & & \downarrow{\scriptstyle f'} \\ Y & \xrightarrow[\eta]{} & Y' \end{array}$$

The composition operation in $\mathscr{C}(2)$ is defined in terms of the composition operation of $\mathscr{C}$, thus:

$$(\xi', \eta') \circ (\xi, \eta) = (\xi' \circ \xi, \eta' \circ \eta).$$

Certain subcategories of the category of pairs are of great importance in our work, but before we discuss these it is desirable to make a few more definitions, for any category $\mathscr{C}$.

Definition (1.6). An object A is initial if for each object X there exists precisely one morphism $A \to X$; in that case the morphisms $X \to A$ are called the cosections of X.

Definition (1.7). An object B is final if for each object X there exists precisely one morphism $X \to B$; in that case the morphisms $B \to X$ are called the sections of X.

Of course initial and final are dual notions here, in the sense that an initial object of $\mathscr{C}$ is a final object of $\mathscr{C}^{\text{op}}$, and vice versa. Initial objects, if they exist, are unique up to equivalence, and similarly with final objects. Both notions are combined in

Definition (1.8). An object O which is both initial and final is a nul object; in that case, for each pair of objects (X, Y), the unique morphism $X \to O \to Y$ is called the nul morphism, and denoted by c.

For example, take the category $\mathscr{S}$ of sets. The empty set $\varnothing$ is an initial object and the singleton (one-element set) $*$ is a final object. There is no nul object.

For another example, take the category $\mathscr{G}$ of groups. The trivial group $*$ is a nul object and, for any pair of groups (G, H), the trivial homomorphism $G \to H$ is the nul morphism.

Given an object A of the category $\mathscr{C}$, the category $\mathscr{C}^A = \hom(A, \ )$ of objects under A is defined as follows. An object under A is a pair consisting of an object X of $\mathscr{C}$ and a morphism $u: A \to X$ of $\mathscr{C}$, called the insertion; in practice X alone is usually sufficient notation. If X, Y are objects under A with insertions u, v then a morphism $f: X \to Y$ of $\mathscr{C}$ is a morphism under A if $fu = v$, as shown below.

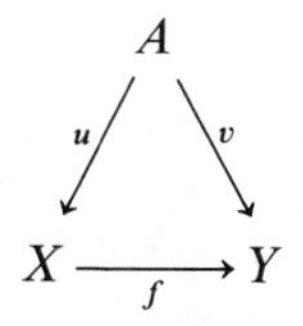

In this way the category $\mathscr{C}^A = \hom(A, \ )$ of objects under A is defined. The equivalences of the category are called equivalences under A, and so on. Notice that if $f: X \to Y$ is a morphism under A then any left inverse of f in the original category $\mathscr{C}$ is automatically a left inverse of f in the category $\mathscr{C}^A$. Notice also that if A itself is regarded as an object under A, with insertion the identity id_A, then A constitutes an initial object of $\mathscr{C}^A$. Moreover if A is already a final object of $\mathscr{C}$ then A constitutes a nul object of $\mathscr{C}^A$. Finally if A is already an initial object of $\mathscr{C}$ then $\mathscr{C}^A$ reduces to $\mathscr{C}$.

For example, take the category $\mathscr{S}$ of sets. The category $\mathscr{S}^*$ of objects under $*$, the singleton, is essentially the category of pointed sets, i.e. sets with basepoint. Specifically a set X under $*$ with insertion u determines, and is determined by, the set X with the basepoint $u(*)$. Also the functions $f\colon X \to Y$ under $*$ are precisely the functions which respect basepoints. We shall therefore refer to $\mathcal{S}^*$ as the category of pointed sets.

Returning to the general case, observe that the categories $\mathscr{C}^A$, for various objects A, may all be regarded as subcategories of the category $\mathscr{C}(2)$ of pairs. This suggests an extension of our terminology as follows. Suppose that we have a commutative diagram of objects and morphisms of $\mathscr{C}$, as shown below.

$$\begin{array}{ccc} A & \xrightarrow{\xi} & A' \\ {\scriptstyle u}\downarrow & & \downarrow{\scriptstyle u'} \\ X & \xrightarrow[\eta]{} & X' \end{array}$$

Then we may refer to η as a morphism under ξ. Of course this is just another way of saying that (ξ, η) is a morphism of $\mathscr{C}(2)$ but in some circumstances it is the more natural expression to use.

Again, given an object B of the category $\mathscr{C}$, the category $\mathscr{C}_B = \hom(\ , B)$ of objects over B is defined as follows. An object over B is a pair consisting of an object X of $\mathscr{C}$ and a morphism $p\colon X \to B$ of $\mathscr{C}$, called the projection; in practice X alone is usually sufficient notation. If X, Y are objects over B with projections p, q then a morphism $f\colon X \to Y$ of $\mathscr{C}$ is a morphism over B if $qf = p$, as shown below.

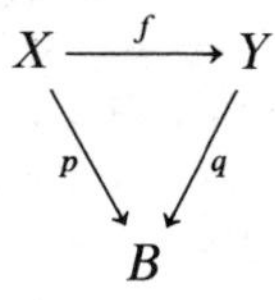

In this way the category $\mathscr{C}_B = \hom(\ , B)$ of objects over B and morphisms over B is defined. The equivalences of the category are called equivalences over B, and so on. Note that if $f\colon X \to Y$ is a morphism over B then any right inverse of f in the original category $\mathscr{C}$ is automatically a right inverse of f in the category $\mathscr{C}_B$. Notice also that if B itself is regarded as an object over B, with projection id_B, then B constitutes a final object of $\mathscr{C}_B$. Moreover if B is already an initial object of $\mathscr{C}$ then B constitutes a nul object of $\mathscr{C}_B$. Finally if B is already a final object of $\mathscr{C}$ then $\mathscr{C}_B$ reduces to $\mathscr{C}$.

Thus, let $\mathscr{C}$ be any category and let B be any object of $\mathscr{C}$. Then B constitutes a nul object of the category

$$\mathscr{C}_B^B = (\mathscr{C}^B)_B = (\mathscr{C}_B)^B$$

of objects "over and under" B. Since we shall be using this category quite a lot let us state explicitly what its objects and morphisms are. An object over and under B is a triple consisting of an object X of $\mathscr{C}$ and a pair of morphisms

$$B \xrightarrow{u} X \xrightarrow{p} B$$

of $\mathscr{C}$ such that $p \circ u = \mathrm{id}_B$; usually X alone is sufficient notation. In particular B is regarded as an object over and under itself, taking id_B to be both insertion and projection. If X, Y are objects over and under B, with projections p, q and insertions u, v, then a morphism $f: X \to Y$ in $\mathscr{C}$ is a morphism over and under B if $f \circ u = v$ and $q \circ f = p$, as shown below.

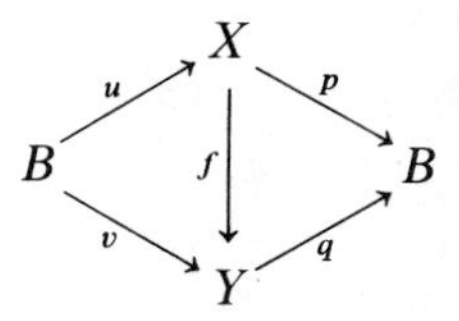

The equivalences of the category are called equivalences over and under B, and so on. The nul object is B itself, and the nul morphism $c: X \to Y$ is the composition

$$X \xrightarrow{p} B \xrightarrow{v} Y.$$

To illustrate these ideas let us examine in detail the case of the category $\mathscr{S}$ of sets. Let X be a set over the given set B with projection $p: X \to B$. The inverse image $p^{-1}b$ of each point $b \in B$ is denoted by X_b and known as the fibre over b. Fibres may be empty since p is not necessarily surjective. Let X, Y be sets over B and let $f: X \to Y$ be a function. If f is over B then f determines, by restriction, a function $f_b: X_b \to Y_b$, for each $b \in B$. Conversely if $fX_b \subset Y_b$, for each $b \in B$, then f is over B. For this reason the term "fibre-preserving function" is often used instead of "function over B" when it is clear which set B is intended.

Again the inverse image $p^{-1}B'$ of each subset $B' \subset B$ is denoted by $X_{B'}$ and known as X restricted to B'; of course $X_{B'}$ may be regarded as a set over B' with projection given by p. Moreover a function $f: X \to Y$ over B determines a function $f_{B'}: X_{B'} \to Y_{B'}$ over B' in the obvious way.

An example which arises frequently is when an equivalence relation $\sim$ is given on a set X. Then B may be taken as the quotient set $X/\sim$ with the natural projection which assigns to each point x its equivalence class $[x]$; such a projection is necessarily surjective. Conversely if X is a set over a given set B and if the projection p is surjective then, by taking the fibres as equivalence classes, an equivalence relation is defined on X such that p induces a bijection between the quotient set and B.

Next consider the category $\mathscr{S}_B^B$ of sets over and under B. The insertion $u: B \to X$ of such a set X assigns to each fibre X_b a basepoint $u(b)$. If X, Y are sets over and under B, a function $f: X \to Y$ is over and under B if it not only satisfies the fibre-preserving condition $fX_b \subset Y_b$, for each $b \in B$, but also preserves basepoints in each fibre. The nul function $c: X \to Y$ sends each fibre X_b to the basepoint in the corresponding fibre Y_b. A function $f: X \to Y$ over and under B determines, for each subset B' of B, a function $f_{B'}: X_{B'} \to Y_{B'}$ over and under B'.

Functors are the next subject for discussion. We begin with the ordinary covariant type. If $\mathscr{C}$ and $\mathscr{C}'$ are categories a functor $\Phi: \mathscr{C} \to \mathscr{C}'$ is a transformation which assigns to each object X of $\mathscr{C}$ an object ΦX of $\mathscr{C}'$, and to each morphism $f: X \to Y$ of $\mathscr{C}$ a morphism $\Phi f: \Phi X \to \Phi Y$ of $\mathscr{C}'$, so that the following two conditions are satisfied.

Condition (1.9). *For each triple (X, Y, Z) of objects of $\mathscr{C}$ the relation*

$$\Phi(g \circ f) = (\Phi g) \circ (\Phi f)$$

holds for all morphisms $f: X \to Y$ and $g: Y \to Z$.

Condition (1.10). *For each object X of $\mathscr{C}$ we have*

$$\Phi(\mathrm{id}_X) = \mathrm{id}_{\Phi X}.$$

For example, suppose that $\mathscr{C}$ is a subcategory of $\mathscr{C}'$. Then one has the inclusion functor $\mathscr{C} \to \mathscr{C}'$ which assigns each object to itself and each morphism to itself. For another example let I, J be indexing sets and let $\alpha: I \to J$ be a function. Then α determines a functor

$$\Phi: \mathscr{C}^J \to \mathscr{C}^I$$

for any category $\mathscr{C}$. Specifically Φ assigns to each J-indexed family $\{X_j\}$ of objects of $\mathscr{C}$ the I-indexed family $\{X'_i\}$ of objects of $\mathscr{C}$, where $X'_i = X_{\alpha(i)}$, and similarly for morphisms.

For another example, if $\mathscr{C}$ is a category with final object B then a functor

$$\sec_B: \mathscr{C} \to \mathscr{S}$$

is given by assigning to each object X of $\mathscr{C}$ the set $\sec_B X = \hom(B, X)$ of sections of X, and similarly with morphisms.

For each category $\mathscr{C}$ one has the identity functor $\mathscr{C} \to \mathscr{C}$. Moreover the composition of functors $\mathscr{C} \to \mathscr{C}'$ and $\mathscr{C}' \to \mathscr{C}''$ is defined in the obvious way as a functor $\mathscr{C} \to \mathscr{C}''$. Thus, aside from logical difficulties, one may imagine a universal category, in which the objects are categories and the morphisms are functors.

Let $\Phi: \mathscr{C} \to \mathscr{C}'$ be a functor. If g is a left (resp. right) inverse of f in $\mathscr{C}$ then Φg is a left (resp. right) inverse of Φf in $\mathscr{C}'$. If f is an equivalence in $\mathscr{C}$ then Φf is an equivalence in $\mathscr{C}'$.

We generally use the notation

$$\Phi_\#: \hom(X, Y) \to \hom(\Phi X, \Phi Y)$$

to denote the transformation of morphisms determined by the functor Φ. Note that

$$\Phi_\#: \mathrm{end}(X) \to \mathrm{end}(\Phi X)$$

is a homomorphism of monoids and that

$$\Phi_\#: \mathrm{aut}(X) \to \mathrm{aut}(\Phi X)$$

is a homomorphism of groups.

A functor from a category $\mathscr{C}$ into itself is called an endofunctor of $\mathscr{C}$. For example, let Φ be an endofunctor of the category $\mathscr{S}$ of sets. In $\mathscr{S}$ every surjective function admits a right inverse; consequently Φ transforms surjections into surjections. Similarly Φ transforms injections into injections, provided the domain is non-empty. The restriction is essential. For let $\theta: A \to B$ be a function, where A and B are sets. Define $\Phi X = A$ when X is empty and $\Phi X = B$ when X is non-empty. Define Φf, where $f: X \to Y$, to be θ when X is empty and Y is non-empty, and to be the identity otherwise. The functor Φ thus defined will not, in general, transform injections into injections.

Let Φ be an endofunctor of the category $\mathscr{S}$. We can "extend" Φ to an endofunctor Φ_B of the category $\mathscr{S}_B$ of sets over a given set B in the natural way thus:

$$\Phi_B(X) = \coprod_{b \in B} \Phi(X_b),$$

for each set X over B, and similarly with morphisms. More generally let $\eta: X' \to X$ be a morphism over a given morphism $\xi: B' \to B$. Then Φ determines a morphism $\Phi_{B'}(X') \to \Phi_B(X)$ over ξ, in the obvious way.

I have already remarked that the categories we shall be dealing with in practice are all concrete categories. For these there is a notion of concrete functor as follows. Let $\mathscr{C}$ and $\mathscr{C}'$ be concrete categories. A concrete functor $\Phi: \mathscr{C} \to \mathscr{C}'$ assigns to each object X of $\mathscr{C}$ an object ΦX of $\mathscr{C}'$ and is accompanied by an endofunctor $|\Phi|$ of $\mathscr{S}$, called the underlying functor, such that the following two conditions are satisfied:

Condition (1.11). *If X is an object of $\mathscr{C}$ then*

$$|\Phi X| = |\Phi||X|.$$

Condition (1.12). *If $f: X \to Y$ is a morphism of $\mathscr{C}$ then $|\Phi| f: \Phi X \to \Phi Y$ is a morphism of $\mathscr{C}'$.*

When these two conditions are satisfied it is clear that Φ constitutes a functor in the previous, more general, sense, with Φf defined to be $|\Phi| f$ for each morphism f. Note that for each concrete category $\mathscr{C}$ the transformation $X \mapsto |X|$ determines a concrete functor $\mathscr{C} \to \mathscr{S}$; this is known as the forgetful functor. Of course every endofunctor of $\mathscr{S}$ is automatically concrete.

So far we have only been dealing with covariant functors. Later on we shall occasionally need to deal with contravariant functors as well. A contravariant functor $\Phi: \mathscr{C} \to \mathscr{C}'$ is simply a covariant functor $\mathscr{C} \to (\mathscr{C}')^{\mathrm{op}}$. Thus Φ assigns to each object X of $\mathscr{C}$ an object ΦX of $\mathscr{C}'$, and to each morphism $f: X \to Y$ of $\mathscr{C}$ a morphism $\Phi f: \Phi Y \to \Phi X$ of $\mathscr{C}'$, such that (1.10) is satisfied but instead of (1.9) we now have

Condition (1.13). *For each triple (X, Y, Z) of objects of $\mathscr{C}$ the relation*

$$\Phi(g \circ f) = (\Phi f) \circ (\Phi g)$$

holds for all morphisms $f: X \to Y$ and $g: Y \to Z$.

For example, let $\mathscr{C}$ be any category. For fixed X and variable Y, hom(X, Y) determines a covariant functor $\mathscr{C} \to \mathscr{S}$, but for fixed Y and variable X, hom(X, Y) determines a contravariant functor $\mathscr{C} \to \mathscr{S}$.

Multiple functors are defined in the obvious way. Take bifunctors $\mathscr{C} \times \mathscr{C} \to \mathscr{C}'$, for example. These are of four types, depending on whether they are covariant or contravariant in the first entry and whether they are covariant or contravariant in the second entry. Thus hom (with $\mathscr{C}' = \mathscr{S}$) is contravariant in the first and covariant in the second. In discussing bifunctors $\Phi: \mathscr{C} \times \mathscr{C} \to \mathscr{C}'$ it is often convenient to use the notation $X\Phi Y$ rather than $\Phi(X, Y)$ and, if this notation is adopted, to use a symbol such as $\Box$ for the bifunctor rather than a greek capital.

For example, in the case of $\mathscr{S}$ we have the bifunctors $\times$ (product) and $+$ (sum). In the case of $\mathscr{S}_B$, where B is a given set, we extend these to bifunctors $\times_B$ (fibre product) and $+_B$ (fibre sum). Specifically let X, Y be sets over B with projections p, q. respectively. The fibre product $X \times_B Y$ is the subset of $X \times Y$ consisting of pairs (x, y) such that $px = qy$, with the projection r given by $r(x, y) = px = qy$. The fibre sum $X +_B Y$ is the set $X + Y$ with the projection given by p on X and by q on Y; we often write $+$ instead of $+_B$. The fibre product and fibre sum of functions are defined similarly.

Definition (1.14). A covariant functor $\Phi: \mathscr{C} \to \mathscr{C}'$ is an equivalence of categories if there exists a covariant functor $\Phi': \mathscr{C}' \to \mathscr{C}$ such that $\Phi' \circ \Phi$ is the identity functor on $\mathscr{C}$ while $\Phi \circ \Phi'$ is the identity functor on $\mathscr{C}'$.

For example, let $\mathscr{C}$ be any category. If A is an initial object of $\mathscr{C}$ then $\mathscr{C}^A$ is equivalent to $\mathscr{C}$. If B is a final object of $\mathscr{C}$ then $\mathscr{C}_B$ is equivalent to $\mathscr{C}$.

Let Φ, $\Psi: \mathscr{C} \to \mathscr{C}'$ be (covariant) functors. A natural transformation $T: \Phi \to \Psi$ assigns to each object X of $\mathscr{C}$ a morphism $TX: \Phi X \to \Psi X$ of $\mathscr{C}'$ such that the following condition is satisfied.

Condition (1.15). *For each morphism $f: X \to Y$ of $\mathscr{C}$ we have the relation*

$$TY \circ \Phi f = \Psi f \circ TX,$$

as in the diagram shown below.

$$\begin{array}{ccc} \Phi X & \xrightarrow{\Phi f} & \Phi Y \\ {\scriptstyle TX}\downarrow & & \downarrow{\scriptstyle TY} \\ \Psi X & \xrightarrow[\Psi f]{} & \Psi Y \end{array}$$

A natural transformation $T: \Phi \to \Psi$ is said to be a natural equivalence if the morphism $TX : \Phi X \to \Psi X$ of $\mathscr{C}'$ is an equivalence of $\mathscr{C}'$ for each object X

of $\mathscr{C}$. In that case an inverse natural equivalence $T^{-1}: \Psi \to \Phi$ is defined, where $T^{-1}X = (TX)^{-1}$, the inverse of the equivalence TX. In situations where the natural equivalence is an obvious one I shall simply write $\Phi \equiv \Psi$ to indicate that the functors Φ and Ψ are naturally equivalent. Examples will be given shortly.

For any category $\mathscr{C}$ we have the switching functor $T: \mathscr{C} \times \mathscr{C} \to \mathscr{C} \times \mathscr{C}$, where for objects

$$T(X, Y) = (Y, X)$$

and similarly for morphisms.

Consider a bifunctor $\square: \mathscr{C} \times \mathscr{C} \to \mathscr{C}$.

Definition (1.16). The bifunctor $\square$ is commutative if the bifunctors

$$\square, \square \circ T: \mathscr{C} \times \mathscr{C} \to \mathscr{C}$$

are naturally equivalent.

Definition (1.17). The bifunctor $\square$ is associative if the trifunctors

$$\square \circ (\square \times \mathrm{id}), \square \circ (\mathrm{id} \times \square): \mathscr{C} \times \mathscr{C} \times \mathscr{C} \to \mathscr{C}$$

are naturally equivalent.

When working in a category $\mathscr{C}$ it is convenient to have terms to describe pairs of morphisms with a common domain or codomain.

Definition (1.18). A triad, with codomain M, is a pair of morphisms

$$X \xrightarrow{\alpha} M \xleftarrow{\beta} Y.$$

Definition (1.19). A cotriad, with domain W, is a pair of morphisms

$$X \xleftarrow{\xi} W \xrightarrow{\eta} Y.$$

For example take the category $\mathscr{S}$ of sets. The set-theoretical sum (disjoint union) $X + Y$ should be regarded as a triad

$$X \xrightarrow{u} X + Y \xleftarrow{v} Y,$$

where u, v are the standard insertions. The cartesian product $X \times Y$ should be regarded as a cotriad

$$X \xleftarrow{p} X \times Y \xrightarrow{q} Y,$$

where p, q are the standard projections.

Of course a triad with codomain M can equally well be regarded as a pair of objects of $\mathscr{C}_M$ while a cotriad with domain W can equally well be regarded as a pair of objects of $\mathscr{C}^W$.

Now consider a commutative diagram of objects and morphisms of $\mathscr{C}$ as shown below.

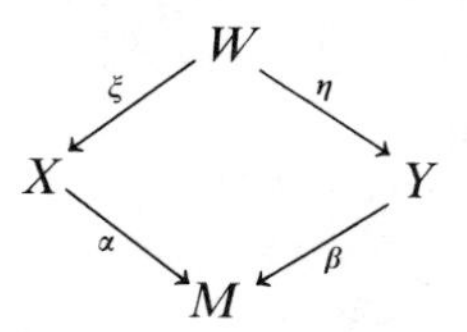

Definition (1.20). The above diagram is cartesian if for each object Z and cotriad

$$X \overset{f}{\leftarrow} Z \overset{g}{\rightarrow} Y$$

with domain Z such that $\alpha f = \beta g$ there exists precisely one morphism $h: Z \to W$ such that $\xi h = f$ and $\eta h = g$.

Definition (1.21). The above diagram is cocartesian if for each object Z and triad

$$X \overset{f}{\rightarrow} Z \overset{g}{\leftarrow} Y$$

with codomain Z such that $\xi f = \eta g$ there exists precisely one morphism $h: M \to Z$ such that $h\alpha = f$ and $h\beta = g$.

The morphism h in (1.20) is called the pull-back of f and g, while the morphism h in (1.21) is called the push-out of f and g.

For example, take the category $\mathscr{S}$ of sets. The product $X \times Y$ and the sum (disjoint union) $X + Y$ are defined for each pair of sets (X, Y), so that the diagram on the left is cartesian while the diagram on the right is cocartesian.

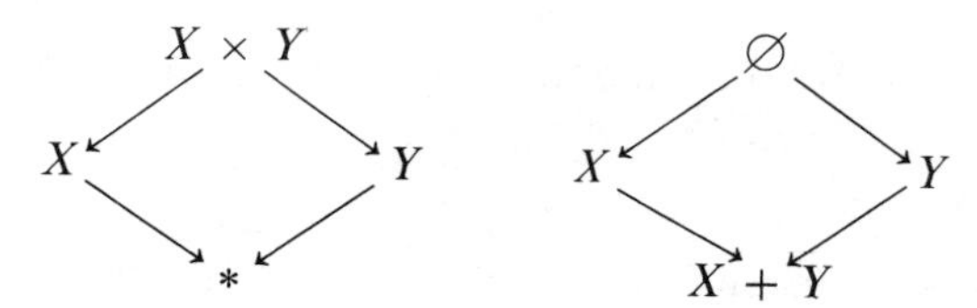

The functions here are the obvious ones. In fact these properties are characteristic of the product and sum, respectively.

Clearly cartesian squares in $\mathscr{C}$ are cocartesian squares in $\mathscr{C}^{op}$, and vice versa. The same square can be both cartesian and cocartesian. For example if $f: X \to Y$ is a morphism then the square shown below has both properties.

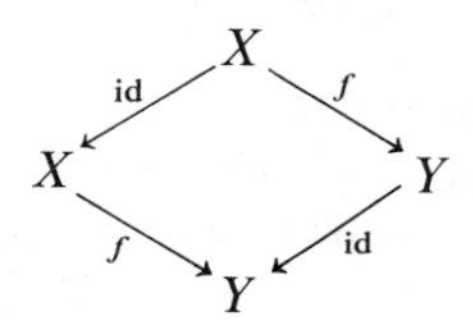

As an illustration of how the cartesian and cocartesian conditions may be used we prove the following result about the commutative square:

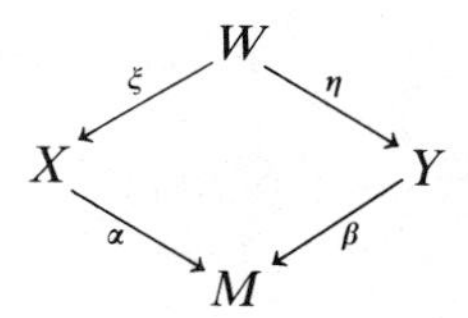

Proposition (1.22). (i) *Suppose that the above square is cartesian. If β is an equivalence then so is ξ.*
(ii) *Suppose that the above square is cocartesian. If ξ is an equivalence then so is so β.*

To prove (i), consider the cotriad

$$X \xleftarrow{\text{id}} X \xrightarrow{\beta^{-1}\alpha} Y,$$

where $\beta^{-1}: M \to Y$ is the inverse of β. Let $\xi': X \to W$ be the pull-back of id_X and $\beta^{-1}\alpha$. Then

$$\xi\xi'\xi = \xi, \qquad \eta\xi'\xi = \beta^{-1}\alpha\xi = \beta^{-1}\beta\eta = \eta.$$

Thus $\xi'\xi$ and id_W are both pull-backs of ξ and η, hence $\xi'\xi = \text{id}_W$, by uniqueness. Therefore ξ is an equivalence, as asserted. This proves (i) and (ii) is just the dual.

Definition (1.23). A pull-back of a triad

$$X \xrightarrow{\alpha} M \xleftarrow{\beta} Y$$

is a cotriad

$$X \xleftarrow{\xi} W \xrightarrow{\eta} Y$$

such that the square thus formed is both commutative and cartesian.

Definition (1.24). A push-out of a cotriad

$$X \xleftarrow{\xi} W \xrightarrow{\eta} Y$$

is a triad

$$X \xrightarrow{\alpha} M \xleftarrow{\beta} Y$$

such that the square thus formed is both commutative and cocartesian.

An important point of principle is that push-outs and pull-backs in the category $\mathscr{C}$ also serve as push-outs and pull-backs in the associated categories $\mathscr{C}^A$ and $\mathscr{C}_B$, for any objects A, B of $\mathscr{C}$. Thus, let

$$X \xrightarrow{\alpha} M \xleftarrow{\beta} Y$$

be a triad in $\mathscr{C}$ with pull-back cotriad

$$X \overset{\xi}{\leftarrow} W \overset{\eta}{\rightarrow} Y$$

in $\mathscr{C}$. Suppose that X, Y and M are objects under A, and that α, β are morphisms under A. Take the pull-back of the insertions of X and Y as an insertion for W. Then W becomes an object under A so that ξ, η are morphisms under A, and the cotriad constitutes a pull-back in $\mathscr{C}^A$. Again, suppose that X, Y, M are objects over B and that α, β are morphisms over B. Make W an object over B so that ξ, η are morphisms over B; then the cotriad constitutes a pull-back in $\mathscr{C}_B$. Similarly with push-outs.

Evidently pull-backs and push-outs, when they exist, are unique up to an appropriate form of equivalence, so that we may usually refer to "the" pull-back and "the" push-out.

Exercise (1.25). Let $f\colon X \to Y$ be a morphism. Consider the triad

$$X \overset{f}{\rightarrow} Y \overset{\text{id}}{\leftarrow} Y$$

and the cotriad

$$X \overset{\text{id}}{\leftarrow} X \overset{f}{\rightarrow} Y.$$

The pull-back of the cotriad is the triad, and the push-out of the triad is the cotriad.

To illustrate these ideas consider first the category $\mathscr{S}$ of sets. Here the pull-back of a triad

$$X \overset{\alpha}{\rightarrow} M \overset{\beta}{\leftarrow} Y$$

is the cotriad

$$X \overset{\xi}{\leftarrow} W \overset{\eta}{\rightarrow} Y,$$

where W is the subset of $X \times Y$ consisting of pairs (x, y) such that $\alpha x = \beta y$, and where

$$\xi(x, y) = x, \qquad \eta(x, y) = y.$$

In other words $W = X \times_M Y$.

Now, still in the category of sets, suppose that we have a cotriad

$$X \overset{\xi}{\leftarrow} W \overset{\eta}{\rightarrow} Y.$$

Since the pull-back in the previous paragraph was a subset of $X \times Y$ we might expect the push-out of the cotriad to be a quotient set of $X + Y$. So consider the triad

$$X \overset{\sigma}{\rightarrow} X + Y \overset{\tau}{\leftarrow} Y.$$

The obvious relation $\sim$ on $X + Y$ to try out is the one where $\sigma x \sim \tau y$ if there exists a point $z \in W$ such that $\xi z = x$ and $\eta z = y$. Unfortunately this is not, in general, an equivalence relation. For example, take $W = X \times Y$ with ξ, η the standard projections. Then $\sigma x \sim \tau y$ for each point $x \in X$ and for each point $y \in Y$, and this is not a transitive relation when X and Y are non-empty. Thus in general it is necessary to form the transitive closure in order to obtain a quotient of $X + Y$ which satisfies the conditions for a push-out.

The procedure of forming the transitive closure can, however, be omitted if one of the insertions is injective. In practice this condition is usually satisfied anyway and so, rather than get involved with transitive closures, let us accept this restriction. It is also worth remarking that if one of the insertions (say v) is surjective then the push-out can be regarded as as quotient of X rather than of $X + Y$.

For example, consider the cotriad

$$X \overset{u}{\leftarrow} A \rightarrow *,$$

where u is injective and A is non-empty. The push-out is the pointed set consisting of (i) the points of $X - uA$ and (ii) the base-point. This is known as the pointed set obtained from X by collapsing with respect to u. In case A is a subset of X and u is the inclusion it is simply known as the pointed set obtained from X by collapsing A, and written X/A.

Let Φ be an endofunctor of the category $\mathscr{S}$. If X is a pointed set the insertion $u: * \rightarrow X$ has a left inverse, hence the transform $\Phi(u): \Phi(*) \rightarrow \Phi(X)$ has a left inverse, and so $\Phi(u)$ is injective. We define $\Phi^*(X)$ to be the pointed set obtained from $\Phi(X)$ by collapsing with respect to $\Phi(u)$. Thus Φ^* associates a pointed set with each pointed set, and similarly with pointed functions. The endofunctor Φ^* of $\mathscr{S}^*$ defined in this way is called the reduction of the given endofunctor Φ of $\mathscr{S}$.

Multiple functors of $\mathscr{S}$ can also be reduced in a similar fashion. For example, consider the topological sum $+$. If X and Y are pointed sets the insertions form a cotriad

$$X \leftarrow * \rightarrow Y.$$

Since the insertions are injective the push-out

$$X \rightarrow X \vee Y \leftarrow Y$$

is defined. Thus $X \vee Y$ is obtained from $X + Y$ by identifying basepoints. We refer to $X \vee Y$ as the wedge of X and Y. The wedge of pointed functions is defined similarly, so that $\vee$ constitutes a binary functor $\mathscr{S}^* \times \mathscr{S}^* \rightarrow \mathscr{S}^*$. It is easy to check that $\vee$ is commutative and associative.

In fact one can proceed a step further. Consider the cotriad

$$X \overset{\xi}{\leftarrow} X \vee Y \overset{\eta}{\rightarrow} Y,$$

where ξ is the push-out of id_X and the nul-function $Y \rightarrow X$ while η is the push-out of the nul-function $X \rightarrow Y$ and id_Y. The pull-back of ξ and η is an injection

$$\zeta: X \vee Y \rightarrow X \times Y.$$

Hence the pointed set $X \wedge Y$ obtained from $X \times Y$ by collapsing with respect to ζ is defined. We refer to $X \wedge Y$ as the smash of X and Y. The smash of pointed functions is defined similarly so that $\wedge$ constitutes a binary functor $\mathscr{S}^* \times \mathscr{S}^* \to \mathscr{S}^*$. It is easy to check that $\wedge$ is also commutative and associative.

All this can be extended to the category $\mathscr{S}_B$ of sets over the given set B. Thus consider a cotriad

$$X \overset{u}{\leftarrow} A \to B$$

where A and X are sets over B with the projection of A surjective, and where u is a function over B. The push-out is the set over and under B consisting of (i) the points of $X - uA$ and (ii) the set B. This is known as the set, over and under B, obtained from X by fibre collapsing with respect to u. In case A is a subset of X and u is the inclusion it is simply known as the set, over and under B, obtained from X by fibre collapsing A, and written $X/_B A$.

Let Φ_B be an endofunctor of the category $\mathscr{S}_B$. If X is a set over and under B the insertion $u: B \to X$ has a left inverse, namely the projection, hence the transform $\Phi_B(u): \Phi_B(B) \to \Phi_B(X)$ has a left inverse, and so $\Phi_B(u)$ is injective. We define $\Phi_B^B(X)$ to be the set over and under B obtained from $\Phi_B(X)$ by collapsing with respect to $\Phi_B(u)$. Thus Φ_B^B associates a set over and under B with every set over and under B, and similarly with functions over and under B. The endofunctor Φ_B^B of $\mathscr{S}_B^B$ defined in this way is called the reduction of the given endofunctor Φ_B of $\mathscr{S}_B$.

Multiple functors of $\mathscr{S}_B$ can also be reduced in a similar fashion. For example, consider the topological sum $+ = +_B$. If X and Y are sets over and under B the insertions form a cotriad

$$X \leftarrow B \to Y.$$

Since the insertions are injective the push-out

$$X \to X \vee_B Y \leftarrow Y$$

is defined. Thus $X \vee_B Y$ is obtained from $X +_B Y$ by identifying basepoints in corresponding fibres. We refer to $X \vee_B Y$ as the fibre wedge of X and Y. The fibre wedge of functions over and under B is defined similarly, so that $\vee_B$ constitutes a binary functor $\mathscr{S}_B^B \times \mathscr{S}_B^B \to \mathscr{S}_B^B$. It is easy to check that $\vee_B$ is commutative and associative.

In fact one can proceed a step further. Consider the cotriad

$$X \overset{\xi}{\leftarrow} X \vee_B Y \overset{\eta}{\to} Y$$

where ξ is the push-out of id_X and the nul-function $Y \to X$ while η is the push-out of the nul-function $X \to Y$ and id_Y. The pull-back of ξ and η is an injection

$$\zeta: X \vee_B Y \to X \times_B Y$$

over B. Hence the set $X \wedge_B Y$ over and under B obtained from $X \times_B Y$ by fibre collapsing with respect to ζ is defined. We refer to $X \wedge_B Y$ as the fibre smash of X and Y. The fibre smash of functions over and under B is defined similarly so that $\wedge_B$ constitutes a binary functor $\mathcal{S}_B^B \times \mathcal{S}_B^B \to \mathcal{S}_B^B$. It is easy to check that $\wedge_B$ is also commutative and associative.

There is yet another kind of terminology which is useful on occasion. Suppose that we have a commutative diagram:

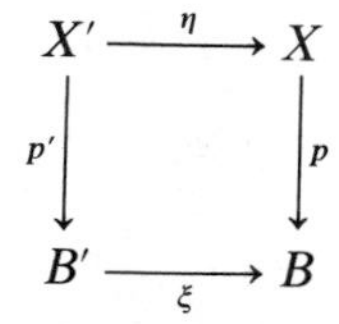

$$\begin{array}{ccc} X' & \xrightarrow{\eta} & X \\ {\scriptstyle p'}\downarrow & & \downarrow{\scriptstyle p} \\ B' & \xrightarrow[\xi]{} & B \end{array}$$

Definition (1.26). The morphism η is a pro-equivalence over ξ if the above diagram is cartesian.

Thus η is a pro-equivalence over ξ if, and only if, for each object Z over B' and morphism $f: Z \to X$ over B there exists precisely one morphism $f': Z \to X'$ over B' such that $\eta f' = f$. For example, each morphism is a pro-equivalence over itself. Quite formally we have

Proposition (1.27). *If $\eta: X' \to X$ is a pro-equivalence over $\xi: B' \to B$ and $\eta': X'' \to X'$ is a pro-equivalence over $\xi': B'' \to B'$ then $\eta \circ \eta': X'' \to X$ is a pro-equivalence over $\xi \circ \xi': B'' \to B$.*

The dual situation arises occasionally. Let us make a change of notation, for reasons of appearance, and consider a commutative diagram of the follow-form,

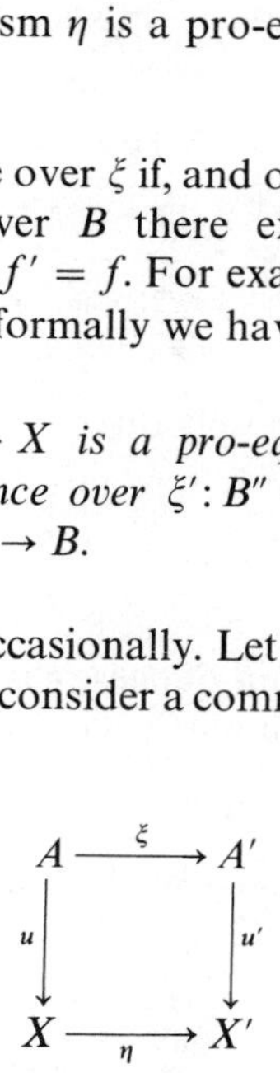

$$\begin{array}{ccc} A & \xrightarrow{\xi} & A' \\ {\scriptstyle u}\downarrow & & \downarrow{\scriptstyle u'} \\ X & \xrightarrow[\eta]{} & X' \end{array}$$

Definition (1.28). The morphism η under ξ is a pro-equivalence under ξ if the above diagram is cocartesian.

Thus η is a pro-equivalence under ξ if, and only if, for each object Z under A' and morphism $f: X \to Z$ under A there exists precisely one morphism $f': X' \to Z$ under A' such that $f'\eta = f$. For example, each morphism is a pro-equivalence under itself. Quite formally we have

Proposition (1.29). *If $\eta: X \to X'$ is a pro-equivalence under $\xi: A \to A'$ and $\eta': X' \to X''$ is a pro-equivalence under $\xi': A' \to A''$ then $\eta' \circ \eta: X \to X''$ is a pro-equivalence under $\xi' \circ \xi: A \to A''$.*

We are now ready to say formally what is meant by the terms product and coproduct, for a given category $\mathscr{C}$. For this purpose consider the functors $L, R: \mathscr{C} \times \mathscr{C} \to \mathscr{C}$ which are given for objects by

$$L(X, Y) = X, \qquad R(X, Y) = Y,$$

and similarly for morphisms. We discuss product functors first.

A product in the category $\mathscr{C}$ consists of a bifunctor

$$\Pi: \mathscr{C} \times \mathscr{C} \to \mathscr{C}$$

together with natural transformations

$$L \xleftarrow{P} \Pi \xrightarrow{Q} R,$$

such that the following condition is satisfied for all pairs of objects (X, Y). Consider the cotriad

$$X \xleftarrow{p} X \Pi Y \xrightarrow{q} Y,$$

where $p = P(X, Y)$, $q = Q(X, Y)$. The condition is that for each object Z of $\mathscr{C}$ and each cotriad

$$X \xleftarrow{f} Z \xrightarrow{g} Y$$

with domain Z there exists precisely one morphism

$$Z \xrightarrow{h} X \Pi Y$$

such that $f = p \circ h$ and $g = q \circ h$.

The morphisms f, g are called the left and right (or first and second) components of h. In view of its uniqueness h is determined by its components and we may write $h = (f, g)$. In particular, take $X = Y = Z$; the morphism

$$(\mathrm{id}, \mathrm{id}): X \to X \Pi X$$

is called the diagonal and denoted by Δ. Again, take $Z = Y \Pi X$; the morphism

$$(q, p): Y \Pi X \to X \Pi Y$$

is called the switching equivalence and denoted by t.

Proposition (1.30). *If $\mathscr{C}$ admits a product functor Π then Π is both commutative and associative.*

Commutativity is given by the switching equivalence, just defined, and associativity is equally obvious. In view of the latter we may omit brackets when writing down multiple products. In fact the whole procedure may be generalized to infinite products in an obvious way. Thus let J be an indexing set: a J-fold product in $\mathscr{C}$ is a functor $\mathscr{C}^J \to \mathscr{C}$ together with natural transformations to each of the projection functors $\mathscr{C}^J \to \mathscr{C}$ satisfying the appropriate conditions. In case J is finite the existence of a product in the previous sense implies the existence of a J-fold product.

A coproduct in the category $\mathscr{C}$ consists of a bifunctor

$$\amalg : \mathscr{C} \times \mathscr{C} \to \mathscr{C}$$

together with natural transformations

$$L \xrightarrow{U} \amalg \xleftarrow{V} R$$

such that the following condition is satisfied for all pairs of objects (X, Y). Consider the triad

$$X \xrightarrow{u} X \amalg Y \xleftarrow{v} Y,$$

where $u = U(X, Y)$ and $v = V(X, Y)$. The condition is that for each object Z of $\mathscr{C}$ and triad

$$X \xrightarrow{f} Z \xleftarrow{g} Y$$

with codomain Z there exists precisely one morphism

$$X \amalg Y \xrightarrow{h} Z$$

such that $f = h \circ u$ and $g = h \circ v$. The morphisms f, g are called the left and right (or first and second) components of h. In view of its uniqueness h is determined by its components and we may write $h = (f, g)$. (In theory this terminology and notation might be confused with that for the product but in practice it is always clear from the context which is meant.) In particular, take $X = Y = Z$; the morphism

$$(\mathrm{id}, \mathrm{id}) : X \amalg X \to X$$

is called the codiagonal and denoted by ∇. Again, take $Z = Y \amalg X$; the morphism

$$(v, u) : X \amalg Y \to Y \amalg X$$

is called the switching equivalence and denoted by t.

Proposition (1.31). *If $\mathscr{C}$ admits a coproduct functor $\amalg$ then $\amalg$ is both commutative and associative.*

Commutativity is given by the switching equivalence, just defined, and associativity is equally obvious. In view of the latter we may omit brackets when writing down multiple coproducts. Infinite coproducts may be defined similarly.

Evidently a product in the category $\mathscr{C}$ also constitutes a product in the category $\mathscr{C}^A$ of objects under a given object A. Likewise a coproduct in the category $\mathscr{C}$ also constitutes a coproduct in the category $\mathscr{C}_B$ of objects over a given object B. However it is not in general true that a coproduct in $\mathscr{C}$ determines a coproduct in $\mathscr{C}^A$ or that a product in $\mathscr{C}$ determines a product in $\mathscr{C}_B$.

Suppose that our category $\mathscr{C}$ admits both a coproduct $\amalg$ and a bifunctor $\square$—possibly, but not necessarily, a product. Then for each triple (X, Y, Z) of objects of $\mathscr{C}$ there is a morphism

$$(X \square Z) \amalg (Y \square Z) \to (X \amalg Y) \square Z,$$

given by $u \square \mathrm{id}$ on the first summand and by $v \square \mathrm{id}$ on the second. When this natural transformation is an equivalence we say that the bifunctor $\square$ is distributive over the coproduct $\amalg$.

For example, the category $\mathscr{S}$ admits the product $\times$ and coproduct $+$, as we have seen, and $\times$ distributes over $+$. Again the category $\mathscr{S}^*$ admits the product $\times$ and coproduct $\vee$, also the binary functor $\wedge$, and $\wedge$ distributes over $\vee$ although $\times$ does not. In the case of $\mathscr{S}_B$ the fibre product $\times_B$ distributes over the fibre sum $+_B$, while in the case of $\mathscr{S}_B^B$ the fibre smash $\wedge_B$ distributes over the fibre wedge $\vee_B$.

Another concept which plays an important part in our work is the following. Let $\mathscr{C}$ be a category and let $\mathscr{C}_B$ be the associated category of objects over the given object B. Suppose that $\mathscr{C}_B$ is equipped with a product, denoted by Π_B. Then for each object B' of $\mathscr{C}$ and morphism $\xi: B' \to B$ of $\mathscr{C}$ a functor

$$\xi^*: \mathscr{C}_B \to \mathscr{C}_{B'}$$

is defined, where $\xi^* X = X \Pi_B B'$ for each object X over B and similarly for morphisms. Here we are regarding B' as an object over B using ξ as projection. With this construction it should be noted that the first projection

$$\xi^* X = X \Pi_B B' \to X$$

is a pro-equivalence over ξ.

Dually let $\mathscr{C}^A$ be the associated category of objects under the given object A. Suppose that $\mathscr{C}^A$ is equipped with a coproduct denoted by $\amalg^A$. Then for each object A' of $\mathscr{C}$ and morphism $\xi: A \to A'$ of $\mathscr{C}$ a functor

$$\xi_*: \mathscr{C}^A \to \mathscr{C}^{A'}$$

is defined, where $\xi_* X = X \amalg^A A'$ for each object X under A and similarly for morphisms. Here we are regarding A' as an object under A using ξ as insertion. With this construction it should be noted that the first insertion

$$X \to X \amalg^A A' = \xi_* X$$

is a pro-equivalence under A.

One of the situations we shall encounter in our work is as follows. Let $\mathscr{C}$ be a category and let $\mathscr{C}'$ be a subcategory having the same objects as $\mathscr{C}$ but not necessarily the same morphisms. Then if X, Y are objects of $\mathscr{C}$ the set $\hom'(X, Y)$ of morphisms of $\mathscr{C}'$ is a subset of the set $\hom(X, Y)$ of morphisms of $\mathscr{C}$. Suppose, however, that each equivalence in $\mathscr{C}$ is also an equivalence in $\mathscr{C}'$. Also suppose that $\mathscr{C}$ admits a product. Then the product of morphisms of $\mathscr{C}'$ is defined as a morphism of $\mathscr{C}$ but not necessarily as a morphism of $\mathscr{C}'$. In that case we can define a subcategory $\mathscr{C}''$ of $\mathscr{C}'$, and hence of $\mathscr{C}$, consisting of the same objects but now the set $\hom''(X, Y)$ of morphisms of $\mathscr{C}''$ consists of the morphisms $f: X \to Y$ of $\mathscr{C}$ with the property that the product

$$f \sqcap \mathrm{id}: X \sqcap T \to Y \sqcap T$$

is a morphism of $\mathscr{C}'$ for each object T. Evidently the product of morphisms of $\mathscr{C}''$ is a morphism of $\mathscr{C}''$. We may describe $\mathscr{C}''$ as the largest subcategory of $\mathscr{C}'$ for which the product of $\mathscr{C}$ provides a product.

Adjoint functors are also important in our work. The following treatment is sufficient for our needs. Let $\mathscr{C}$ be a category and let

$$\Phi, \Psi: \mathscr{C} \times \mathscr{C} \to \mathscr{C}$$

be bifunctors, where Φ is covariant in both entries while Ψ is contravariant in the first entry and covariant in the second. Consider the trifunctors

$$\mathscr{C} \times \mathscr{C} \times \mathscr{C} \to \mathscr{S}$$

given for each triple of objects (X, Y, Z) by

$$\hom(\Phi(X, Y), Z), \qquad \hom(X, \Psi(Y, Z)),$$

respectively, and similarly for morphisms. Thus both trifunctors are contravariant in the first and second entries, covariant in the third.

Definition (1.32). If there exists a natural equivalence between the above trifunctors then the bifunctors Φ and Ψ are adjoint, and any such natural equivalence is an adjunction.

Strictly speaking, Ψ is a right adjoint of Φ and Φ is a left adjoint of Ψ. It is easy to see that adjoints, if they exist, are unique up to natural equivalence. For let Ψ, Ψ' be right adjoints of the same functor Φ. Then

$$\hom(X, \Psi(Y, Z)) \equiv \hom(X, \Psi'(Y, Z)),$$

from which it follows (taking $X = \Psi(Y, Z)$ and $X = \Psi'(Y, Z)$ in turn) that

$$\Psi(Y, Z) \equiv \Psi'(Y, Z).$$

Thus we are justified in referring to "the adjoint", rather than "an adjoint".

An adjunction between the trifunctors defined by Φ and Ψ assigns to each morphism

$$\phi: \Phi(X, Y) \to Z$$

a morphism

$$\psi: X \to \Psi(Y, Z)$$

and vice versa. We refer to ψ as the (right) adjoint of ϕ, and to ϕ as the (left) adjoint of ψ. More colloquially, we say that ψ is obtained from ϕ by taking Y across to the right, and ϕ is obtained from ψ by taking Y across to the left.

For the prototype of the situation we have been discussing, consider the category $\mathscr{S}$ itself. Take

$$\Phi(X, Y) = X \times Y, \quad \Psi(Y, Z) = \hom(Y, Z)$$

and similarly for functions. Now functions $\phi: X \times Y \to Z$ correspond precisely to functions $\psi: X \to \hom(Y, Z)$, where

$$\psi(x)(y) = \phi(x, y) \qquad (x \in X, y \in Y).$$

In this way an adjunction between Φ and Ψ is defined.

More generally, consider the category $\mathscr{S}_B$ of sets over a given set B. Here the adjoint of the fibre product $\times_B$ is the extension $\hom_B$ of hom given, according to our standard procedure, by

$$\hom_B(X, Y) = \coprod_{b \in B} \hom(X_b, Y_b)$$

for each pair X, Y of sets over B and similarly for functions over B. This should not be confused with the set $\mathrm{HOM}_B(X, Y)$ of functions $X \to Y$ over B. In fact each function ϕ over B determines a section of $\hom_B(X, Y)$, by assigning to each point $b \in B$ the function $\phi_b: X_b \to Y_b$. Evidently the transformation

$$\mathrm{HOM}_B(X, Y) \to \sec_B \hom_B(X, Y)$$

thus defined is an equivalence.

Finally, consider the category $\mathscr{S}_B^B$ of sets over and under B. Here the adjoint of the fibre smash product $\wedge_B$ is the bifunctor $\hom_B^B$ given by

$$\hom_B^B(X, Y) = \coprod_{b \in B} \hom^*(X_b, Y_b),$$

where $\hom^*(X_b, Y_b) \subset \hom(X_b, Y_b)$ denotes the subset of pointed functions. Here again we find an equivalence between the pointed set $\mathrm{HOM}_B^B(X, Y)$ of functions over and under B and the pointed set $\sec_B \hom_B^B(X, Y)$.

Let $\mathscr{C}$ be a category which admits both a nul object O and a product functor Π. Then certain other categories can be derived from $\mathscr{C}$ as follows.

Definition (1.33). A multiplication on an object G of $\mathscr{C}$ is a morphism $m: G \,\Pi\, G \to G$. The pair (G, m) is called a binary system.

For present purposes system, rather than binary system, is sufficient terminology. Also G is often sufficient notation, rather than (G, m).

Definition (1.34). The system (G, m) is commutative if $mt = m$, where t denotes the switching equivalence as shown below.

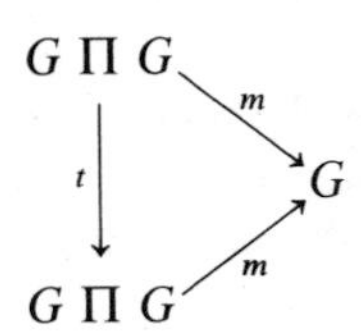

Definition (1.35). The system (G, m) is associative if $m \circ (\mathrm{id}\ \Pi\ m) = m \circ (m\ \Pi\ \mathrm{id})$, as shown below.

$$\begin{array}{ccc} G \,\Pi\, G \,\Pi\, G & \xrightarrow{m\,\Pi\,\mathrm{id}} & G \,\Pi\, G \\ {\scriptstyle \mathrm{id}\,\Pi\, m}\downarrow & & \downarrow{\scriptstyle m} \\ G \,\Pi\, G & \xrightarrow[m]{} & G \end{array}$$

Definition (1.36). The system G is unital if both the compositions

$$G \xrightarrow{\Delta} G \,\Pi\, G \xrightarrow[c\,\Pi\,\mathrm{id}]{\mathrm{id}\,\Pi\, c} G \,\Pi\, G \xrightarrow{m} G$$

equal the identity.

Here, as elsewhere, c denotes the nul morphism $G \to O \to G$. In this context c is often referred to as the neutral morphism.

Definition (1.37). A morphism $v: G \to G$ is an inversion for the system G if both the compositions

$$G \xrightarrow{\Delta} G \,\Pi\, G \xrightarrow[v\,\Pi\,\mathrm{id}]{\mathrm{id}\,\Pi\, v} G \,\Pi\, G \xrightarrow{m} G$$

equal the neutral morphism.

Definition (1.38). An associative system G which satisfies the unital condition is a semigroup object of $\mathscr{C}$. If in addition G admits an inversion then G is a group object of $\mathscr{C}$; in that case the inversion is uniquely determined.

Definition (1.39). Let $G = (G, m)$ and $G' = (G', m')$ be systems of the category $\mathscr{C}$. A morphism $\phi: G \to G'$ is a morphism of $\mathscr{C}$ such that $m' \circ (\phi\ \Pi\ \phi) = \phi \circ m$, as shown below.

$$\begin{array}{ccc} G \,\Pi\, G & \xrightarrow{\phi\,\Pi\,\phi} & G' \,\Pi\, G' \\ {\scriptstyle m}\downarrow & & \downarrow{\scriptstyle m'} \\ G & \xrightarrow[\phi]{} & G' \end{array}$$

With the above definition of morphism the binary systems of the category $\mathscr{C}$ form a category. The equivalences in the new category are called isomorphisms, to prevent confusion with the equivalences of $\mathscr{C}$ itself. The category of binary systems contains the categories of semigroup and group objects as full subcategories. Moreover, each of these categories inherits from $\mathscr{C}$ both the nul object O and the product Π.

Of course all this can be dualized. For this purpose we assume that $\mathscr{C}$ has a nul object O, as before, but now a coproduct $\amalg$ rather than a product Π. This ensures that the preceding definitions can be applied in the case of the opposite category $\mathscr{C}^{\mathrm{op}}$. A cobinary system of $\mathscr{C}$ is defined to be a binary system of $\mathscr{C}^{\text{op}}$, and so forth. Note that if $\mathscr{C}$ is a category with nul object and product then hom(X,), for each object X of $\mathscr{C}$, constitutes a functor from the category of binary systems of $\mathscr{C}$ to the category of binary systems of $\mathscr{S}^*$. Also that if $\mathscr{C}$ is a category with nul object and coproduct then hom(, Y), for each object Y of $\mathscr{C}$, constitutes a functor from the category of cobinary systems of $\mathscr{C}$ to the category of binary systems of $\mathscr{S}^*$. When $\mathscr{C}$ is a category with nul object O, with product Π and with coproduct $\amalg$ we have

Proposition (1.40). *Let X be a unital cobinary system and let Y be a unital binary system. Then the binary operations $\amalg$ and Π on the set* hom(X, Y) *coincide; moreover they are both commutative and associative.*

To see this let $p_i: Y \Pi Y \to Y$ $(i = 1, 2)$ be the standard projections and let $s_j: X \to X \amalg X$ $(j = 1, 2)$ be the standard insertions. Given a morphism $h: X \amalg X \to Y \Pi Y$ write

$$h_{ij} = p_i \circ h \circ s_j: X \to Y \qquad (i, j = 1, 2).$$

Let $m: Y \Pi Y \to Y$ be the binary operation on Y and let $n: X \to X \amalg X$ be the cobinary operation on X. Then

$$mh = p_1 h \Pi p_2 h, \qquad hn = hs_1 \amalg hs_2.$$

Hence it follows at once that

$$(h_{11} \Pi h_{21}) \amalg (h_{12} \Pi h_{22}) = (mh)n = m(hn) = (h_{11} \amalg h_{12}) \Pi (h_{21} \amalg h_{22}).$$

The equality of the induced operations is obtained by putting $h_{12} = c = h_{21}$. Commutativity is obtained by putting $h_{11} = c = h_{22}$ and associativity by putting $h_{21} = c$.

Now let $\mathscr{C}$ be a category with final object O and let $\mathscr{C}^O = \text{hom}(O, \ )$ be the associated category of objects and morphisms under O. Assume that $\mathscr{C}$, and hence $\mathscr{C}^O$, admits a product functor Π. Since O is a nul object of $\mathscr{C}^O$ the category of group objects of $\mathscr{C}^O$ is defined, as above. We shall now construct the category G–$\mathscr{C}$ of G-objects and G-morphisms of $\mathscr{C}$, where G is a given group object of $\mathscr{C}^O$.

By an action of the group object G on an object X of $\mathscr{C}$ I mean a morphism

$$k: G \Pi X \to X$$

satisfying the following two conditions:

Condition (1.41). *The composition*

$$X \xrightarrow{\Delta} X \Pi X \xrightarrow{c \Pi \text{id}} G \Pi X \xrightarrow{k} X$$

is equal to the identity.

Condition (1.42). *The relation*

$$k(m \Pi \text{id}) = k(\text{id} \Pi k)$$

holds, as shown below.

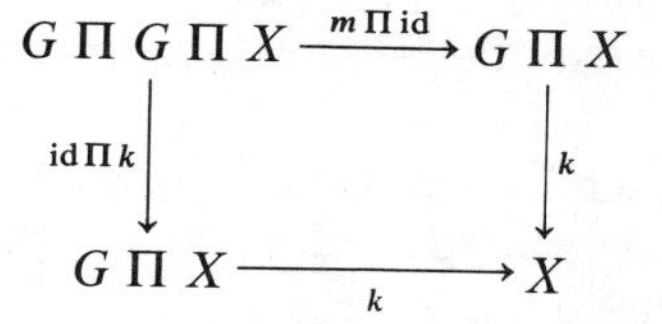

For example, take $X = G$ and take $k = m$. This is called the translation action of G on itself. For another example take $X = G$ again but take k to be the composition:

$$G \sqcap G \xrightarrow{\Delta \sqcap \mathrm{id}} G \sqcap G \sqcap G \xrightarrow{\mathrm{id} \sqcap t} G \sqcap G \sqcap G$$

$$\xrightarrow{\mathrm{id} \sqcap \mathrm{id} \sqcap v} G \sqcap G \sqcap G \xrightarrow{m \sqcap \mathrm{id}} G \sqcap G \xrightarrow{m} G;$$

here t denotes the switching equivalence and v the inversion of G.

In case $\mathscr{C} = \mathscr{S}$, the category of sets, a group object G is just a group in the ordinary sense and an action of G on a set X determines, and is determined by, a homomorphism of G into the group $\mathrm{aut}(X)$ of equivalences of X with itself.

Still in case $\mathscr{C} = \mathscr{S}$, other examples of a group G acting on a set arise in the case of a subgroup H of G. Then G acts by translation on the factor set G/H of cosets. If, further, H is normal in G then G also acts by conjugation on the factor group G/H.

An object X of $\mathscr{C}$ with an action of the group object G of $\mathscr{C}^O$ is called a G-object. Note that any object X of $\mathscr{C}$ may be regarded as a G-object by taking the action to be the second projection $G \sqcap X \to X$; this is known as the trivial action.

Definition (1.43). Let X, Y be G-objects of $\mathscr{C}$ with actions k, l, respectively. A morphism $\phi: X \to Y$ of $\mathscr{C}$ is a G-morphism (or is a morphism of G-objects) if $\phi \circ k = l \circ (\mathrm{id} \sqcap \phi)$, as shown below.

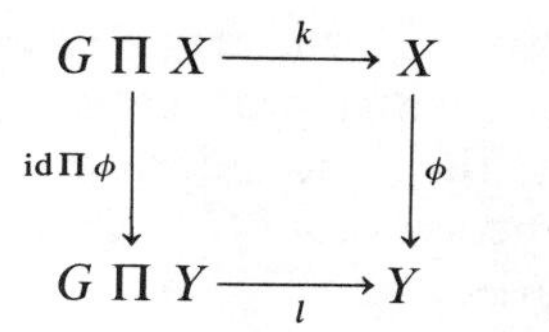

Thus the category G–$\mathscr{C}$ of G-objects and G-morphisms is defined. Note that if $\phi: X \to Y$ is a G-morphism then a left inverse $\psi: Y \to X$ of ϕ as a morphism of $\mathscr{C}$ is automatically a left inverse of ϕ as a G-morphism. In particular if ϕ is an equivalence as a morphism of $\mathscr{C}$ then ϕ is a G-equivalence.

Note that $\mathrm{hom}(A, \ )$, for each object A of $\mathscr{C}$, constitutes a functor from the category of G-objects of $\mathscr{C}$ to the category of $\mathrm{hom}(A, G)$-objects of $\mathscr{S}$.

The category G–$\mathscr{C}$ inherits many of the properties of the original category $\mathscr{C}$. For example G–$\mathscr{C}$ is concrete if $\mathscr{C}$ is concrete: the underlying set of the system (G, m) being the same as the underlying set of G. Again the category G–$\mathscr{C}$ inherits the final object O of $\mathscr{C}$, also the product Π of $\mathscr{C}$. Finally a coproduct $\amalg$ of $\mathscr{C}$, if it exists, determines a coproduct of G–$\mathscr{C}$.

Example (1.44). The G-set X determines a groupoid in which the objects are elements of X and, for elements x, y of X, the morphisms from x to y are the elements g of G such that $gx = y$.

In case $\mathscr{C} = \mathscr{S}$, the category of sets, one has two particularly important functors from G–$\mathscr{S}$ to $\mathscr{S}$. One is the fixed point functor, which assigns to each G-set X the subset $X^G \subset X$ of elements which are fixed under the action of G, and similarly with G-functions. The other is the orbit functor, which assigns to each G-set X the orbit set X/G, i.e. the quotient set of X with respect to the equivalence relation determined by the action of G, and similarly with G-functions. Both these functors play an important part in our work.

Of course one may proceed as above in the case of the associated category $\mathscr{C}^O$, identifying $(\mathscr{C}^O)^O$ with $\mathscr{C}^O$. In that case we obtain the category G–$\mathscr{C}^O$ of G-objects and G-maps over and under O. When $\mathscr{C} = \mathscr{S}$, in particular, this is just the category of pointed G-sets.

Notice that if G, G' are group objects of $\mathscr{C}$ then a homomorphism $\alpha: G' \to G$ determines a functor

$$\alpha_{\#}: G\text{–}\mathscr{C} \to G'\text{–}\mathscr{C}.$$

Specifically if $k: G \,\Pi\, X \to X$ is the action of G on the object X then $k(\alpha \,\Pi\, \mathrm{id}): G' \,\Pi\, X \to X$ is the action of G' on the object X.

Given an endofunctor Φ of $\mathscr{S}$ there is a standard procedure for converting an action of a group G on a set X into an action of G on ΦX. Essentially this is just the procedure which converts a J-indexed family of functions $f_j: X \to Y$ into the J-indexed family of functions $\Phi(f_j): \Phi X \to \Phi Y$. Specifically, if $g_{\#}$ is the action of the element g of G on the G-set X then the action of g on ΦX is defined to be $\Phi(g_{\#})$. The endofunctor of G–$\mathscr{S}$ thus defined can also be denoted by Φ. Note that Φ transforms trivial G-sets into trivial G-sets. Hence if X/G is regarded as a G-set ,with trivial action, so that the natural projection $X \to X/G$ is equivariant, then the transform $\Phi X \to \Phi(X/G)$, being also equivariant, induces a function $(\Phi X)/G \to \Phi(X/G)$. In general this function is neither injective nor surjective.

So far the actions of the group object G we have been discussing have been "on the left" of the object X. Actions "on the right" can also be discussed as follows. An action of the group object G on the right of the object X is a morphism $l: X \,\Pi\, G \to X$ satisfying the following two conditions.

Condition (1.45). *The composite*

$$X \xrightarrow{\Delta} X \,\Pi\, X \xrightarrow{\mathrm{id}\,\Pi\, c} X \,\Pi\, G \xrightarrow{l} X$$

is equal to the identity.

Condition (1.46). *The relation*

$$l(\mathrm{id} \sqcap m) = l(l \sqcap \mathrm{id})$$

holds, as shown below.

$$\begin{array}{ccc} X \sqcap G \sqcap G & \xrightarrow{\mathrm{id} \sqcap m} & X \sqcap G \\ {\scriptstyle l \sqcap \mathrm{id}} \downarrow & & \downarrow {\scriptstyle l} \\ X \sqcap G & \xrightarrow[l]{} & X \end{array}$$

Evidently an action k of G on the left of X determines an action l of G on the right of X, and vice versa, through the formula

(1.47) $$k \circ (v \sqcap \mathrm{id}) = l \circ t,$$

as shown below, where v is the inversion in the group object G.

$$\begin{array}{ccc} G \sqcap X & \xrightarrow{t} & X \sqcap G \\ {\scriptstyle v \sqcap \mathrm{id}} \downarrow & & \downarrow {\scriptstyle l} \\ G \sqcap X & \xrightarrow[k]{} & X \end{array}$$

This leads straight on to

Definition (1.48). Suppose that the group object G acts on the left of the object X while the group object H acts on the right of X. The left action k commutes with the right action l if $l(k \sqcap \mathrm{id}) = k(\mathrm{id} \sqcap l)$, as shown below.

$$\begin{array}{ccc} G \sqcap X \sqcap H & \xrightarrow{\mathrm{id} \sqcap l} & G \sqcap X \\ {\scriptstyle k \sqcap \mathrm{id}} \downarrow & & \downarrow {\scriptstyle k} \\ X \sqcap H & \xrightarrow[l]{} & X \end{array}$$

For example, take $\mathscr{C} = \mathscr{S}$, the category of sets. Suppose that the group G acts on the left of the set X while the group H acts on the right of X, and that the actions commute as in (1.48). Then there are induced actions of G on the left of the right orbit set X/H and of H on the right of the left orbit set $G\backslash X$, such that the identity of X induces an equivalence between $G\backslash(X/H)$ and $(G\backslash X)/H$.

Another construction we need to consider is the balanced or mixed product, which associates to each right G-set X and left G-set Y the set

$$X \times_G Y = (X \times Y)/G;$$

here the action of G on $X \times Y$ is given by

$$g(x, y) = (xg, g^{-1}y) \qquad (g \in G, x \in X, y \in Y).$$

The following three results are standard.

Proposition (1.49). *Let X, Y, Z be sets. Suppose that the group G acts on the right of X and the left of Y, while the group H acts on the right of Y and the left of Z. Also suppose that the action of G on the left of Y commutes with the action of H on the right of Y. Then G acts on the left of $Y \times_H Z$, and H acts on the right of $X \times_G Y$, so that*

$$(X \times_G Y) \times_H Z \equiv X \times_G (Y \times_H Z).$$

Proposition (1.50). *Suppose that the group G acts on the right of the set X and on the left of the set Y. Also suppose that the group H acts on the right of Y, and that the actions of G and H on Y commute with each other. Then there is an induced action of H on the right of $X \times_G Y$ such that*

$$(X \times_G Y)/H \equiv X \times_G (Y/H).$$

Proposition (1.51). *Let H be a subgroup of the group G. Then for each H-set X and G-set Y the H-functions of X into $\alpha^{\#} Y$ correspond precisely to the G-functions of $G/H \times_H X$ into Y, where $\alpha\colon H \subset G$.*

REFERENCES

S. MacLane. *Categories for the Working Mathematician.* Springer-Verlag, New York, 1971.

CHAPTER 2

The Axioms of Topology

In this chapter I have tried to summarize that part of general topology which is directly relevant to what we are going to do later. Of course there is hardly anything which cannot be found in the standard texts. However by selecting only the relevant material and omitting everything else I believe it is possible to give a clearer picture of this part of the subject, at least from the point of view I have adopted.

In topology the elements of the sets involved are always called points. Given a set X we shall be studying families of subsets of X satisfying various conditions. In principle such families are always indexed. However in many situations the indexing set can remain in the background and the notation used does not need to mention it. We shall be particularly concerned with families of subsets which are closed under the operations of union $\cup$ and/or intersection $\cap$. Here we distinguish between finite unions and unrestricted unions, and similarly with intersections. Formally the definitions are as follows.

Definition (2.1). Let X be a set and, for some indexing set J, let $\{X_j\}$ $(j \in J)$ be a family of subsets of X. The family is closed under the operation of union (resp. intersection) if for each subset $I \subset J$ the union $\bigcup_{j \in I} X_j$ (resp. the intersection $\bigcap_{j \in I} X_j$) is a member of the family. The family is closed under the operation of finite union (resp. finite intersection) if the appropriate condition is satisfied for each finite subset $I \subset J$.

Definition (2.2). Let $\{X_j\}$ $(j \in J)$ be a family of subsets of the set X. The family is point-finite if for each point $x \in X$ there exist at most a finite number of indices j such that $x \in X_j$.

Note that it is the indices which matter here, not the subsets as such: the same subset may well be given by different indices.

Definition (2.3). Let $\{X_j\}$ $(j \in J)$ be a family of subsets of the set X. The family covers X (or is a covering of X) if each point of X belongs to at least one member of the family. More generally the family covers a given subset A of X if each point of A belongs to at least one member of the family.

The following result, which we shall not be using, may be proved with the aid of Zorn's lemma.

Exercise (2.4). Let $\{X_j\}$ $(j \in J)$ be a point-finite covering of X. Then there exists an irreducible subcovering of X, i.e. a subcovering such that, whenever any member is removed, the set X is no longer covered.

Here subcovering means, of course, a subfamily which covers X.

After these preliminaries we are ready to make the fundamental definition of topology.

Definition (2.5). A topology for the set X is a family of subsets of X satisfying the following three conditions. First X and $\varnothing$ are members of the family: we call these the special subsets. Secondly the family is closed under the operation of (unrestricted) union. Thirdly the family is closed under the operation of finite intersection.

The members of such a family are called open sets: their complements are called closed sets. It should be emphasized that a subset may be both open and closed; for example, the special subsets have this property.

Of course the three axioms for a topology can equally well be stated in terms of closed rather than open sets, as follows. First the special subsets are closed. Secondly the intersections of closed sets are closed. Thirdly the finite unions of closed sets are closed.

A set with more than one point will admit more than one topology. We say that a topology Γ on X is a refinement of a topology Γ' on X if Γ' is a subfamily of Γ. In this situation we also say that Γ is finer than Γ' and that Γ' is coarser than Γ. These terms are even used (somewhat improperly) to include the case when $\Gamma = \Gamma'$

Definition (2.6). The discrete topology on the set X is the topology in which each subset of X is an open set (and hence each subset is a closed set). The indiscrete topology on X is the topology in which only the special subsets are open (and hence only the special subsets are closed).

Thus the discrete topology is the finest possible while the indiscrete topology is the coarsest possible.

Definition (2.7). Let Γ be a family of subsets of the set X. The topology generated by Γ is the coarsest topology in which all the members of Γ are open sets.

In this situation Γ is called a subbasis for the topology. Contrary to standard practice we do not insist that Γ should cover X. The open sets of the topology are (unrestricted) unions of finite intersections of members of Γ, supplemented by the special subsets. When Γ satisfies the following condition the open sets are unions of members of Γ, supplemented by the special subsets, it is unnecessary to take finite intersections first.

Condition (2.8). *The intersection of two (and hence of any finite number of) members of Γ is a union of members of Γ.*

This is called the basis condition; when it is satisfied Γ is said to be a basis for the topology which it generates.

A set X together with a topology Γ is called a topological space, or simply a space. Usually X alone is sufficient notation. We have already mentioned discrete and indiscrete spaces. Metric spaces provide an important class of spaces where the topology is given by a basis. Thus let X be a metric space with metric d. For each point x of X and positive real number ε the open ball $B_\varepsilon(x)$ of radius ε around x consists of the points $x' \in X$ such that $d(x, x') < \varepsilon$. The family of open balls, for all points $x \in X$ and all $\varepsilon > 0$, satisfies the basis condition and thus defines a topology on X, called the metric topology. Of course this basis is not the only one for the metric topology: another consists of the open balls $B_\varepsilon(x)$ for $x \in X$ and rational $\varepsilon > 0$.

Exercise (2.9). Let X be a set with metric d given by $d(x, x') = 1$ whenever $x \neq x'$. Show that the metric topology determined by d is the discrete topology. Can the indiscrete topology be obtained in this way with a suitable choice of metric?

Let X be a space. By a neighbourhood of a point x of X we mean an open set containing x. Further, by a neighbourhood of a subset A of X we mean an open set containing A. Thus X is a neighbourhood of each of its points and of each of its subsets. If X has the indiscrete topology then X is the only neighbourhood of any point or of any non-empty subset. If X is a metric space then $B_\varepsilon(x)$, for $\varepsilon > 0$, is a neighbourhood of the point x in the metric topology.

Definition (2.10). Let X be a space and let A be a subset of X. The interior $\mathscr{I}nt\ A$ of A is the union of all the open sets contained in A. The closure $\mathscr{C}l\ A$ of A is the intersection of all the closed sets containing A.

Thus we may describe $\mathscr{Int}\, A$ as the largest open set contained in A and $\mathscr{Cl}\, A$ as the smallest closed set containing A. Note that

$$X - \mathscr{Int}\, A = \mathscr{Cl}\,(X - A).$$

Exercise (2.11). Let A be a subset of the space X. Show that $x \in \mathscr{Cl}\, A$ if and only if each neighbourhood of x meets A. Also show that $\mathscr{Cl}(\mathscr{Cl}\, A) = \mathscr{Cl}\, A$.

The closure operator $\mathscr{Cl}$, in the space X, assigns to each subset X' of X a subset $\mathscr{Cl}\, X'$ of X so as to satisfy four conditions:

(i) $X' \subset \mathscr{Cl}\, X'$, for each subset X';
(ii) $\mathscr{Cl}\, \varnothing = \varnothing$;
(iii) $\mathscr{Cl}(X_1 \cup X_2) = \mathscr{Cl}\, X_1 \cup \mathscr{Cl}\, X_2$, for each pair of subsets X_1, X_2;
(iv) $\mathscr{Cl}(\mathscr{Cl}\, X') = \mathscr{Cl}\, X'$, for each subset X'.

Suppose that we are given an operator $\mathscr{Cl}_*: \mathscr{P}(X) \to \mathscr{P}(X)$ satisfying these four conditions. Then $\mathscr{Cl}_*$ determines a topology on X in which a set X' is closed if and only if $\mathscr{Cl}_*\, X' = X'$. Moreover the given operator is then the closure operator in this topology. Thus another approach to the basic definition of topology is to begin with an operator satisfying the four conditions: this is called the Kuratowski approach.

Definition (2.12). Let $\{X_j\}$ $(j \in J)$ be a family of subsets of the space X. The family is locally finite if each point of X has a neighbourhood U such that U meets X_j for at most a finite number of indices j.

Obviously locally finite implies point-finite but the converse is false. For example, the family of one-point subsets of a space X is always point finite but not, in general, locally finite (e.g. take X to be the real numbers $\mathbb{R}$, with the standard topology). A family may be locally finite even though each of its members meets infinitely many other members. In the discrete space N of natural numbers the family $\{A_n\}$ $(n = 0, 1, \ldots)$, where A_n consists of the integers greater than n, provides an illustration of this.

Proposition (2.13). *Let $\{X_j\}$ $(j \in J)$ be a locally finite family of subsets of the space X. Then the family $\{\mathscr{Cl}\, X_j\}$ of closures is also locally finite, and the union $\bigcup_{j \in J} \mathscr{Cl}\, X_j$ is closed in X.*

The first assertion is almost obvious, since if U is a neighbourhood which does not meet X_j then U does not meet $\mathscr{Cl}\, X_j$. To prove the second assertion, write $X' = \bigcup_{j \in J} \mathscr{Cl}\, X_j$. For each point $x \notin X'$ there exists a neighborhood U meeting at most finitely many of the $\mathscr{Cl}\, X_j$. The intersection of the complements of these $\mathscr{Cl}\, X_j$ with U is a neighbourhood of x which does not meet X'. Therefore X' is closed, as asserted.

Next, let $\phi: X \to Y$ be a function, where X and Y are sets. For each subset X' of X the direct image $\phi X'$ is defined as the subset of points $y \in Y$ such that

$y = \phi x$ for some point $x \in X'$. For each subset Y' of Y the inverse image $\phi^{-1}Y'$ is defined as the subset of points $x \in X$ such that $\phi x \in Y'$. The inverse images of the points of Y are often called the fibres of ϕ. The relations

$$X' \subset \phi^{-1}\phi X', \qquad \phi\phi^{-1}Y' \subset Y'$$

always hold. If ϕ is surjective then $Y' = \phi\phi^{-1}Y'$ for all Y' while if ϕ is injective then $X' = \phi^{-1}\phi X'$ for all X'. We refer to $\phi^{-1}\phi X'$ as the saturation of X' and denote this by sat X'. Subsets X' of X which satisfy the condition sat $X' = X'$ are called saturated subsets.

The inverse image behaves well in relation to the operations of union and intersection. Specifically if $\{Y_j\}$ $(j \in J)$ is a family of subsets of Y then

(2.14)
$$\left.\begin{aligned} &\text{(a)}\quad \phi^{-1}\Big(\bigcup_{j \in J} Y_j\Big) = \bigcup_{j \in J} (\phi^{-1}Y_j), \\ &\text{(b)}\quad \phi^{-1}\Big(\bigcap_{j \in J} Y_j\Big) = \bigcap_{j \in J} (\phi^{-1}Y_j). \end{aligned}\right\}$$

The direct image behaves well with respect to the operation of union but not so well with respect to the operation of intersection. Specifically if $\{X_j\}$ $(j \in J)$ is a family of subsets of X then

(2.15)
$$\left.\begin{aligned} &\text{(a)}\quad \phi\Big(\bigcup_{j \in J} X_j\Big) = \bigcup_{j \in J} (\phi X_j), \\ &\text{(b)}\quad \phi\Big(\bigcap_{j \in J} X_j\Big) \subset \bigcap_{j \in J} (\phi X_j). \end{aligned}\right\}$$

The failure of equality to hold in (2.15(b)) is a point of fundamental importance. There is, however, one situation in which the direct image of an intersection is the intersection of the direct images, namely when at most one of the sets fails to be saturated. We shall be using this observation frequently.

As we have remarked in the previous chapter, there are various ways of constructing a category in which the objects are spaces, depending on the morphisms chosen. Undoubtedly the most important of these is that where the morphisms are continuous functions.

Definition (2.16). Let $\phi: X \to Y$ be a function, where X and Y are spaces. Then ϕ is continuous if and only if for each open set V of Y the inverse image $\phi^{-1}V$ is an open set of X. Equivalently, ϕ is continuous if for each closed set F of Y the inverse image $\phi^{-1}F$ is a closed set of X.

Every function $\phi: X \to Y$ is continuous if either X is discrete or Y is indiscrete. In particular the functions $\varnothing \to X$ and $X \to *$ are continuous for every space X.

Clearly the identity function id_X on a space X is continuous. Also if X, Y, Z are spaces and $\phi: X \to Y$, $\psi: Y \to Z$ are continuous functions then the composition $\psi\phi: X \to Z$ is continuous. Thus a category is defined in which the objects are spaces and the morphisms are continuous functions. We denote this category by $\mathscr{T}$ and, in context, usually refer to continuous functions simply as maps.

Equivalences in the category $\mathscr{T}$ are called homeomorphisms; spaces which are equivalent in this sense are said to be homeomorphic. A property of spaces which, if it holds for one space, must also hold for all homeomorphic spaces, is said to be a topological property.

It is important to appreciate that a continuous bijection $\phi: X \to Y$ is not necessarily a homeomorphism; the inverse function $\phi^{-1}: Y \to X$ may fail to be continuous. To illustrate this point take Y to have the same underlying set as X but a coarser topology. Then the identity function is a continuous bijection but not a homeomorphism.

An example of a topological property is that of being a homogeneous space.

Definition (2.17). The space X is homogeneous if for each pair x, x' of points of X there exists a homeomorphism $X \to X$ which sends x into x'.

For example discrete spaces and indiscrete spaces are homogeneous. Also the euclidean space $\mathbb{R}^n$ ($n = 0, 1, \ldots$) is homogeneous. For examples of non-homogeneous spaces take X to be a finite set with the topology generated by some, but not all, of the one-point subsets.

Proposition (2.18). *Let $\phi: X \to Y$ be a function, where X and Y are spaces. Suppose that the topology of Y is generated by a family Γ of subsets of Y. Then ϕ is continuous if (and only if) $\phi^{-1}V$ is open in X for each member V of Γ.*

This follows at once from (2.14) above. It applies, in particular, when Γ satisfies the basis condition. This is useful, for example, in relation to functions with values in a metric space.

Recall that a space is connected if it contains no subset other than the special subsets which is both open and closed.

Proposition (2.19). *Either the space X is connected or there exists a continuous surjection $\phi: X \to D$, where D is the two-point discrete space.*

It is convenient, and entails no real loss of generality, to take $D = \{0, 1\} \subset \mathbb{R}$. Either X is connected or X contains a non-special open and closed set U, say. Given U, define ϕ by $\phi(U) = 0$, $\phi(X - U) = 1$. Given ϕ, define $U = \phi^{-1}(0)$.

Corollary (2.20). *The image of a connected space under a continuous surjection is connected.*

Recall that a subset X of the real numbers $\mathbb{R}$ is an interval if, whenever $x, y \in X$, with $x < y$, then $z \in X$ for any $z \in \mathbb{R}$ such that $x < z < y$. A basic result of real number theory is that intervals are connected. For let $\phi: X \to D$ be continuous. Let $x, y \in X$, with $x < y$, and suppose that $\phi x = 0$. Consider the real number

$$s = \sup\{z \mid x \leq z \leq y \text{ and } \phi z = 0\}.$$

It is easy to show that $\phi s = 0$ and that a contradiction arises if $s < y$. Thus $\phi y = 0$ and so $\phi x = \phi y$. Hence ϕ is constant, hence not surjective, and so X is connected.

Continuity is relatively easy to handle, due to the relations (2.14), since the definition is in terms of inverse images. Our next definition is in terms of direct images, and quite different properties emerge.

Definition (2.21). Let $\phi: X \to Y$ be a function, where X and Y are spaces. If the direct image of each open set of X is an open set of Y then ϕ is open. If the direct image of each closed set of X is a closed set of Y then ϕ is closed.

The first point to notice is that the condition for a function to be open is not equivalent to the condition for a function to be closed, whereas for continuity the definition in terms of open sets is equivalent to the definition in terms of closed sets.

Clearly every function $\phi: X \to Y$ is both open and closed if Y is discrete. In practice the conditions in (2.21) are usually imposed on top of continuity, when we use the terms open map or closed map, as the case may be. Note that for each space X the maps $\varnothing \to X$ and $X \to *$ are both open and closed.

Exercise (2.22). Let $\phi: X \to Y$ be a function, where X and Y are spaces. Show that ϕ is a closed map if, and only if, the condition

$$\phi(\mathscr{Cl}\, X') = \mathscr{Cl}(\phi X')$$

holds for each subset X' of X.

Evidently the identity map id_X on a space X is both open and closed. Also if X, Y, Z are spaces and $\phi: X \to Y$, $\psi: Y \to Z$ are open (resp. closed) maps then the composition $\psi\phi: X \to Z$ is an open (resp. closed) map. Thus subcategories of $\mathscr{T}$ can be defined in which the morphisms are (i) open maps or (ii) closed maps or (iii) maps which are both open and closed. In each case the equivalences in the subcategory are the same as those in $\mathscr{T}$ itself.

Exercise (2.23). Let $\phi: \mathbb{R} \to \mathbb{R}$ be a polynomial map. Then ϕ is closed.

Proposition (2.24). *Let $\phi: X \to Y$ be a function where X and Y are spaces. Let Γ be a basis for the topology of X. Then ϕ is open if (and only if) the direct image ϕU is open in Y for each member U of Γ.*

The proof is obvious. It must be emphasized, however, that it is not sufficient for Γ to be a subbasis for the topology, since one would again encounter the problem of dealing with direct images of intersections. For the same reason there is no counterpart of (2.24) for closed, rather than open, functions. However when the function is injective we have

Proposition (2.25). *Let $\phi: X \to Y$ be an injection, where X and Y are spaces. Let Γ be a subbasis for the topology of X. Then ϕ is open (resp. closed) if and only if the direct image of each member of Γ is open (resp. of the complement of each member of Γ is closed).*

Proposition (2.26). *Let X, Y, Z be spaces. Let $\phi: X \to Y$ and $\psi: Y \to Z$ be continuous functions such that the composition $\psi\phi: X \to Z$ is open (resp. closed). If ψ is injective then ϕ is open (resp. closed). If ϕ is surjective then ψ is open (resp. closed).*

I leave the proofs of these last two propositions to serve as exercises.

Proposition (2.27). *Let $\phi: X \to Y$ be a function, where X and Y are spaces. Suppose that for each point y of Y and neighbourhood U of $\phi^{-1}y$ there exists a neighbourhood V of y such that $\phi^{-1}V \subset U$. Then ϕ is closed. Conversely, suppose that ϕ is closed. Then for each subset H of Y and neighbourhood U of $\phi^{-1}H$ there exists a neighbourhood V of H such that $\phi^{-1}V \subset U$.*

The first part is obvious. To prove the second part, take V to be $Y - \phi(X - U)$. Then $H \subset V$ since $\phi^{-1}H \subset U$, and V is open, since ϕ is closed. Since

$$\phi^{-1}V = X - \phi^{-1}(\phi(X - U)) \subset X - (X - U) \subset U$$

the conclusion follows at once. Taking H to be a point we see that (2.27) provides a characterization of closed functions; we shall be using this quite frequently.

Initial and final topologies play an important part in what is to follow. These topologies arise with reference to a function $\phi: X \to Y$, where X and Y are sets, If Y, on the one hand, is given a topology then the initial topology for X is the coarsest topology which makes ϕ continuous. If X, on the other hand, is given a topology then the final topology for Y is the finest topology which makes ϕ continuous. More explicitly the definitions are as follows.

Definition (2.28). Let $\phi: X \to Y$ be a function, where X is a set and Y is a space. The initial topology on X, determined by ϕ, is the topology in which the open (resp. closed) sets of X are the inverse images of the open (resp. closed) sets of Y.

With the initial topology one sees that for each space X' and function $\xi: X' \to X$ the continuity of ξ is implied by the continuity of $\phi\xi$. Moreover this condition characterizes the initial topology.

Definition (2.29). Let $\phi: X \to Y$ be a function, where X is a space and Y is a set. The final topology on Y, determined by ϕ, is the topology in which the open (resp. closed) sets of Y are the direct images of the saturated open (resp. closed) sets of X.

Another way to express this is to say that a subset of Y is open (resp. closed) if and only if its inverse image in X is open (resp. closed). With the final topology one sees that for each space Y' and function $\eta: Y \to Y'$ the continuity of η is implied by the continuity of $\eta\phi$. Moreover this condition characterizes the final topology.

Note that the discrete topology on the set X is the final topology determined by the function $\varnothing \to X$, while the indiscrete topology is the initial topology determined by the function $X \to *$.

Both initial and final topologies satisfy a transitivity condition. Thus let X, Y, Z be sets and let $\phi: X \to Y$, $\psi: Y \to Z$ be functions. Suppose, on the one hand, that Z has a topology. Then the initial topology on X determined by $\psi\phi$ coincides with the result of first giving Y the initial topology determined by ψ and then giving X the initial topology determined by ϕ. Suppose, on the other hand, that X has a topology. Then the final topology on Z determined by $\psi\phi$ coincides with the result of first giving Y the final topology determined by ϕ and then giving Z the final topology determined by ψ.

However it is necessary to exercise care when mixing initial and final topologies. The following result is relevant.

Proposition (2.30). *Let X, Z be spaces and let Y be a set. Let $\phi: X \to Y$ be a surjection and let $\psi: Y \to Z$ be an injection. Suppose that the composition $\psi\phi: X \to Z$ is continuous and either open or closed. Then the final topology on Y determined by ϕ coincides with the initial topology on Y determined by ψ.*

For if $W \subset Z$ is open then $\phi^{-1}\psi^{-1}W \subset X$ is open, by continuity of $\psi\phi$. Conversely if $\phi^{-1}V$ is open, where $V \subset Y$, then $V = \psi^{-1}\psi V$ and $\psi V = \psi\phi\phi^{-1}V$ is open, provided $\psi\phi$ is open. The argument when $\psi\phi$ is closed is the same except that closed sets are used instead of open sets throughout.

Special terminology is used in case the function $\phi: X \to Y$ is injective or surjective. If ϕ is injective the initial topology on X is called the induced topology; if X has this topology then ϕ is called an embedding. If ϕ is surjective then the final topology on Y is called the quotient topology; if Y has this topology then ϕ is called a quotient map. Embeddings and quotient maps are not necessarily either open or closed. However an open or closed injection is necessarily an embedding, while an open or closed surjection is necessarily a quotient map. Notice also that if $\phi: X \to Y$ is a map which admits a left inverse then ϕ is an embedding while if ϕ admits a right inverse then ϕ is a quotient map.

Let X be a space and let X' be a subset of X with inclusion $u: X' \to X$. In the induced topology for X' the open (resp. closed) sets of X' are the intersections with X' of the open (resp. closed) sets of X. If X' is open (resp. closed) in X then the embedding u is open (resp. closed). In general, however, one must be careful to distinguish between subsets of X' which are open (resp. closed) in X' and subsets of X' which are open (resp. closed) in X; the latter include the former but not vice versa. In what follows it is always assumed, unless the contrary is stated, that subsets of a space receive the induced topology. The terms subset and subspace are therefore used interchangeably. Note that, by transitivity of the initial topology, if X' is a subspace of X and X'' is a subspace of X' then X'' is a subspace of X.

One of the main uses of the quotient topology is when we have an equivalence relation $\sim$ on the space X. In that case the set $X/\sim$ of equivalence classes, with the quotient topology determined by the natural projection, is called the quotient space of X with respect to $\sim$. In general the natural projection is neither open nor closed. Transitivity holds for quotient maps, of course, but one must exercise care in situations where the induced topology and the quotient topology occur together. Thus suppose that X' is a subset of X and $\sim$ is an equivalence relation on X which determines, by restriction, an equivalence relation on X'. The result of giving X' the induced topology and then giving $X'/\sim$ the quotient topology is not, in general, the same as the result of giving $X/\sim$ the quotient topology and then giving $X'/\sim$ the induced topology.

We shall also encounter multiple forms of the initial and final topologies but there are some other results which need to be considered first.

Proposition (2.31). *Let $\phi: X \to Y$ be a function, where X and Y are spaces. If ϕ is continuous then so is the function $X' \to \phi X'$ determined by ϕ for each subset $X' \subset X$. If ϕ is open or closed then so is the function $\phi^{-1}Y' \to Y'$ determined by ϕ for each subset $Y' \subset Y$.*

The proof is left as an exercise. We now seek some results which go in the other direction. For these it is necessary to consider coverings of certain kinds. In practice these are usually open coverings or finite closed coverings. However the coverings to which the results apply are more general than these.

Proposition (2.32). *Let $\{X_j\}$ be a family of subsets of the space X such that either*

(i) *$\{\mathcal{I}nt\, X_j\}$ is an open covering of X, or*
(ii) *$\{X_j\}$ is a locally finite closed covering of X.*

Then a subset A of X is open (resp. closed) if and only if $A \cap X_j$ is open (resp. closed) in X_j for each index j.

Proposition (2.33). *Let $\{X_j\}$ be a family of subsets of the space X such that either*

(i) *$\{\mathscr{Int}\, X_j\}$ is an open covering of X, or*
(ii) *$\{X_j\}$ is a locally finite closed covering of X.*

Then a function $\phi: X \to Y$, for any space Y, is continuous if the restriction $\phi \mid X_j$ is continuous for each index j.

Proposition (2.34). *Let $\{Y_j\}$ be a family of subsets of the space Y such that either*

(i) *$\{\mathscr{Int}\, Y_j\}$ is an open covering of Y, or*
(ii) *$\{Y_j\}$ is a locally finite closed covering of Y.*

Then a function $\phi: X \to Y$, for any space X, is open (resp. closed) if the function $\phi^{-1} Y_j \to Y_j$ determined by ϕ is open (resp. closed) for each index j.

Let us digress somewhat to prove a few more results about connectedness which will be useful later, beginning with

Example (2.35). Let E be a connected subspace of the space X. Then the closure $\mathscr{Cl}\, E$ is also connected. More generally F is connected for any set F such that $E \subset F \subset \mathscr{Cl}\, E$.

For let $\phi: F \to D$ be a continuous function, where D is discrete. Since E is connected and since $\phi \mid E$ is continuous it follows at once that $\phi \mid E$ is constant with value $a \in D$, say. Suppose, to obtain a contradiction, that $\phi(x) \neq a$ for some $x \in F$. Since $\phi(x)$ is open in D the subset $\phi^{-1}\phi(x)$ is open in F. Therefore $\phi^{-1}\phi(x) = F \cap U$ for some open U of X. Now $x \in U$ and $x \in \mathscr{Cl}\, E$. Hence there exists a point x', say, of $E \cap U$. Since $x' \in E$ we have $\phi(x') = a$. But this gives a contradiction, since $x' \in E \cap U \subset F \cap U = \phi^{-1}\phi(x)$ and $\phi(x) \neq a$. This proves the result.

Proposition (2.36). *Let $\{X_j\}$ $(j \in J)$ be a covering of the space X by connected subsets X_j. Suppose that every member of the covering meets every other member. Then X is connected.*

For let $\phi: X \to D$ be a continuous function, where D is discrete. Then $\phi \mid X_j$ is continuous and so constant at $a_j \in D$, say. But $a_j = a_k$ for each pair of indices j, k, since X_j meets X_k. Therefore ϕ is constant and the conclusion follows.

These last two results show that for each connected subset A of the space X there exists a maximal connected subset containing A, i.e. the union of all the connected subsets which contain A. This maximal subset is called the (connectedness) component containing A. The components containing the individual points of X are called the components of X, Evidently these are closed in X. Moreover two components either coincide or are disjoint. Hence the family of components constitutes a partition of X. In the space $\mathbb{Q}$ of rationals, for example, the components are just the individual points, even though $\mathbb{Q}$ does not have the discrete topology.

Exercise (2.37). Let $\phi: X \to Y$ be continuous. Then the image of each component of X is contained in a component of Y.

There is a local form of connectedness which is also important.

Definition (2.38). The space X is locally connected if the connected open sets of X form a basis for the topology of X.

In other words the condition is that for each point x of X and each neighbourhood V of x there exists a connected neighbourhood U of x such that $U \subset V$. Discrete spaces are locally connected, for example. On the other hand the rationals $\mathbb{Q}$ and the irrationals $\mathbb{R} - \mathbb{Q}$ are not locally connected.

Exercise (2.39). Let $\phi: X \to Y$ be a continuous open surjection. Show that Y is connected and locally connected if X is connected and locally connected.

Clearly any open subset of a locally connected space is again locally connected.

Exercise (2.40). The space X is locally connected if and only if the components of each open set of X are also open.

Exercise (2.41). Let $\phi: X \to Y$ be a quotient map. If X is locally connected then Y is locally connected.

If A is a subspace of the space X a left inverse $X \to A$ of the inclusion $A \to X$ is called a retraction. We say that A is a retract of X if there exists a retraction $X \to A$. Obviously every space is a retract of itself. Also each point of a space is a retract of that space. On the other hand the point-pair $\dot{I}$ is not a retract of the unit interval I, for reasons of connectivity.

Exercise (2.42). Let $J^{n-1} \subset \dot{I}^n$ be one face of the n-cube I^n. Show that $\dot{I}^n - \mathscr{I}nt\, J^{n-1}$ is a retract of I^n.

A subspace A of a space X is said to be a neighbourhood retract if there exists a neighbourhood U of A in X of which A is a retract. Obviously this is so when A is a retract of X. On the other hand the subspace

$$\mathbb{R}_0 = \{0\} \cup \left\{\frac{1}{n} \mid n = 1, 2, \ldots\right\}$$

of the real line $\mathbb{R}$ is not a neighbourhood retract. For suppose, to obtain a contradiction, that there exists a neighbourhood U of $\mathbb{R}_0$ in $\mathbb{R}$ and a retraction $r: U \to A$. Then $[0, \varepsilon) \subset U$, for some $\varepsilon > 0$. Now $r(x) = 0$ for $0 \leq x < \varepsilon$, since $[0, \varepsilon]$ is connected while the component of 0 in $\mathbb{R}_0$ is just 0. Taking $x = 1/n$, where $n > \varepsilon^{-1}$, we obtain a contradiction.

Next let us turn to the subject of topological sums and products. In practice it is sums and products of just a pair of spaces which one is dealing with most of the time. We shall therefore mainly consider these; the general case is very similar.

Let X_1, X_2 be spaces. At the set-theoretic level the sum $X_1 + X_2$ has been defined in the previous chapter, together with the standard insertions

$$X_1 \underset{\sigma_1}{\rightarrow} X_1 + X_2 \underset{\sigma_2}{\leftarrow} X_2.$$

Note that the image of σ_1 does not meet the image of σ_2. An arbitrary subset of $X_1 + X_2$ is of the form $A_1 + A_2$, where A_i is a subset of X_i $(i = 1, 2)$. We topologize $X_1 + X_2$ so that the open (resp. closed) sets are the sets $A_1 + A_2$, where A_i is an open (resp. closed) subset of X_i $(i = 1, 2)$. With this topology each of the insertions σ_i is an open and closed embedding. Moreover for any space X a function

$$\phi: X_1 + X_2 \to X$$

is continuous if and only if each of the functions $\phi\sigma_i$ is continuous. Thus the conditions for a coproduct in the category $\mathscr{T}$ are satisfied. Of course one may regard the procedure used here as an example of a multiple form of the final topology.

The topological product is not quite so straight-forward. At the set-theoretic level the product $X_1 \times X_2$ has been defined in the previous chapter together with the standard projections

$$X_1 \underset{\pi_1}{\leftarrow} X_1 \times X_2 \underset{\pi_2}{\rightarrow} X_2.$$

However it is by no means true that every subset of $X_1 \times X_2$ is of the form $A_1 \times A_2$, where A_i is a subset of X_i $(i = 1, 2)$. Let us give $X_1 \times X_2$ the coarsest topology which makes each of the functions π_i continuous. This is the topology generated by the family of subsets of the form $U_1 \times X_2$ or $X_1 \times U_2$, where U_i is open in X_i $(i = 1, 2)$. It is also the topology generated by the basis consisting of subsets of the form

$$(U_1 \times X_2) \cap (X_1 \times U_2) = U_1 \times U_2,$$

where U_i is open in X_i $(i = 1, 2)$. (Of course this does not mean that each open set of $X_1 \times X_2$ is of this form). For any space X a function $\phi: X \to X_1 \times X_2$ is continuous if and only if each of the functions $\pi_i\phi$ is continuous. Thus the conditions for a product in the category $\mathscr{T}$ are satisfied.

With this product topology, as it is called, the topological product $X_1 \times X_2$ is defined. Clearly each projection π_i is an open map, but it is not in general true that each projection is a closed map. For example, take

$X_1 = X_2 = \mathbb{R}$, the real numbers. Consider the closed subset H of the real plane $\mathbb{R} \times \mathbb{R}$ consisting of the points (x_1, x_2) such that $x_1 x_2 = 1$. The image of H is the subset $\mathbb{R} - \{0\}$ of $\mathbb{R}$, which is not closed.

In dealing with the product topology one tries, as far as possible, to avoid returning to the definition all the time; instead one uses its characteristic properties, as in

Proposition (2.43). *Let X and Y be spaces and let $\phi: X \to Y$ be continuous. Then the function*

$$\Gamma_\phi: X \to X \times Y,$$

given by $\Gamma_\phi(x) = (x, \phi x)$, is an embedding.

The function Γ_ϕ is known as the graph function and its direct image is called the graph of ϕ. Evidently Γ_ϕ maps bijectively onto the graph, and is continuous. Also an inverse $\Gamma_\phi X \to X$ is given by the first projection, so that Γ_ϕ maps homeomorphically onto the graph, as asserted.

Proposition (2.44). *For all spaces X_1, X_2 and Y,*

$$(X_1 \times Y) + (X_2 \times Y) \equiv (X_1 + X_2) \times Y.$$

Here we use the symbol $\equiv$, as in the previous chapter, to denote an obvious equivalence. In this case the map is given by $\sigma_1 \times \mathrm{id}_Y$ on the first summand and by $\sigma_2 \times \mathrm{id}_Y$ on the second. The map is clearly bijective and open, since $\sigma_1 \times \mathrm{id}_Y$ and $\sigma_2 \times \mathrm{id}_Y$ are open, and hence is a homeomorphism. We may refer to (2.44) as the distributive law.

The definitions of topological sum and product in the general case are on similar lines. Thus if $\{X_j\}$ $(j \in J)$ is a collection of spaces then the topological sum

$$\coprod_{j \in J} X_j$$

is the set-theoretic sum with the topology in which a subset

$$\coprod_{j \in J} A_j$$

is open (resp. closed) if and only if each A_j is open (resp. closed) in X_j. The distributive law (2.44) generalizes in the obvious way. Similarly the topological product

$$\prod_{j \in J} X_j$$

is defined to be the set-theoretic product with the topology generated by the family of subsets

$$\prod_{j \in J} A_j,$$

where A_j is open in X_j for all values of j and where $A_j \neq X_j$ for at most one value of j. Thus a basis for the topology consists of the subsets

$$\prod_{j \in J} A_j$$

where A_j is open in X_j for all values of j and where $A_j \neq X_j$ for at most a finite number of values of j. It is important to observe that

$$\prod_{j \in J} A_j$$

is not necessarily open without this last restriction. This means, for example, that the product of an infinite collection of discrete spaces is not, as a rule, discrete.

The functorial behaviour of the topological product may be summarized as follows. Let

$$\xi_j \colon X_j \to Y_j \qquad (j \in J)$$

be a collection of maps, where X_j and Y_j are spaces. Then the product map

$$\prod_{j \in J} \xi_j \colon \prod_{j \in J} X_j \to \prod_{j \in J} Y_j$$

is defined, and is open whenever each ξ_j is open. Moreover the product is an embedding (resp. closed embedding) when each of the ξ_j is an embedding (resp. closed embedding).

Pull-backs present no difficulty in our category. Thus let

$$X \xrightarrow{p} B \xleftarrow{q} Y$$

be a triad. At the set-theoretic level the pull-back $X \times_B Y$ is defined as the subset of $X \times Y$ consisting of pairs (x, y) such that $px = qy$. We give $X \times Y$ the product topology, of course, and then give $X \times_B Y$ the induced topology. The projections

$$X \leftarrow X \times_B Y \rightarrow Y$$

are then continuous and the diagram shown below is cartesian in $\mathscr{T}$.

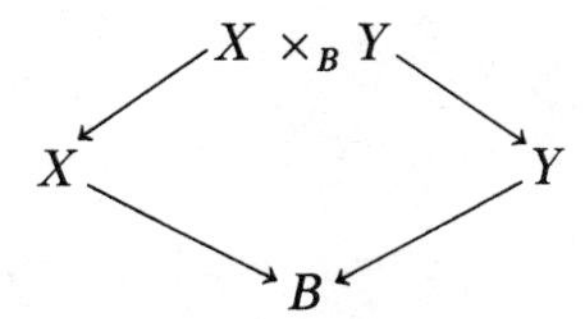

Thus the cotriad constitutes the pull-back of the given triad.

For the push-out, let

$$X \xleftarrow{u} A \xrightarrow{v} Y$$

be a cotriad where one of the maps u, v (say u) is injective. At the set-theoretic level the push-out $X +^A Y$ is defined as a quotient set of $X + Y$. We give

$X + Y$ the sum topology and then $X +^A Y$ the quotient topology. Then the insertions

$$X \to X +^A Y \leftarrow Y$$

are continuous, and the triad thus formed constitutes the push-out of the given triad in $\mathscr{T}$, so that the diagram shown below is cocartesian.

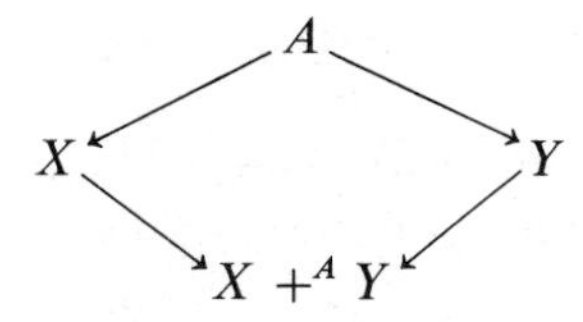

Proposition (2.45). *Suppose that u is an open (resp. closed) embedding and that v is an open (resp. closed) map. Then the quotient map $X + Y \to X +^A Y$ is open (resp. closed).*

The saturation of a subset $X' + Y'$ of $X + Y$ is the union of $X' + Y'$ with

$$uv^{-1}Y' + vu^{-1}X'$$

Suppose that u and v are open. If X' and Y' are open then $u^{-1}X'$ and $v^{-1}Y'$ are open, by continuity, and so $uv^{-1}Y'$ and $vu^{-1}X'$ are open. Hence the saturation of $X' + Y'$ is open. This proves (2.45) in the open case; the other case is similar.

In case A is a subset of X, and u the inclusion, the push-out is sometimes called "the space obtained from X by attaching Y by means of v".

Returning to the general case, it should be noted that if v is surjective, as well as u injective, then the push-out of the cotriad

$$X \overset{u}{\leftarrow} A \overset{v}{\rightarrow} Y$$

may be regarded as a quotient space of X.

In case X and Y are both subsets of a space Z there is a continuous bijection to $X \cup Y$ from the push-out of the cotriad

$$X \leftarrow X \cap Y \rightarrow Y$$

of inclusions. The bijection is a homeomorphism if, and only if, $X \cap Y$ is closed in Z.

Finally a few words about the functorial behaviour of push-outs. Suppose that we have a commutative diagram as shown below, where u and u' are injective.

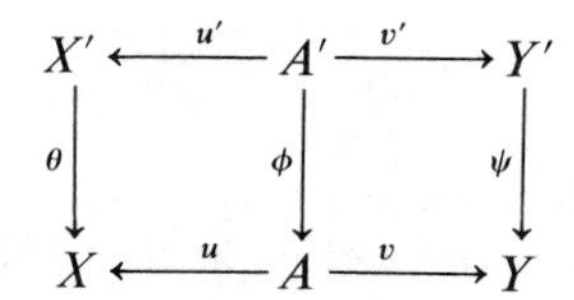

Then θ, ϕ, ψ together determine a map ξ making the following diagram commutative.

$$\begin{array}{ccc} X' + Y' & \xrightarrow{\pi'} & X' +^{A'} Y' \\ {\scriptstyle \theta+\psi}\downarrow & & \downarrow{\scriptstyle \xi} \\ X + Y & \xrightarrow[\pi]{} & X +^{A} Y \end{array}$$

Proposition (2.46). *Suppose that u is an open (resp. closed) embedding, and that θ, ψ and v are open (resp. closed) maps. Then ξ is an open (resp. closed) map.*

For (to take the "open" alternative first) if θ, ψ are open then so is $\theta + \psi$. If π is also open then so is $\pi(\theta + \psi) = \xi\pi'$. Also π' is surjective and so ξ is open. The "closed" alternative is similar. Suppose moreover that θ and ψ are quotient maps. Then $\theta + \psi$ is a quotient map, hence $\pi(\theta + \psi) = \xi\pi'$ is a quotient map, and so ξ is a quotient map.

The category $\mathscr{T}$ admits some interesting and important endofunctors. We discuss these chiefly in the unary case, since the generalization to the multiple case is straightforward. The endofunctors which are most useful for our purposes are those which satisfy the following continuity condition.

Condition (2.47). *The endofunctor Φ is continuous if for each map $f\colon B \times X \to Y$ the corresponding function $\hat{f}\colon B \times \Phi X \to \Phi Y$ is continuous, where $\hat{f}_b = \Phi(f_b)$ $(b \in B)$.*

In the first instance we allow B, X, Y to be any spaces and f to be any map. Later, however, we shall need to impose some restrictions on the spaces concerned, although not on the maps.

For example, the functor $T\times$ is continuous for each space T. However, consider the following endofunctor Φ, which plays an important part in our work. Let T be a space and let T_0 be a subset of T. For each space X define ΦX to be the push-out of the cotriad

$$T \times X \leftarrow T_0 \times X \rightarrow T_0$$

and similarly for maps. Thus ΦX is a quotient space of $T \times X + T_0$, although for non-empty X we may regard ΦX as a quotient space of $T \times X$ alone. In general this functor Φ is not continuous. However if the pair (T, T_0) satisfies certain conditions, to be considered later in this chapter, the continuity condition is satisfied. Fortunately these conditions are satisfied in the cases which concern us most, namely when $T = I = [0, 1] \subset \mathbb{R}$ and either $T_0 = \{0\}$ or $T_0 = \dot{I} = \{0, 1\}$. For the pair $(I, \{0\})$ the functor Φ is called the cone, and denoted by Γ. For the pair $(I, \dot{I})$ the functor Φ is called the suspension and denoted by Σ.

There is, however, an alternative way to proceed, based on the use of initial rather than final topologies. Returning to the general case, let ΦX denote the push-out set of the cotriad

$$T \times X \leftarrow T_0 \times X \rightarrow T_0.$$

Then ΦX comes equipped with two "coordinate" functions. The first is the function $\Phi X \rightarrow \Phi * = T$, given by $(t, x) \mapsto t$; we denote this by t. The second is the function $t^{-1}(T - T_0) \rightarrow X$, given by the second projection; we denote this by x. We now give the push-out set the coarsest topology which makes these two coordinate functions continuous; in other words the topology generated by the family of subsets $t^{-1}U$ for all open $U \subset T$, and by the family of subsets $x^{-1}V$ for all open $V \subset X$. In general this topology is coarser than the one previously used, but the endofunctor is now continuous.

As it happens the example which is most important in our work is that of a bifunctor, called the join and denoted by $*$. This too exists in a fine version and a coarse version. Thus let X and Y be spaces. In the fine version $X * Y$ is defined as the quotient space of

$$I \times X \times Y + X + Y$$

with respect to the relations

$$(0, x, y) \sim x, \qquad (1, x, y) \sim y.$$

With this topology the join operation is commutative, in the sense that $X * Y$ is homeomorphic to $Y * X$ under the transformation given by $(t, x, y) \mapsto (1 - t, y, x)$. However the join operation is not associative with the fine topology. Notice that the cone ΓX can be identified with $X * \{0\}$ and the suspension ΣX with $X * \{0, 1\}$, still using the fine topology of course.

Exercise (2.48). Show that if $\phi_i\colon X_i \rightarrow Y_i$ $(i = 0, 1)$ is a closed embedding, then so is

$$\phi_0 * \phi_1 : X_0 * X_1 \rightarrow Y_0 * Y_1,$$

in the fine topology.

At the set theoretic level the set $X * Y$ comes equipped with three coordinate functions:

$$t: X * Y \rightarrow I, \qquad x: t^{-1}[0, 1) \rightarrow X, \qquad y: t^{-1}(0, 1] \rightarrow Y,$$

defined in the obvious way. In the coarse version we give $X * Y$ the coarsest topology which makes these three functions continuous. In other words $X * Y$ has the topology generated by three families of subsets; the inverse images under t of the open sets of I, the inverse images under x of the open sets of X, and the inverse images under y of the open sets of Y.

Proposition (2.49). *For all spaces X and Y there is a natural homeomorphism*

$$(\Gamma(X * Y), X * Y) \rightarrow (\Gamma X \times \Gamma Y, \Gamma X \times Y \cup X \times \Gamma Y)$$

in the coarse topology.

Here the transformation from domain to codomain is given by

$$(s, (t, x, y)) \mapsto ((S, x), (T, y)),$$

where $s, t \in I$, $x \in X$, $y \in Y$ and where

$$S = s, \qquad T = 1 - 2t(1 - s) \qquad (t \leq \tfrac{1}{2}),$$

$$S = 1 - 2(1 - s)(1 - t), \qquad T = s \qquad (t \geq \tfrac{1}{2});$$

while the transformation from codomain to domain is given by

$$((S, x), (T, y)) \mapsto (s, (t, x, y)),$$

where $S, T \in I$, $x \in X$, $y \in Y$ and where

$$s = S; \qquad t = (1 - T)/2(1 - S) \qquad (S \leq T),$$

$$s = T, \qquad t = 1 - (1 - S)/2(1 - T) \qquad (S \geq T).$$

These transformations are easily seen to be continuous by checking the coordinate functions.

Proposition (2.50). *The join operation, with coarse topology, is associative. Specifically there is a natural homeomorphism*

$$(X * Y) * Z \equiv X * (Y * Z)$$

for all spaces X, Y, Z.

I leave the proof of (2.50) to serve as an exercise. This result enables us to suppress brackets when discussing multiple coarse joins.

Returning now to the basic theory let us recall that the composition of open maps is open and the composition of closed maps is closed. Also that the sum of open maps is open and the sum of closed maps is closed. However the product of closed maps is not necessarily closed, although the product of open maps is open. Putting it another way we may say that the category of open maps admits a coproduct and a product, while the category of closed maps admits a coproduct but not a product.

What we are going to do next is to study the subcategory of $\mathscr{T}$ consisting of spaces and maps satisfying the following stronger condition.

Definition (2.51). The map $\phi: X \to Y$ is compact if the map

$$\phi \times \mathrm{id}_T : X \times T \to Y \times T$$

is closed for each space T.

The usual term is proper map rather than compact map but it will emerge in due course why I depart from this. It follows at once from the definition and what we have already proved that the category of spaces and compact maps admits both coproduct $+$ and product $\times$, and that equivalences in $\mathscr{T}$ are also equivalences in the subcategory. The following results are straight forward consequences of the corresponding results for closed maps.

Proposition (2.52). *The continuous injection $\phi: X \to Y$ is compact if and only if ϕ is a closed embedding.*

Proposition (2.53). *Let $\phi: X \to Y$ and $\psi: Y \to Z$ be maps such that $\psi\phi: X \to Z$ is compact. Then ϕ is compact if ψ is injective and ψ is compact if ϕ is surjective.*

Proposition (2.54). *Let $\{X_i\}$ be a finite covering of the space X. Let $\phi: X \to Y$ be a map such that $\phi | X_i$ is compact for each index i. Then ϕ is compact.*

Proposition (2.55). *Let $\phi: X \to Y$ be a compact map. Then the map $\phi': \phi^{-1}Y' \to Y'$ determined by ϕ is compact for each subset Y' of Y.*

Proposition (2.56). *Let $\phi: X \to Y$ be a map. Let $\{Y_j\}$ $(j \in J)$ be a family of subsets of Y such that either*

(i) *the family $\{\mathscr{I}nt\, Y_j\}$ covers Y, or*
(ii) *the family $\{Y_j\}$ is a locally finite closed covering of Y.*

Suppose that the map $\phi_j: \phi^{-1}Y_j \to Y_j$ determined by ϕ is compact for each index j. Then ϕ is compact.

Obviously the map $\varnothing \to X$ is compact for all spaces X. However compactness of the map $X \to *$ imposes a condition on X. When X is discrete, for for example, the map is compact if and only if X is finite. This leads us to

Definition (2.57). The space X is compact if the map $X \to *$ is compact.

From this definition, and what we have already proved, it follows at once that a closed subspace of a compact space is compact, that the continuous image of a compact space is compact, and that the sum and product of two compact spaces are compact. However we need to establish the equivalence of (2.57) with the more familiar

Definition (2.58). The space X is compact if each open covering of X admits a finite subcovering.

First suppose that X satisfies the condition in (2.58). Consider the projection $\pi: X \times T \to T$, for any space T. Let E be a closed set of $X \times T$ and let t be a point of T such that $t \notin \pi E$. Since the sets $X \times \{t\}$ and E are disjoint there exists, for each point $x \in X$, a neighbourhood $U_x \times V_x$ of (x, t) which does not meet E. Since $X \times \{t\}$ is homeomorphic to X we can extract from the open covering $\{U_x \times V_x\}$ $(x \in X)$ of $X \times \{t\}$ a finite subcovering. Then the intersection of the factors $\{V_x\}$ from the subcovering is a neighbourhood of t which does not meet πE. Therefore $T - \pi E$ is open and so πE is closed. Thus X satisfies the condition in (2.57).

Conversely suppose that X satisfies the condition in (2.57). Let $\{U_j\}$ $(j \in J)$ be an open covering of X. We need to establish the existence of a finite subcovering. With no real loss of generality we may suppose that $\{U_j\}$ is closed under finite unions, so that what we have to establish is that $U_j = X$ for some j. So suppose, to obtain a contradiction, that $U_j \neq X$ for all $j \in J$.

Consider the set $X' = X + *$ with the topology given by a basis consisting of (i) the complements $X' - U_j$ in X' of the U_j and (ii) the intersections $U \cap (X - U_j)$ for all open sets U of X and each of the U_j. It is easy to check that the basis is closed under finite intersection. Also X is not closed in X' since no $U_j = X$. There is no suggestion that the inclusion $X \to X'$ is continuous.

Consider next the graph D of the inclusion, as a subset of $X \times X'$, and write $\mathscr{Cl}\ D = C$. Since $\pi: X \times X' \to X'$ is closed, by assumption, we have $\pi C = \mathscr{Cl}\ \pi D = \mathscr{Cl}\ X$, by (2.22). Now $\mathscr{Cl}\ X \neq X$, since X is not closed, and so πC contains $*$. Therefore there exists a point x of X such that $(x, *) \in C$.

We complete the proof by showing that this point x does not lie in any of the U_j. For if U is a neighbourhood of x in X then the product $U \times (X' - U_j)$ is a neighbourhood of $(x, *)$ in $X \times X'$. Since $C = \mathscr{Cl}\ D$ and $(x, *) \in C$ this neighbourhood meets D. Therefore U meets $X' - U_j$ and so meets $X - U_j$. Since this is true for each neighbourhood U of x we conclude that $x \notin \mathscr{Int}\ U_j = U_j$. This establishes a contradiction and completes the proof that the two definitions of compactness are equivalent.

One of the fundamental results of the subject is that any bounded closed interval in $\mathbb{R}$, and hence any bounded closed set in $\mathbb{R}^n$, is compact. It is sufficient to prove this in the case $I = [0, 1]$.

Proposition (2.59). *The unit interval $I \subset \mathbb{R}$ is compact.*

To prove (2.59), let Γ be an open covering of I. For each point $x \in I$ we can choose an open interval U_x containing x, such that U_x is contained in a member of Γ. These open intervals form an open covering $\{U_x\}$ $(x \in I)$ of I which refines Γ. If we can extract a finite subcovering from $\{U_x\}$ then we can extract a finite subcovering from Γ. So we may assume, without real loss of generality, that each member of Γ is an open interval.

Let $\phi: I \to \dot{I}$ be the function such that $\phi(x) = 0$ if $[0, x]$ can be covered by a finite subfamily of Γ, and $\phi(x) = 1$ otherwise. Then $\phi(0) = 0$, and we need to show $\phi(1) = 0$. We begin by showing that ϕ is constant on each member U of Γ.

If $\phi(x) = 0$ for some point x of U then $[0, x]$ can be covered by a finite subfamily of Γ. Hence $[0, y]$ can be covered by a finite subfamily of Γ for each point y of U (i.e. the previous subfamily augmented by U). So either $\phi = 0$ throughout U or $\phi = 1$ throughout U. Thus ϕ is constant, and hence continuous, on each member of Γ, and so is continuous on I. Hence ϕ is constant on I, since I is connected. Since $\phi(0) = 0$ the constant is 0, hence $\phi(1) = 0$ and (2.59) is proved.

It is convenient to slip in at this stage a more technical result we shall need later.

Proposition (2.60). *Let C be a compact subset of the space X and let D be a compact subset of the space Y. Let W be a neighbourhood of $C \times D$ in $X \times Y$. Then neighbourhoods U of C in X and V of D in Y exist such that $U \times V \subset W$.*

For let x, y be arbitrary points of C, D, respectively. There exist neighbourhoods $U_{x,y}$ of x in X and $V_{x,y}$ of y in Y such that $U_{x,y} \times V_{x,y} \subset W$. For fixed x and varying $y \in D$ the family $\{V_{x,y}\}$ forms an open covering of the compact D. Extract a finite subcovering indexed by $y_1, \ldots, y_n$, say. Then the intersection

$$U_x = U_{x,y_1} \cap \cdots \cap U_{x,y_n}$$

is a neighbourhood of x, and the union

$$V_x = V_{x,y_1} \cup \cdots \cup V_{x,y_n}$$

is a neighbourhood of D such that $U_x \times V_x \subset W$. Now for varying $x \in C$ the sets U_x form an open covering of the compact C. Extract a finite subcovering indexed by $x_1, \ldots, x_m$, say. Then the union

$$U = U_{x_1} \cup \cdots \cup U_{x_m}$$

is a neighbourhood of C and the intersection

$$V = V_{x_1} \cap \cdots \cap V_{x_m}$$

is a neighbourhood of D such that $U \times V \subset W$, as required.

Returning now to the subject of compact maps, the following provides a useful criterion in practice.

Proposition (2.61). *The closed map $\phi: X \to Y$ is compact if and only if the fibre $\phi^{-1}y$ is compact for each point $y \in Y$.*

For suppose that the fibres are compact. Consider the product

$$\phi \times \mathrm{id}_T : X \times T \to Y \times T$$

for any space T. Let (y, t) be a point of $Y \times T$ and let W be a neighbourhood of $\phi^{-1}y \times \{t\}$ in $X \times T$. For each point $x \in \phi^{-1}y$ there exist neighbourhoods U_x of x in X and V_x of t in T such that $U_x \times V_x \subset W$. From the open covering $\{U_x\}$ $(x \in \phi^{-1}y)$ of the compact $\phi^{-1}y$ we can extract a finite subcovering indexed by $x_1, \ldots, x_m$, say. Then the union

$$U = U_{x_1} \cup \cdots \cup U_{x_m}$$

is a neighbourhood of $\phi^{-1}y$ in X. Since ϕ is closed there exists, by (2.27), a neighbourhood N of y in Y such that $\phi^{-1}N \subset U$. Moreover the intersection

$$V = V_{x_1} \cap \cdots \cap V_{x_m}$$

is a neighbourhood of t in T. Thus $N \times V$ is a neighbourhood of (y, t) in $Y \times T$ such that $(\phi \times \mathrm{id}_T)^{-1}(N \times V) \subset \phi^{-1}N \times V \subset U \times V \subset W$. Hence $\phi \times \mathrm{id}_T$ is closed and so ϕ is compact. The converse is an immediate consequence of (2.55) and so the proof of (2.61) is complete.

Example (2.62). The euclidean norm

$$D: \mathbb{R}^n \to \mathbb{R} \qquad (n = 0, 1, \ldots)$$

is a compact map.

For the fibre over the point $t > 0$ is the sphere $S_t(0) \subset \mathbb{R}^n$, which is closed and bounded, and so compact. To show that D is a closed map, let U be a neighbourhood of $S_t(0)$. For each point x of $S_t(0)$ there exists an $\varepsilon_x > 0$ such that the open ball $B_{\varepsilon_x}(x)$ is contained in U. Extract a finite subcovering of this covering of $S_t(0)$, indexed by $x_1, \ldots, x_n$, say. Then $\varepsilon > 0$, where

$$\varepsilon = \min(\varepsilon_{x_1}, \ldots, \varepsilon_{x_n}),$$

and so $(t - \varepsilon, t + \varepsilon)$ is a neighbourhood of t such that $D^{-1}(t - \varepsilon, t + \varepsilon) \subset U$. Thus D is closed, by (2.27), and so compact.

Next let us discuss push-outs and similar constructions. Consider a cotriad

$$X \overset{u}{\leftarrow} A \overset{v}{\rightarrow} Y$$

where, as usual, we suppose u to be injective.

Proposition (2.63). *Suppose that u and v are compact. Then the natural projection*

$$\pi: X + Y \to X +^A Y$$

is also compact.

For clearly the fibres of π are equivalent either to fibres of u or fibres of v, and so are compact. Also u and v are closed, hence π is closed by (2.45).

Definition (2.64). An endofunctor Φ of the category $\mathscr{T}$ is compact if Φ transforms compact spaces into compact spaces and compact maps into compact maps.

Evidently it is sufficient if the condition on maps is satisfied and if, in addition, the transform $\Phi *$ of the singleton $*$ is compact.

For example, the functor $T\times$ which transforms each space X into $T \times X$, and similarly for maps, is a compact functor whenever T is compact.

Proposition (2.65). *Let T_0 be a closed subspace of the compact space T. Let Φ be the endofunctor which transforms each space X into the push-out of the cotriad*

$$T \times X \leftarrow T_0 \times X \rightarrow T_0,$$

and similarly for maps. Then Φ is compact.

In particular, the suspension Σ and the cone Γ are compact. A similar argument shows that the join $*$ is a compact bifunctor, in the obvious sense.

To prove (2.65) we first observe that $\Phi *$ is equivalent to T, and so is compact. Let $\phi: X \to Y$ be a compact map, and consider the diagram below, where the horizontals are determined by ϕ, in the obvious way, and the verticals are quotient maps.

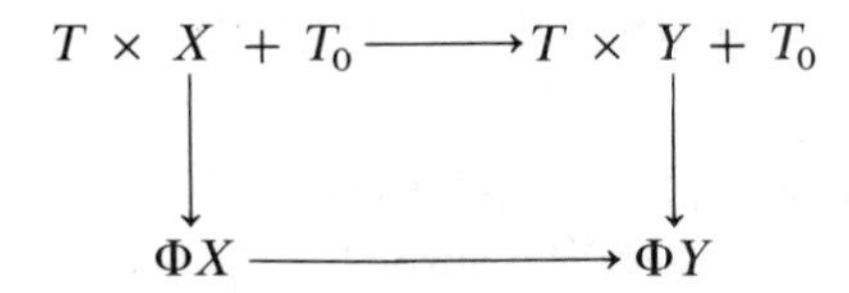

Now the verticals are surjective and closed, as we saw in (2.45). Also the top horizontal is a sum of closed maps and so is closed. Hence the bottom horizontal, which is $\Phi(\phi)$, is also closed. Finally the fibres of $\Phi(\phi)$ are either point-spaces or are products with T of the corresponding fibres of ϕ. Since the fibres of ϕ are compact so also are the fibres of $\Phi(\phi)$. Therefore $\Phi(\phi)$ is compact as asserted.

There is also a local form of compactness which plays an important part in our work.

Definition (2.66). The map $\phi: X \to Y$ is locally compact if for each point x of X there exists a neighbourhood U such that $\phi | \mathscr{Cl}\, U$ is compact. The space X is locally compact if the map $X \to *$ is locally compact, i.e. if each point x of X admits a neighbourhood U such that $\mathscr{Cl}\, U$ is compact.

For example, discrete spaces are locally compact, since in that case $U = \{x\}$ is a neighbourhood of x which meets the requirements. For another example, the euclidean spaces $\mathbb{R}^n$ $(n = 0, 1, \ldots)$ are locally compact, since the open ball of radius 1, say, around each point is a neighbourhood with compact closure: the closed ball with the same radius.

The basic results for local compactness are analogous to those for compactness. Instead of writing them out at length we shall simply refer to them, when necessary, as (e.g.) the local version of (2.53).

Subsets E, F of a space X are said to be separated if there exist neighbourhoods U of E and V of F which are disjoint. Of the various separation axioms in the literature the ones we shall be using most are the Hausdorff axiom and the regularity axiom.

Definition (2.67). The space X is Hausdorff if for each distinct pair (x, x') of points of X there exist neighbourhoods U of x and U' of x' which are disjoint.

For example, discrete spaces are Hausdorff, since each point is a neighbourhood of itself. Metric spaces provide another example. For if $d(x, x') = \varepsilon$, where d is the metric, then $\varepsilon > 0$, since $x \neq x'$, and we can take $U = B_\eta(x)$, $U' = B_\eta(x')$, where $\eta = \varepsilon/2$.

Clearly every subset of a Hausdorff space is also Hausdorff. Also the sum and product of two Hausdorff spaces are Hausdorff.

An alternative way to characterize Hausdorff spaces is provided by

Proposition (2.68). *The space X is Hausdorff if and only if the diagonal embedding*

$$\Delta: X \to X \times X$$

is closed.

For if Δ is closed then each point (x, x') of $X \times X - \Delta X$ has a neighbourhood W which does not meet ΔX and therefore has a neighbourhood $U \times U' \subset W$ which does not meet ΔX; then U does not meet U'. The argument in the reverse direction is similar.

Corollary (2.69). *Let Y be a Hausdorff space. Then for each space X and each pair of maps $\phi, \phi': X \to Y$ the points $x \in X$ such that $\phi x = \phi' x$ form a closed subset of X.*

For consider the respective diagonal maps:

$$X \xrightarrow{\Delta} X \times X \xrightarrow{\phi \times \phi'} Y \times Y \xleftarrow{\Delta} Y.$$

The subset of X we are concerned with is precisely $\Delta^{-1}(\phi \times \phi')^{-1}(\Delta Y)$. Since ΔY is closed, by (2.68), the conclusion follows immediately.

Corollary (2.70). *Let Y be a Hausdorff space. Then for each space X and each map $\phi: X \to Y$ the graph*

$$\Gamma_\phi: X \to X \times Y$$

is a closed embedding.

This follows similarly from consideration of the commutative diagram shown below.

$$\begin{array}{ccc} X & \xrightarrow{\Gamma} & X \times Y \\ \downarrow{\scriptstyle \phi} & & \downarrow{\scriptstyle \phi \times \mathrm{id}} \\ Y & \xrightarrow[\Delta]{} & Y \times Y \end{array}$$

For $\Gamma X = (\phi \times \mathrm{id})^{-1}\Delta Y$, and ΔY is closed.

Our next result shows that not only points but, more generally, compact sets can be separated in a Hausdorff space.

Proposition (2.71). *Let C, D be disjoint compact subsets of the Hausdorff space X. Then there exist disjoint neighbourhoods U of C and V of D.*

This follows at once from (2.60) with $X = Y$ and $W = X \times X - \Delta X$.

Proposition (2.72). *Let $\phi: X \to Y$ be a compact surjection. If X is Hausdorff then so is Y.*

For since ϕ is a compact surjection so is $\phi \times \phi$, in the diagram shown below.

$$\begin{array}{ccc} X & \xrightarrow{\Delta} & X \times X \\ {\scriptstyle\phi}\downarrow & & \downarrow{\scriptstyle\phi\times\phi} \\ Y & \xrightarrow[\Delta]{} & Y \times Y \end{array}$$

By (2.68) the diagonal ΔX is closed, since X is Hausdorff, hence $(\phi \times \phi)\Delta X = \Delta\phi X$ is closed. But $\Delta\phi X = \Delta Y$, since ϕ is surjective, and so Y is Hausdorff, by (2.68) again.

Proposition (2.73). *Let $\phi: X \to Y$ be a map, where X is compact and Y is Hausdorff. Then ϕ is compact.*

For $\phi = \pi_2 \Gamma_\phi$, as shown below, where Γ_ϕ is the graph function:

$$X \xrightarrow{\Gamma_\phi} X \times Y \xrightarrow{\pi_2} Y.$$

Since both π_2 and Γ_ϕ are compact, so is ϕ.

It follows, in particular, that compact subspaces of a Hausdorff space are closed; for example points are closed in a Hausdorff space. Thus in a compact Hausdorff space the compact subspaces are precisely the closed subspaces. We also have

Example (2.74). Let X be a subset of $\mathbb{R}^n$, where $n \geq 0$. Then X is compact if and only if X is closed and bounded.

This result assures us of a plentiful supply of compact spaces. We have already observed, in relation to (2.59), that closed and bounded subsets of $\mathbb{R}^n$ are compact. To prove the converse, suppose that X is compact. Then X is closed, since $\mathbb{R}^n$ is Hausdorff. Consider, for some point $x \in X$, the covering of X by open balls $B_\varepsilon(x)$ for all $\varepsilon > 0$. Extract a finite subcovering indexed by $\varepsilon_1, \ldots, \varepsilon_n$, say. The finite set $\{\varepsilon_1, \ldots, \varepsilon_n\}$ is bounded by δ, say, and so X itself is bounded by 2δ.

Proposition (2.75). *Let T_0 be a closed subset of the Hausdorff space T. Then for each Hausdorff space X the push-out of the cotriad*

$$T \times X \leftarrow T_0 \times X \rightarrow T_0$$

is also a Hausdorff space.

We prove this for the coarse topology since that implies the result for any finer topology. The coarse topology is generated by the inverse images of the open sets of T under the coordinate function t and the inverse images of the open sets of X under the coordinate function x. Since both these functions separate points the result follows at once.

This shows that if X is a Hausdorff space then so are the suspension ΣX and the cone ΓX. A similar argument shows that the join $X * Y$ of Hausdorff spaces is Hausdorff in the coarse topology and hence in any finer topology.

Taken together (2.65) and (2.75) show that if T_0 is a closed subset of the compact Hausdorff space T then the push-out of the cotriad

$$T \times X \leftarrow T_0 \times X \rightarrow T_0$$

is compact Hausdorff for each compact Hausdorff space X. Moreover it follows, using (2.73), that the coarse and the fine topologies for the push-out then coincide. In particular this is true for the cone and the suspension of a compact Hausdorff space and, more generally, for the join of compact Hausdorff spaces.

Example (2.76). The cone $\Gamma(S^{n-1})$ on the sphere S^{n-1} is homeomorphic to the closed ball B^n, while the suspension $\Sigma(S^{n-1})$ is homeomorphic to the sphere S^n.

Points of B^n may be represented in the form tx, where $x \in S^{n-1}$ and $0 \leq t \leq 1$. Thus a map $\lambda: I \times S^{n-1} \rightarrow B^n$ is given by $\lambda(t, x) = tx$, and λ induces a continuous bijection $\Gamma(S^{n-1}) \rightarrow B^n$. Since $\Gamma(S^{n-1})$ is compact and B^n is Hausdorff the bijection is a homeomorphism. The argument in the case of $\Sigma(S^{n-1})$, based on the use of polar coordinates, is similar and will be left to serve as an exercise. Similarly it can be shown that the join $S^{p-1} * S^{q-1}$ of S^{p-1} and S^{q-1} is homeomorphic to S^{p+q-1}.

The other separation axiom we shall be using is that of regularity. This is neither weaker nor stronger than the Hausdorff axiom.

Definition (2.77). The space X is regular if for each point x of X and each neighbourhood V of x there exists a neighbourhood U of x such that $\mathscr{Cl}\, U \subset V$.

It is important to notice that if the condition holds for all V in a subbasis for the topology of X then it holds for all V and so X is regular.

The condition for regularity can of course be reformulated in terms of the complement of the original neighbourhood V. Then the condition is that for each point x of X and each closed set E which does not contain x there exist neighbourhoods U of x and W of E which are disjoint.

Compact Hausdorff spaces are regular, by (2.71). Also indiscrete spaces are regular. For if X is indiscrete the only neighbourhood of a point is X itself, and $\mathscr{Cl}\ X = X$. Since indiscrete spaces are not, in general, Hausdorff this shows that regularity does not imply the Hausdorff condition. Nor is the opposite implication true. For example consider the set $\mathbb{R}$ of real numbers, with the topology generated by the open intervals together with the set $\mathbb{Q}$ of rationals. This is finer than the metric topology and so is Hausdorff. However $\mathbb{Q}$ is a neighbourhood of unity, in this topology, but there exists no neighbourhood U of 1 such that $\mathscr{Cl}\ U \subset \mathbb{Q}$. Thus $\mathbb{R}$, with this topology, is non-regular.

Incidentally this non-regular topology for $\mathbb{R}$ enables us to give an example where the product of quotient maps fails to be a quotient map. For let $\mathbb{P}$ be the space obtained from $\mathbb{R}$, with non-regular topology, by collapsing the closed set $\mathbb{R} - \mathbb{Q}$ to the point $*$. Obviously $*$ and $\pi(1)$ cannot be separated, and so the diagonal $\Delta\mathbb{P}$ is not closed in $\mathbb{P} \times \mathbb{P}$. However the inverse image

$$(\pi \times \pi)^{-1}(\Delta\mathbb{P}) = \Delta\mathbb{R} \cup ((\mathbb{R} - \mathbb{Q}) \times (\mathbb{R} - \mathbb{Q}))$$

is closed in $\mathbb{R} \times \mathbb{R}$, and so $\pi \times \pi$ is not a quotient map.

Metric spaces are regular. For if $B_\varepsilon(x)$ is an ε-neighbourhood of x, where $\varepsilon > 0$, then $B_{\varepsilon/2}(x)$ is neighbourhood of x with closure contained in $B_\varepsilon(x)$.

The points of a regular space X are not necessarily closed. However the closures of the one-point subsets of X form a partition of X and the projection of X onto the quotient space thus defined is both open and closed. Furthermore the quotient space is both Hausdorff and regular. The reader may care to check these statements as an exercise.

In general a subset of a regular space need not be regular. However closed subsets are regular, from the definition of the induced topology. Also the sum and product of two regular spaces are regular.

Next we prove what is generally known as the shrinking theorem, for regular spaces. We begin with a special case.

Proposition (2.78). *Let C be a compact subset of the regular space X, and let V be a neighbourhood of C in X. Then there exists a neighbourhood U of C in X such that $\mathscr{Cl}\ U \subset V$.*

Since V is a neighbourhood of each point x of C there exist, by regularity, disjoint neighbourhoods U_x of x and W_x of $X - V$. The open sets $\{U_x\}$ $(x \in C)$ cover the compact C; extract a finite subcovering indexed by $x_1, \ldots, x_m$, say. Then the union

$$U = U_{x_1} \cup \cdots \cup U_{x_m}$$

is a neighbourhood of C and the intersection

$$W = W_{x_1} \cap \cdots \cap W_{x_m}$$

is a neighbourhood of $X - V$. These neighbourhoods do not intersect and so we have $\mathscr{Cl}\ U \subset V$, as required. We at once generalize this to

Proposition (2.79). *Let C be a compact subset of the regular space X. Let $\{V_j\}$ $(j = 1, \ldots, n)$ be a finite open covering of C. Then there exists an open covering $\{U_j\}$ $(j = 1, \ldots, n)$ of C such that $\mathscr{Cl}\ U_j \subset V_j$ for each index j.*

The method of proof is step-by-step. Write $V = V_2 \cup \cdots \cup V_n$; then $X - V$ is closed in X. Hence $C \cap (X - V)$ is closed in C and so is compact. Applying the previous proposition to the neighbourhood V_1 of $C \cap (X - V)$, we obtain a neighbourhood U of $C \cap (X - V)$ such that $\mathscr{Cl}\ U \subset V_1$. Now $C \cap V$ and $C \cap (X - V)$ cover C, hence V and U cover C. Thus $U_1 = U$ is a shrinking of V_1. We continue by repeating the argument for $\{U_1, V_2, \ldots, V_n\}$, so as to shrink V_2, and so on. Thus the result is obtained.

Proposition (2.80). *Let $\phi: X \to Y$ be a compact surjection. If X is regular then so is Y.*

To see this, let V be a neighbourhood of a given point y of Y. Since $\phi^{-1}y$ is compact and X is regular there exists, by (2.78), a neighbourhood U of $\phi^{-1}y$ such that $\mathscr{Cl}\ U \subset \phi^{-1}V$. Since ϕ is closed there exists, by (2.27), a neighbourhood W of y such that $\phi^{-1}W \subset U$. Then $\mathscr{Cl}\ W \subset V$, which proves the result. We deduce

Corollary (2.81). *Let T_0 be a closed subset of the compact regular space T. Then for each compact regular space X the push-out of the cotriad*

$$T \times X \leftarrow T_0 \times X \rightarrow T_0$$

is compact regular.

In particular the suspension ΣX and the cone ΓX of a compact regular space X are compact regular. Similarly one can show that the join of two compact regular spaces is compact regular.

Proposition (2.82). *Let C be a compact subset of the locally compact regular space X, and let V be a neighbourhood of C in X. Then there exists a neighbourhood U of C in X such that $\mathscr{Cl}\ U \subset V$ and such that $\mathscr{Cl}\ U$ is compact.*

First take the special case where C is a point, say the point x, and V is a neighbourhood of x. Since X is locally compact there exists a neighbourhood bourhood W of x such that $\mathscr{Cl}\ W$ is compact. Since X is regular, so is $\mathscr{Cl}\ W$.

Now $V \cap \mathscr{Cl}\, W$ is a neighbourhood of x in $\mathscr{Cl}\, W$ and so there exists a neighbourhood N of x in $\mathscr{Cl}\, W$ such that, in $\mathscr{Cl}\, W$, $\mathscr{Cl}\, N \subset V \cap \mathscr{Cl}\, W$. So $N \cap W$ is a neighbourhood U of x in X such that $\mathscr{Cl}\, U \subset V$ and such that $\mathscr{Cl}\, U$ is compact, as required.

In the general case, therefore, each point x of C admits a neighbourhood U_x such that $\mathscr{Cl}\, U_x$ is compact and such that $\mathscr{Cl}\, U_x \subset V$. The sets $\{U_x\}$ $(x \in C)$ constitute an open covering of the compact C; extract a finite subcovering indexed by $x_1, \ldots, x_n$, say. Then the union

$$U = U_{x_1} \cup \cdots \cup U_{x_n}$$

is a neighbourhood of C such that $\mathscr{Cl}\, U \subset V$, as required.

We turn now to the problem of constructing an adjoint functor for the product functor in the category $\mathscr{T}$. As will be seen, certain separation and compactness conditions need to be imposed on the spaces concerned. Various ways to get round these restrictions have been proposed in the literature but for present purposes it is hardly necessary to go into these.

Consider, for spaces X and Y, the set map(X, Y) of maps $X \to Y$. For each pair of subsets $C \subset X$ and $V \subset Y$ let (C, V) denote the set of maps $\phi\colon X \to Y$ such that $\phi C \subset V$. With this notation we have the relation

(2.83)
$$\bigcap_{j \in J} (C_j, V) = \left(\bigcup_{j \in J} C, V\right)$$

for an indexed family $\{C_j\}$ $(j \in J)$ of subsets of X, and the relation

(2.84)
$$\bigcap_{j \in J} (C, V_j) = \left(C, \bigcap_{j \in J} V_j\right)$$

for an indexed family $\{V_j\}$ $(j \in J)$ of subsets of Y. These relations will be used later.

Definition (2.85). The compact-open topology for the set map(X, Y) is the topology generated by the family of subsets (C, V), for all compact subsets C of X and all open sets V of Y.

The subbasic sets (C, V) are called the compact-open sets; with this topology map(X, Y) is called the space of maps of X into Y.

If X is discrete, for example, then map(X, Y) is equivalent to the Xth power of Y, for all Y. In particular map$(*, Y)$ is equivalent to Y. If Y is indiscrete then so is map(X, Y), for all X.

Let X, Y, Z be spaces. From (2.53) it is clear that precomposition with a map $\theta\colon X \to Y$ determines a map

$$\theta^*\colon \text{map}(Y, Z) \to \text{map}(X, Z),$$

while postcomposition with a map $\phi: Y \to Z$ determines a map

$$\phi_*: \mathrm{map}(X, Y) \to \mathrm{map}(X, Z).$$

Proposition (2.86). (i) *If θ is a compact surjection then θ^* is an embedding.* (ii) *If ϕ is an embedding then ϕ_* is an embedding.*

(i) Let θ be a compact surjection. If C is a compact subset of Y then $\theta^{-1}C$ is a compact subset of X, by (2.55). Since θ^* is injective we have $(C, V) = \theta^{*-1}(\theta^{-1}C, V)$ for each open set V of Z. Since $(\theta^{-1}C, V)$ is open this proves that θ^* is an embedding.

(ii) Let ϕ be an embedding. Then ϕ_* is injective. Also for each compact-open subset (C, U) of map(X, Y) there exists an open set V of Z such that $U = \phi^{-1}V$ and so (C, V) is a compact-open subset of map(X, Z). Since $(C, U) = \phi_*^{-1}(C, V)$ this shows that ϕ_* is an embedding.

Proposition (2.87). *Let X_1, X_2 be spaces and let*

$$X_1 \xrightarrow{\sigma_1} X_1 + X_2 \xleftarrow{\sigma_2} X_2$$

be the standard insertions. Then the map

$$\mathrm{map}(X_1 + X_2, Y) \to \mathrm{map}(X_1, Y) \times \mathrm{map}(X_2, Y)$$

given by σ_1^*, σ_2^* is a homeomorphism for all spaces Y.

This follows at once from the definition of the compact-open topology since the compact subsets of $X_1 + X_2$ are the sums $C_1 + C_2$, where $C_i \subset X_i$ $(i = 1, 2)$ is compact.

Proposition (2.88). *If Y is regular then so is* map(X, Y) *for all spaces X.*

For let $\phi: X \to Y$ be a map, regarded as a point of map(X, Y), and let (C, V) be a compact-open neighbourhood of ϕ. Then V is a neighbourhood of the compact ϕC and so, by (2.78), there exists a neighbourhood U of ϕC such that $\mathscr{Cl}\ U \subset V$. Then $\phi \in (C, U)$ and $\mathscr{Cl}(C, U) \subset (C, V)$. To see the latter, let $\psi: X \to Y$ be a map such that ψ is not contained in (C, V). Then $(C \cap \psi^{-1}(Y - V), Y - \mathscr{Cl}\ U)$ is a neighbourhood of ψ which does not meet (C, U), whence the result.

Proposition (2.89). *If Y is Hausdorff then so is map(X, Y) for all spaces X.*

For let ϕ, $\phi': X \to Y$ be distinct maps. Then $\phi x \neq \phi' x$ for some point $x \in X$. Since Y is Hausdorff there exist neighbourhoods U of ϕx and U' of $\phi' x$ which are disjoint. Then (x, U) and (x, U') are neighbourhoods of ϕ and ϕ' which are disjoint.

In what follows we prefer to assume regularity rather than the Hausdorff property where there is a choice. This is because the regularity results generalize more satisfactorily to the fibrewise theory, as we shall see in the next chapter.

After these preliminaries we can begin to show that map(,) has the properties of an adjoint functor to $\times$. The first step is

Proposition (2.90). *Let X, Y, Z be spaces. If $h: X \times Y \to Z$ is continuous then so is the function $\hat{h}: X \to$ map(Y, Z) given by*

$$\hat{h}(x)(y) = h(x, y) \qquad (x \in X, y \in Y).$$

For let (C, V) be a compact-open subset of map(Y, Z) and let $x \in \hat{h}^{-1}(C, V)$. Then $h^{-1}V$ is a neighbourhood in $X \times Y$ of the compact subset $\{x\} \times C$. Hence by (2.27) there exists a neighbourhood U of x in X such that $U \times C \subset h^{-1}V$. Then $U \subset \hat{h}^{-1}(C, V)$ and so $\hat{h}$ is continuous.

Exercise (2.91). The function

$$\alpha: R^{n+1} \to \text{map}(R, R) \qquad (n = 0, 1, \ldots)$$

assigns to each $(n + 1)$-vector $(a_0, a_1, \ldots, a_n)$ the polynomial

$$a_0 + a_1 t + \cdots + a_n t^n$$

in the real variable t. Show that α is continuous.

Proposition (2.92). *Let X be regular. Let Y be a space with topology generated by a family Γ of subsets. Then the topology of* map(X, Y) *is generated by the family of compact-open subsets (C, V), where $C \subset X$ is compact and $V \in \Gamma$.*

In view of (2.84) we may assume, without real loss of generality, that Γ forms a basis for the topology of Y and not merely a subbasis. So let (C, V) be an arbitrary compact-open subset of map(X, Y), and let $\phi: X \to Y$ be a map such that $\phi \in (C, V)$. Then $\phi C \subset V$, so $\phi^{-1}V$ is a neighbourhood of the compact C. Now V is the union of a family $\{V_j\}$ $(j \in J)$ of basic open sets, so that the family $\{\phi^{-1}V_j\}$ of inverse images forms an open covering of C. Extract a finite subcovering indexed by $j_1, \ldots, j_n$, say. Since X is regular we can shrink this to an open covering $\{U_j\}$ $(j = j_1, \ldots, j_n)$ of C such that $\mathscr{Cl}\, U_j \subset V_j$ for each index j. Each of the sets $\phi^{-1}V_j$ is a neighbourhood of the corresponding set $C \cap \mathscr{Cl}\, U_j$, which is closed in C and so compact. Therefore each of the sets $(C \cap \mathscr{Cl}\, U_j, V_j)$ is a compact-open neighbourhood of ϕ. Since

$$\bigcap_{i=1}^{n} (C \cap \mathscr{Cl}\, U_{j_i}, V_{j_i}) \subset \left(C, \bigcup_{i=1}^{n} V_{j_i}\right) = (C, V)$$

this proves (2.92). As an immediate application we have

Proposition (2.93). *Let Y_1, Y_2 be spaces and let*

$$Y_1 \underset{\pi_1}{\leftarrow} Y_1 \times Y_2 \underset{\pi_2}{\rightarrow} Y_2$$

be the standard projections. Then the map

$$\text{map}(X, Y_1 \times Y_2) \to \text{map}(X, Y_1) \times \text{map}(X, Y_2)$$

given by $(\pi_1)_*$, $(\pi_2)_*$ is a homeomorphism for all regular spaces X.

The compact-open topology is by no means the only significant topology for spaces of maps, and there are many results which show that it is consistent with topologies defined in other ways, such as

Example (2.94). Let U, V be a real vector spaces of finite dimension. Then the euclidean topology on $\hom(U, V)$ agrees with the induced topology obtained from $\text{map}(U, V)$.

In view of the product theorem (2.93) and its counterpart in linear algebra it is sufficient to prove this in case $V = \mathbb{R}$. Choose an inner product $\langle\ ,\ \rangle$ on U and consider the isomorphism $\xi: U \to \hom(U, \mathbb{R})$ given by

$$\xi(x)(y) = \langle x, y\rangle \qquad (x, y \in U).$$

We prove (2.94) by showing that the composition $\sigma\xi: U \to \text{map}(U, \mathbb{R})$ is an embedding, where $\sigma: \hom(U, \mathbb{R}) \to \text{map}(U, \mathbb{R})$ is the injection; from this it will follow at once that σ is an embedding, as asserted. So consider the open ball $B_\varepsilon(x) \subset U$, where $x \in U$ and $\varepsilon > 0$. We have to show that this is the inverse image of an open set of $\text{map}(U, \mathbb{R})$. By homogeneity it is sufficient to consider the case $x = 0$. However $B_\varepsilon(0)$ is the inverse image under $\sigma\xi$ of the compact-open subset

$$(\mathscr{Cl}\ B_1(0), (-\varepsilon, \varepsilon))$$

of $\text{map}(U, \mathbb{R})$. Thus $\sigma\xi$, and hence σ, is an embedding.

Proposition (2.95). *Let X, Y be regular. Then for all spaces Z the compact-open topology on $\text{map}(X \times Y, Z)$ is generated by the family of subsets $(E \times F, V)$ where $E \subset X$ and $F \subset Y$ are compact and where $V \subset Z$ is open.*

For let C be a compact subset of $X \times Y$ and let N be a neighbourhood of C. Since X and Y are regular, so is $X \times Y$. Hence there exists, for each point c of C, a neighbourhood with closure contained in N. Hence, by the product topology, there exists a neighbourhood of the form $U_c \times V_c$, where $U_c \subset X$ and $V_c \subset Y$ are open, such that $\mathscr{Cl}\ U_c \times \mathscr{Cl}\ V_c \subset N$. Since C is compact we can extract from the open covering $\{U_c \times V_c\}$ $(c \in C)$ of C a finite subcovering indexed by $c_1, \ldots, c_n$, say.

Now let C_X, C_Y be the projections of C in X, Y, respectively, and write

$$E_i = C_X \cap \mathscr{Cl}\, U_{c_i}, \qquad F_i = C_Y \cap \mathscr{Cl}\, V_{c_i} \qquad (i = 1, \ldots, n).$$

Then E_i and F_i are compact, since C_X and C_Y are compact, and we have

$$C \subset \bigcup_{i=1}^{n} (E_i \times F_i) \subset N.$$

So let (C, W) be an arbitrary compact-open subset of $\mathrm{map}(X \times Y, Z)$. If $\phi \in (C, W)$ then $N = \phi^{-1}W$ is a neighbourhood of C. So, by the above, there exist compact subsets $E_i \subset X$, $F_i \subset Y$ $(i = 1, \ldots, n)$ such that

$$\phi \in \bigcap_{i=1}^{n} (E_i \times F_i, W) = \left(\bigcup_{i=1}^{n} (E_i \times F_i), W\right) \subset (C, W).$$

This proves (2.95) and we at once deduce

Corollary (2.96). *Let X_i $(i = 1, 2)$ be regular spaces and let Y_i $(i = 1, 2)$ be spaces. Then the function*

$$\mathrm{map}(X_1, Y_1) \times \mathrm{map}(X_2, Y_2) \to \mathrm{map}(X_1 \times X_2, Y_1 \times Y_2)$$

given by the product functor is an embedding.

Proposition (2.97). *Let T_0 be a closed subset of the compact regular space T. For each space X let ΦX be the push-out of the cotriad*

$$T \times X \leftarrow T_0 \times X \to T_0,$$

and similarly for maps. Then the function

$$\Phi_\#: \mathrm{map}(X, Y) \to \mathrm{map}(\Phi X, \Phi Y)$$

is continuous for regular X and all Y.

To see this consider the commutative diagram shown below, where T stands for the functor $T\times$ and $p: T \to \Phi$ is the obvious natural transformation.

$$\begin{array}{ccc} \mathrm{map}(X, Y) & \xrightarrow{T_\#} & \mathrm{map}(TX, TY) \\ \downarrow{\scriptstyle \Phi_\#} & & \downarrow{\scriptstyle p_*} \\ \mathrm{map}(\Phi X, \Phi Y) & \xrightarrow[p^*]{} & \mathrm{map}(TX, \Phi Y) \end{array}$$

Since T, and hence T_0, is compact regular the natural projection $p: TX \to \Phi X$ is a compact surjection, hence p^* is an embedding by (2.86) above. Also $T_\#$ is continuous, as we have seen and p_* is continuous from first principles. Therefore $p^*\Phi_\# = p_* T_\#$ is continuous, and so $\Phi_\#$ is continuous as asserted.

Composition of maps determines a function

$$\mathrm{map}(Y, Z) \times \mathrm{map}(X, Y) \to \mathrm{map}(X, Z).$$

In particular (take $X = *$ and replace Y, Z by X, Y, respectively) evaluation determines a function

$$\text{map}(X, Y) \times X \to Y.$$

Although these functions are not continuous in general, we have

Proposition (2.98). *If Y is locally compact regular the composition function*

$$\text{map}(Y, Z) \times \text{map}(X, Y) \to \text{map}(X, Z)$$

is continuous for all spaces X, Z.

Corollary (2.99). *If X is locally compact regular the evaluation function*

$$\text{map}(X, Y) \times X \to Y$$

is continuous for all spaces Y.

To prove (2.98) let (C, V) be a compact-open subset of $\text{map}(X, Z)$. Let $\theta: X \to Y$ and $\phi: Y \to Z$ be maps such that $\phi\theta \in (C, V)$. Then $\phi^{-1}V$ is a neighbourhood of the compact subset θC of Y. Since Y is locally compact regular we can shrink $\phi^{-1}V$ to a neighbourhood U of θC such that $\mathscr{C}\ell\ U \subset \phi^{-1}V$ and $\mathscr{C}\ell\ U$ is compact. Then composition sends $(C, U) \times (\mathscr{C}\ell\ U, V)$ into (C, V). Since $\theta \in (C, U)$ and $\phi \in (\mathscr{C}\ell\ U, V)$ this proves (2.98) and hence (2.99).

Corollary (2.100). *Let $\hat{h}: X \to \text{map}(Y, Z)$ be continuous, where Y is locally compact regular. Then $h: X \times Y \to Z$ is continuous, where*

$$h(x, y) = \hat{h}(x)(y) \qquad (x \in X, y \in Y).$$

This constitutes a converse of (2.90), subject to the restriction on Y. It is usual to refer to $\hat{h}$ as the adjoint of h, when the functions are related as above. For the proof of (2.100) it is only necessary to observe that h may be expressed as the composition

$$X \times Y \to \text{map}(Y, Z) \times Y \to Z$$

of $\hat{h} \times \text{id}_Y$ and the evaluation map.

Another application of these results is to prove

Proposition (2.101). *Suppose that $\phi: X \to Y$ is a quotient map then so is $\phi \times \text{id}_T: X \times T \to Y \times T$ for each locally compact regular T.*

For let Z be a space and let f, g be functions making the diagram on the left commutative.

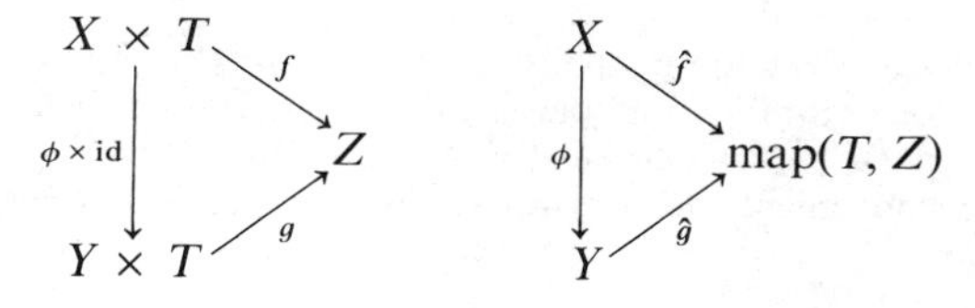

Suppose that f is continuous. Then the adjoints $\hat{f}$, $\hat{g}$ are defined, as shown on the right, and $\hat{f}$ is continuous. Therefore $\hat{g}$ is continuous, since ϕ is a quotient map, and so g is continuous, by (2.100). Therefore $\phi \times \text{id}$ is a quotient map.

Of course this result can perfectly well be proved directly, but the proof given is somewhat easier. As an application the reader may wish to demonstrate.

Exercise (2.102). Show that if either:
(i) X and Z are locally compact regular; or
(ii) X and Y (or Y and Z) are compact regular then

$$(X * Y) * Z \equiv X * (Y * Z),$$

in the fine topology.

I conclude the chapter with the exponential law, which justifies the use of the term adjoint in this context.

Proposition (2.103). *Let X, Y, Z be spaces and let*

$$\xi\colon \text{map}(X \times Y, Z) \to \text{map}(X, \text{map}(Y, Z))$$

be the injection defined by taking adjoints as in (2.90). *If X is regular then ξ is continuous. If Y is also regular then ξ is an open embedding. If in addition Y is locally compact then ξ is a homeomorphism.*

First observe that if $E \subset X$, $F \subset Y$ and $V \subset Z$ then

$$\xi(E \times F, V) = (E, (F, V)), \qquad \xi^{-1}(E, (F, V)) = (E \times F, V).$$

When X is regular the topology of $\text{map}(X, \text{map}(Y, Z))$ is generated by compact-open sets of the form $(E, (F, V))$, where E, F are compact and V is open. The inverse image of this subset is $(E \times F, V)$, which is also compact-open, and so ξ is continuous. When Y is also regular the topology of map $(X \times Y, Z)$ is generated by compact-open sets of the form $(E \times F, V)$, where E and F are compact and V is open. The direct image of $(E \times F, V)$ is $(E, (F, V))$, which is also compact-open, and so ξ is an open embedding. The final assertion follows at once from (2.100).

References

N. Bourbaki. *Topologie Générale*. Hermann, Paris, 1965.

D. E. Cohen. Products and carrier theory. *Proc. London Math. Soc.* (3), **26** (1957), 219–248.

J. Dugundji. *Topology*. Allyn and Bacon, Boston, 1966.

R. H. Fox. On topologies for function spaces. *Bull. Amer. Math. Soc.* **51** (1945), 429–432.

J. L. Kelley. *General Topology*. Van Nostrand, New York, 1955.

N. E. Steenrod. A convenient category for topological spaces. *Michigan Math. J.* **14** (1967), 133–152.

CHAPTER 3

Spaces Under and Spaces Over

Following the programme outlined in the first chapter we now turn to the category $\mathscr{T}(2)$ of pairs associated with the category $\mathscr{T}$ of spaces and maps. We begin by discussing the category of spaces under a given space, then turn to the category of spaces over a given space, and finally consider the category of spaces over and under a given space.

Recall that a space under a given space A is a pair consisting of a space X and a map $u: A \to X$, called the insertion. Usually X alone is sufficient notation. Thus A is regarded as a space under itself with insertion the identity map.

Let X, Y be spaces under A with insertions u, v, respectively. A map $\phi: X \to Y$ under A is a map in the ordinary sense such that $\phi u = v$; the set of such maps is denoted by $\mathrm{MAP}^A(X, Y)$. With this definition of morphism the category $\mathscr{T}^A$ of spaces under A is defined. We know from the generalities of the first chapter that A constitutes an initial object of the category. We also know that the product $\times$ in $\mathscr{T}$ determines a product $\times^A$ in $\mathscr{T}^A$; usually it is sufficient to write $\times$ instead of $\times^A$. We also know that the pull-back in $\mathscr{T}$ of a triad in $\mathscr{T}^A$ is automatically a pull-back in $\mathscr{T}^A$, and that the push-out in $\mathscr{T}$ of a cotriad in $\mathscr{T}^A$ is automatically a push-out in $\mathscr{T}^A$. Note that a functor $A +: \mathscr{T} \to \mathscr{T}^A$ is given by $(A+)T = A + T$, and similarly with maps.

In the category $\mathscr{T}^A$ a special role is played by those spaces under A for which the insertion is injective. Thus let X, Y be spaces under A with insertions u, v, respectively, and suppose that u is injective. Then the push-out $X +^A Y$ of the cotriad

$$X \overset{u}{\leftarrow} A \overset{v}{\rightarrow} Y$$

is defined. In particular $A +^A Y$ is defined, for each space Y under A, and is naturally equivalent to Y. In case both u and v are injective we usually write

$X \vee^A Y$ instead of $X +^A Y$ and refer to this as the wedge-sum under A. The wedge-sum comes equipped with insertions

$$X \to X \vee^A Y \leftarrow Y$$

induced by the corresponding insertions

$$X \to X + Y \leftarrow Y.$$

With these insertions the wedge-sum constitutes a co-product for the category $\mathscr{T}^A$. It is easy to check that the wedge-sum is commutative and associative, in the sense of Chapter 1.

When $A = *$, the point-space, the insertion is automatically injective. In that case spaces under $*$ are called pointed spaces and maps under $*$ are called pointed maps. The wedge-sum of pointed spaces X, Y is denoted by $X \vee Y$.

Let X, X' be spaces and let $\phi: X' \to X$ be a continuous injection. The push-out of the contriad

$$X \overset{\phi}{\leftarrow} X' \to *$$

is called the space obtained from X by collapsing with respect to ϕ. Note that the push-out has a natural basepoint, which is closed when ϕ is closed. Provided X' is non-empty the push-out may be regarded as a quotient space of X alone; when X' is empty the push-out is just the disjoint union $X + *$. In practice X' is usually a subspace of X and ϕ the inclusion; then the push-out is called the space obtained from X by collapsing X' and is denoted by X/X'.

Exercise 3.1. Let X, Y be spaces and let X' be a closed subspace of X. Either suppose that X' is compact or that Y is locally compact regular. Then

$$\frac{X * Y}{X' * Y} \equiv \left(\frac{X}{X'}\right) * Y$$

in the fine topology.

Let X, Y be pointed spaces and let $c: X \to Y$, $c: Y \to X$ be the nul-maps. Consider the continuous injection

$$\zeta: X \vee Y \to X \times Y,$$

which is given by (id, c) on X, by (c, id) on Y. The pointed space obtained from $X \times Y$ by collapsing with respect to ζ is called the smash product of X and Y, and denoted by $X \wedge Y$. The smash product of pointed maps is defined similarly, so that the smash product constitutes a bifunctor $\mathscr{T}^* \times \mathscr{T}^* \to \mathscr{T}^*$.

Pointed spaces where the basepoint is a closed set play a special role, as in

Proposition (3.2). *Let X, Y, Z be spaces with closed basepoints. Then*

$$(X \wedge Z) \vee (Y \wedge Z) \equiv (X \vee Y) \wedge Z.$$

Suppose that either:
(i) *X and Z are locally compact regular; or that*
(ii) *X and Y (or Y and Z) are compact regular. Then*

$$(X \wedge Y) \wedge Z \equiv X \wedge (Y \wedge Z).$$

To prove the first part, consider the diagram shown below, where the maps are the obvious ones.

$$\begin{array}{ccc} X \times Z + Y \times Z & \longrightarrow & (X + Y) \times Z \\ \downarrow & & \downarrow \\ (X \wedge Z) \vee (Y \wedge Z) & \longrightarrow & (X \vee Y) \wedge Z \end{array}$$

The left vertical is a quotient map, from first principles. The right vertical is a quotient map by (2.63). The top horizontal is a homeomorphism by the distributive law. The bottom horizontal is a bijection and hence, by commutativity of the diagram, a homeomorphism.

To prove the second part, under either hypothesis, consider the commutative diagram shown below, where the vertical maps are the obvious ones and the bottom horizontal is induced by the identity at the top.

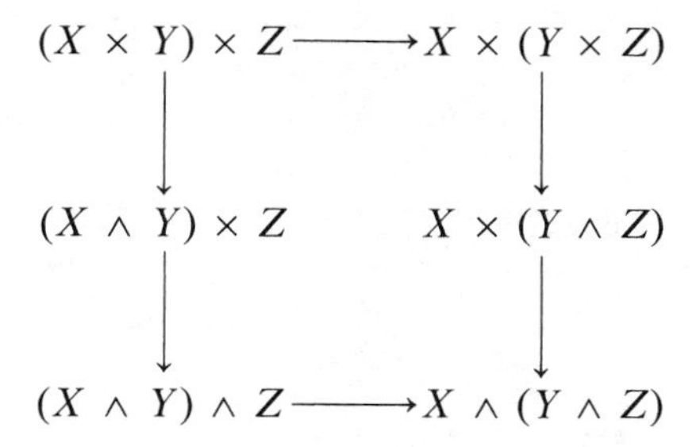

Under hypothesis (i) the first vertical on the left is a quotient map, by (2.101), while the second is a quotient map by definition. Thus the composition is a quotient map. Similarly the composition on the right is a quotient map. It follows at once that the bijection on the bottom is a homeomorphism. Under hypothesis (ii), with X and Y compact regular, the projection $X \times Y \to X \wedge Y$ is compact, by (2.63), and so the first vertical on the left is compact and so a quotient map. Hence the left vertical is a quotient map, moreover so is the right vertical as before. The result now follows at once.

We refer to these two results as the distributive law and the associative law for the smash product. In fact the assumption that the basepoints are closed is not strictly necessary for these results.

The proof of our next result is similar to that of (3.2) and will therefore be left to serve as an exercise.

Proposition (3.3). *Let X, Y be spaces and let X' be a closed subspace of X. If Y is locally compact regular or if X' is compact then*

$$\frac{X \wedge Y}{X' \wedge Y} \equiv \left(\frac{X}{X'}\right) \wedge Y.$$

Suppose that we have an endofunctor Φ of $\mathscr{T}$. For each pointed space X the pointed map $* \to X$ has a left inverse, hence the transform $\Phi* \to \Phi X$ has a left inverse. We define $\Phi^* X$ to be the push-out of the cotriad

$$\Phi X \leftarrow \Phi* \to *,$$

and similarly for pointed maps. The endofunctor Φ^* of $\mathscr{T}^*$ thus defined is called the reduction of the original endofunctor Φ.

For example, let T_0 be a closed subset of the locally compact regular space T. Define ΦX, for each space X, to be the push-out of the cotriad

$$T \times X \leftarrow T_0 \times X \to T_0,$$

and similarly for maps. Using (3.3) and transitivity of quotients it follows at once that in this case $\Phi^* X$, for each space X with closed basepoint, is equivalent to the smash product $(T/T_0) \wedge X$, and similarly for pointed maps. In particular the reduced cone functor Γ^* is equivalent to $I \wedge$ and the reduced suspension functor Σ^* to $S \wedge$, where I has basepoint 0 and $S = I/\dot{I}$ has basepoint $\dot{I}/\dot{I}$.

Another relationship involving the smash product may be deduced from (2.49) above. First observe that the suspension Σ may be regarded as a functor $\mathscr{T} \to \mathscr{T}^*$, rather than as an endofunctor of $\mathscr{T}$. Specifically we embed the given space X in the cone ΓX, through the correspondence $x \mapsto (1, x)$, and then take ΣX to be the pointed space $\Gamma X/X$; of course this is equivalent to the previous definition of suspension in the fine topology. Note that the basepoint of ΣX is always closed. Now (2.49) shows that for all spaces X, Y we have a homeomorphism

$$(\Gamma(X * Y), X * Y) \to (\Gamma X \times \Gamma Y, \Gamma X \times Y \cup X \times \Gamma Y)$$

and hence a homeomorphism

$$\frac{\Gamma(X * Y)}{X * Y} \to \frac{\Gamma X \times \Gamma Y}{\Gamma X \times Y \cup X \times \Gamma Y}.$$

We therefore conclude that

(3.4) $$\Sigma(X * Y) \equiv (\Sigma X) \wedge (\Sigma Y),$$

for all spaces X, Y. For example, $S^p \wedge S^q \equiv S^{p+q}$.

We turn now to the process known as compactification. This should be regarded as a functor from the category of locally compact Hausdorff spaces and compact maps to the category of compact Hausdorff pointed spaces and pointed maps. As we shall see this leads, inter alia, to a better understanding of the relation between the sum $+$ and product $\times$ in $\mathscr{T}$ and the wedge $\vee$ and smash $\wedge$ in $\mathscr{T}^*$.

Let X be a locally compact Hausdorff space. Then a compact Hausdorff space X^+ can be constructed as follows. Take X^+, as a set, to be the disjoint union $X/\varnothing = X + *$, and give X^+ the topology with the following basis. The basic open sets of X^+ are of two kinds. The first kind are the open sets of X, regarded as subsets of X^+. The second kind are the complements in X^+ of the compact subsets of X. Note that the inclusion $X \to X^+$ is an open embedding.

I assert that X^+ is Hausdorff. Certainly each distinct pair of points in X can be separated in X and so can be separated in X^+. It remains to be shown that each point x of X can be separated from $*$ in X^+. Since X is locally compact there exists a neighborhood U of x such that $\mathscr{Cl}\, U$ is compact. Then U and $X^+ - \mathscr{Cl}\, U$ are disjoint neighbourhoods of x and $*$, respectively. Thus X^+ is Hausdorff, as asserted.

I also assert that X^+ is compact. To see this, let $\{U_j\}$ $(j \in J)$ be an open covering of X^+. Choose a member U of the covering which contains $*$. Then $X^+ - U = C$ is compact. Regarding $\{U_j\}$ as an open covering of C we extract a finite subcovering and then supplement it by U. The result is a finite subcovering of X^+, and so the second assertion is proved.

The compact Hausdorff space X^+ is called the compactification of X. We regard X^+ as a pointed space with basepoint $*$. It is easy to check that when X itself is compact Hausdorff the compactification $X^+ = X + *$, the disjoint union in the topological sense.

Example (3.5). The compactification $\mathbb{R}^+$ of the real line $\mathbb{R}$ is equivalent to the circle S.

To see this we regard $\mathbb{R}$ as the tangent to S at the south pole. For each point $x \in \mathbb{R}$ the line joining x to the north pole meets the complement of that pole in a point $f(x)$. Extend f to a function $g\colon \mathbb{R}^+ \to S$ by sending $*$ to the north pole. Then g is bijective and is easily shown to be continuous. Since $\mathbb{R}^+$ is compact and S is Hausdorff it follows from (2.73) that g is a homeomorphism.

Let X, Y be locally compact Hausdorff spaces. Clearly each function $\phi\colon X \to Y$ determines a pointed function $\phi^+\colon X^+ \to Y^+$, and vice versa. Suppose that ϕ is continuous. I assert that ϕ^+ is continuous if and only if ϕ is compact. For suppose that ϕ is compact. If $C \subset Y$ is compact then $\phi^{-1}C \subset X$ is compact, by (2.55). Hence if $Y^+ - C$ is a neighbourhood of $*$ in Y^+ then $X^+ - \phi^{-1}C$ is a neighbourhood of $*$ in X^+. This proves continuity of ϕ^+ at $*$; continuity away from $*$ is obvious.

Conversely suppose that ϕ^+ is continuous. Then $\phi^{-1}(Y^+ - C) = X^+ - \phi^{-1}C$ is open for each compact C, and hence $\phi^{-1}C$ is compact. So let $\{U_j\}$ $(j \in J)$ be an open covering of Y such that $\mathscr{Cl}\, U_j$ is compact for each index j. Then each of the maps $\phi^{-1}\, \mathscr{Cl}\, U_j \to \mathscr{Cl}\, U_j$ determined by ϕ is closed and so compact, by (2.61). Therefore ϕ itself is compact, as asserted.

Exercise (3.6). Let X be locally compact Hausdorff. Let $\phi: X' \to X$ be a closed embedding. Then X' is locally compact Hausdorff. Also $\phi^+: X'^+ \to X^+$ is a closed embedding. Finally X/X' is equivalent to X^+/X'^+, as a pointed space.

Proposition (3.7). *Let X, Y be locally compact Hausdorff. Then*

$$X^+ \vee Y^+ \equiv (X + Y)^+, \qquad X^+ \wedge Y^+ \equiv (X \times Y)^+,$$

as pointed spaces.

Thus, using (3.4) and (3.5), the second part of (3.7) shows that $(R^n)^+ \equiv S^n$. More generally we have

Corollary (3.8). *Let X be locally compact Hausdorff. Then*

$$(X \times R^n)^+ \equiv X \wedge S^n,$$

as pointed spaces.

To prove the first part of (3.7) consider the triad

$$X \to X + Y \leftarrow Y.$$

Since the insertions are compact they determine a triad

$$X^+ \to (X + Y)^+ \leftarrow Y^+.$$

The wedge-sum

$$X^+ \vee Y^+ \to (X + Y)^+$$

of these maps is a continuous bijection and hence an equivalence of pointed spaces, since the domain is compact and the codomain is Hausdorff.

To prove the second part of (3.7) consider the bijection

$$\xi: (X \times Y)^+ \to X^+ \wedge Y^+$$

which is given by the identity on $X \times Y$ and sends basepoint to basepoint. Clearly ξ sends $U \times V$, where $U \subset X$ and $V \subset Y$ are open, into $U \times V$, which is open in the codomain. Also $\xi | (X \times Y)$ is continuous and so sends the compact subset K of $X \times Y$ into the compact subset $\xi K \subset X^+ \wedge Y^+$. Moreover $X^+ \wedge Y^+$ is Hausdorff, by (2.72), since X^+ and Y^+ are Hausdorff and $X^+ \times Y^+ \to X^+ \wedge Y^+$ is compact. Therefore ξK is closed, and hence the image of the complement of K in $(X \times Y)^+$ is open. Thus ξ is continuous. However $(X \times Y)^+$ is compact and $X^+ \wedge Y^+$ is Hausdorff, as we have seen, so that ξ is an equivalence, as asserted.

Let us break off our discussion of pointed spaces at this stage, since we have gone far enough to show how it works, and go on to consider spaces over a base. Later we shall be ready to discuss sectioned spaces over a base. When the base is a point, these are just pointed spaces. The results we have just been discussing will all be generalized and we shall then continue the discussion at a more general level rather than pursue the special case any further now.

We turn now to the category of spaces over a given space. As we shall see this has features for which there are no counterparts in the category of spaces under a given space. In fact one can develop an extensive theory of general topology over a space, and what we are going to do may be regarded as part of that theory.

The space we work over is called the base and usually denoted by B. A space over B consists of a space X and a map $p: X \to B$, called the projection. Usually X alone is sufficient notation. Thus B is regarded as a space over itself with projection the identity map. For example, let X be a space with an equivalence relation $\sim$; then X may be regarded as a space over the quotient space $X/\sim$ using the natural projection.

Let X, Y be spaces over B with projections p, q respectively. By a map $\phi: X \to Y$ over B we mean a map in the ordinary sense such that $q\phi = p$; the set of such maps is denoted by $\mathrm{MAP}_B(X, Y)$. With this definition of morphism the category $\mathscr{T}_B$ of spaces over B is defined. We know from the generalities of the first chapter that B constitutes a final object of the category. We also know that the sum $+$ in $\mathscr{T}$ determines a sum $+_B$ in $\mathscr{T}_B$; usually it is sufficient to write $+$ instead of $+_B$. We also know that the push-out in $\mathscr{T}$ of a cotriad in $\mathscr{T}_B$ is automatically a push-out in $\mathscr{T}_B$, and that the pull-back in $\mathscr{T}$ of a triad in $\mathscr{T}_B$ is automatically a pull-back in $\mathscr{T}_B$. Note that a functor $B\times : \mathscr{T} \to \mathscr{T}_B$ is given by $(B\times)T = B \times T$, and similarly for maps.

If X, Y are spaces over B with projections p, q respectively then the fibre product $X \times_B Y$ is defined as the subspace of $X \times Y$ consisting of pairs (x, y) such that $px = qy$, with projection r given by $r(x, y) = px = qy$. In this way a bifunctor $\times_B: \mathscr{T}_B \times \mathscr{T}_B \to \mathscr{T}_B$ is defined which satisfies the conditions for a product functor as stated in Chapter 1. The product is commutative and associative, in the sense of that chapter, and is distributive over the sum.

If X is a space over B then for each space B' and map $\xi: B' \to B$ the fibre product $B' \times_B X$ is defined, and can be regarded as a space over B' under the first projection. Here B' is regarded as a space over B using ξ as projection. We denote $B' \times_B X$ by ξ^*X, as a space over B'. In this way a functor $\xi^*: \mathscr{T}_B \to \mathscr{T}_{B'}$ is defined, and from the laws of distributivity and associativity we at once obtain

Proposition (3.9). *If X, Y are spaces over B then*

$$\xi^*(X +_B Y) \equiv \xi^*X +_{B'} \xi^*Y, \qquad \xi^*(X \times_B Y) \equiv \xi^*X \times_{B'} \xi^*Y,$$

for all spaces B' and maps $\xi: B' \to B$.

Now suppose that we are given the space B and a set X over B. In general there will be a choice of topologies for X, each of which makes the projection $p: X \to B$ continuous. All these topologies have in common that $p^{-1}U = X_U$

is open for each open U of B. We refer to such sets as the special subsets of X, as a set over B. I must emphasize that if, in some topology for X, a subset X_H happens to be open although H is not open in B then X_H is not regarded as a special subset.

The special subsets of X constitute a topology for X which we may call the indiscrete topology over B. This topology, of course, is just the initial topology determined by p. It has the property that for each space X' over B, each function $X' \to X$ over B is continuous when X has the indiscrete topology over B. Note that if X has the indiscrete topology over B then so does any subspace of X. Also that if X has the indiscrete topology over B then ξ^*X has the indiscrete topology over B', for each space B' and map $\xi: B' \to B$. Finally the fibre product of two indiscrete spaces over B is again an indiscrete space over B.

Again let X be a set over the space B, and let Γ be a family of subsets of X. If we give X the topology generated by Γ, in the sense of the previous chapter, there is no guarantee that p will be continuous. The topology we need to use in this situation is that generated by Γ together with the family of special subsets. Let us describe this as the topology generated by Γ after supplementation, and say that Γ is a subbasis for that topology after supplementation. Similarly if Γ satisfies the basis condition (2.8) we say that Γ is a basis for the topology after supplementation. In that case each non-special open set of X is a union of open sets of the form $V \cap X_W$, where V is a member of Γ and W is open in B.

An example may help to make this clearer. Consider the space $B \times T$ over B, where T is any space. If Γ_0 is a family of subsets of T which satisfies the basis condition for T then the corresponding family

$$\Gamma = \{B \times U \mid U \in \Gamma_0\}$$

satisfies the basis condition for $B \times T$. But Γ is not a basis for the product topology without supplementation.

The above example may be regarded as a special case of

Proposition (3.10). *Let X be a space over B and let Γ be a basis (resp. subbasis) for the topology of X after supplementation. Let B' be a space and let $\xi: B' \to B$ be a map. Let $\eta: \xi^*X \to X$ be the canonical map over ξ. Then the pull-back $\eta^{-1}\Gamma$ is a basis (resp. subbasis) for the topology of ξ^*X, after supplementation.*

The proofs of (3.10), and of the next few results, are entirely straightforward and will therefore be left to serve as exercises.

Proposition (3.11). *Let $\phi: X \to Y$ be a function over B, where X and Y are spaces over B. Let Γ be a subbasis for the topology of Y after supplementation. Then ϕ is continuous if (and only if) $\phi^{-1}V$ is open for each member V of Γ.*

Proposition (3.12). *Let $\phi: X \to Y$ be a function over B, where X and Y are spaces over B. Let Γ be a basis for the topology of X after supplementation. Then ϕ is open if (and only if) ϕU is open for each member U of Γ.*

Proposition (3.13). *Let $\phi: X \to Y$ be an injection over B, where X and Y are spaces over B. Let Γ be a subbasis for the topology of X after supplementation. Then ϕ is open (resp. closed) if and only if the direct image of each member of Γ is open (resp. of the complement of each member of Γ is closed).*

Thus provided we confine ourselves to fibre-preserving functions, the results of the previous chapter about bases and subbases generalize in this way.

Proposition (3.14). *Let $\phi: X \to Y$ be a function over B, where X and Y are spaces over B. If ϕ is open (resp. closed) then the restriction $\phi': X_{B'} \to Y_{B'}$ is open (resp. closed) for each subset B' of B.*

Proposition (3.15). *Let $\phi: X \to Y$ be a function over B, where X and Y are spaces over B. Let $\{B_j\}$ $(j \in J)$ be a family of subsets of B such that either*

(i) *$\{\mathcal{I}nt\ B_j\}$ is an open covering of B, or*
(ii) *$\{B_j\}$ is a locally finite closed covering of B.*

If each of the functions $\phi_j: X_{B_j} \to Y_{B_j}$ determined by ϕ is open (resp. closed) then ϕ is open (resp. closed).

For under (i), since $Y_{\mathcal{I}nt\ B'} \subset \mathcal{I}nt\ Y_{B'}$, it follows that $\{\mathcal{I}nt\ Y_{B_j}\}$ is an open covering of Y, and under (ii) that $\{Y_{B_j}\}$ is a locally finite closed covering of Y. So (3.15) is a corollary of (2.34).

If X is a set over B and X has a topology we can always refine the topology so as to make the projection continuous. Specifically the open sets in the new topology are the unions

$$\bigcup_{j \in J} (V_j \cap X_{W_j})$$

for arbitrary J, where each V_j is an open set of X in the old topology, and each W_j is an open set of B. To distinguish between the two topologies, we use $\check{X}$ when we refer to the old topology and $\hat{X}$ when we refer to the new one.

Of course the open sets of the old topology satisfy the basis condition, in a particularly strong way, and the new topology is generated by this basis after supplementation. Thus the fibres of X have the same topology whether the old or the new topology is used

Proposition (3.16). *Let $\phi: X \to Y$ be a map over B, where X is a set over B and Y is a space over B. Suppose that ϕ is compact, for some topology $\check{X}$ for the set X. Then ϕ is compact for the refined topology $\hat{X}$.*

The first step is to show that each fibre $\phi^{-1}y$ $(y \in Y)$ is compact in the new topology as well as the old. Suppose therefore that we have a covering of $\phi^{-1}y$ by sets $\{U_k\}$ $(k \in K)$ which are open in the new topology. We have to extract a finite subcovering. Now each U_k is the union of a family

$$\{V_{j,k} \cap X_{W_{j,k}}\} \qquad (j \in J_k),$$

where each $V_{j,k}$ is open in the old topology and each $W_{j,k}$ is open in B. Without real loss of generality we may assume that each $W_{j,k}$ contains the point $qy = b$, say. Now the family

$$\left\{\bigcup_{j \in J_k} V_{j,k}\right\} \qquad (k \in K)$$

is a covering of $\phi^{-1}y$ by sets which are open in the old topology. Since $\phi^{-1}y$ is compact in the old topology we can extract a finite subcovering, indexed by a finite subset $L \subset K$. Since each set $X_{W_{j,l}}$ $(l \in L)$ contains X_b and hence $\phi^{-1}y$ it follows at once that $\{U_l\}$ $(l \in L)$ covers $\phi^{-1}y$. Thus we have extracted a finite subcovering of the original covering as required.

It remains to be shown that the function $\phi\colon \hat{X} \to Y$ is closed. So let U be a neighbourhood of $\phi^{-1}y$ in the new topology. We have to show that there exists a neighbourhood N of y in Y such that $\phi^{-1}N \subset U$. Now U is the union of the members of a covering

$$\{V_j \cap X_{W_j}\} \qquad (j \in J)$$

of $\phi^{-1}y$, where each V_j is open in the old topology and each W_j is open in B. Since $\phi^{-1}y$ is compact, as we have seen, we can assume without real loss of generality that J is finite. We can also assume, without real loss of generality, that each W_j contains the point $qy = b$. Now

$$V = \bigcup_{j \in J} V_j$$

is a neighbourhood of $\phi^{-1}y$ in the old topology and so $N' = Y - \phi(X - V)$ is a neighbourhood of y such that $\phi^{-1}N' \subset V$. Also

$$W = \bigcap_{j \in J} W_j$$

is a neighbourhood of b, and so $N = N' \cap Y_W$ is a neighbourhood of y such that $\phi^{-1}N \subset U$ as required.

In the first chapter it was explained how an endofunctor Φ of $\mathscr{S}$ can be extended, in a natural way, to an endofunctor $\Phi(2)$ of $\mathscr{S}(2)$. We recall that for a set X over a given set B the transform of X under $\Phi(2)$ is denoted by $\Phi_B(X)$, and similarly with functions over B. The procedure can be made topological as follows.

For each space X over the given space B let $\Phi_B(X)$ denote the set

$$\coprod_{b \in B} \Phi(X_b)$$

over B with the topology induced by the function

$$\lambda\colon \Phi_B(X) \to B \times \Phi(X).$$

Here λ_1 is the projection and λ_2 is given on $\Phi(X_b)$ by $\Phi(u_b)$, where $u_b: X_b \subset X$. In other words $\Phi_B(X)$ has the topology induced by λ_2 after supplementation. It follows at once that if $\eta: X' \to X$ is a map over $\xi: B' \to B$ then the function $\Phi(2)(\eta): \Phi_{B'}(x') \to \Phi_B(x)$ over ξ, defined as in the first chapter, is continuous.

Now suppose that Φ is a continuous functor in the sense of Chapter 2. Then for each space T the identity on $B \times T$ is transformed into a map $B \times \Phi(T) \to \Phi(B \times T)$, with the same image as

$$\lambda_2: \Phi_B(B \times T) \to \Phi(B \times T).$$

Suppose further that Φ transforms injections into injections, as is always the case in the applications. Then it follows at once that $\Phi_B(B \times T)$ is equivalent to $B \times \Phi(T)$, as a space over B.

Another important property of continuous endofunctors is the following. Let X, Y be spaces over B, and let A be a space. Regarding $A \times X$ as a space over B, through the second projection, let $f: A \times X \to Y$ be a map over B. Then the function

$$\hat{f}: A \times \Phi_B(X) \to \Phi_B(Y),$$

defined in the obvious way, is continuous. I leave the proof of this as an exercise for the reader.

Not every endofunctor of $\mathscr{T}_B$ originates as an endofunctor of $\mathscr{T}$, of course. Thus let T be a space over B and let T_0 be a subset of T. Define $\Phi_B(X)$, for any space X over B, to be the push-out of the cotriad

$$T \times_B X \leftarrow T_0 \times_B X \rightarrow T_0,$$

and similarly for maps over B. In case $(T, T_0) = B \times (T', T'_0)$ where T' is a space and T'_0 is a subset of T', we may interpret $\Phi_B(X)$ as the extension to $\mathscr{T}_B$ of the endofunctor Φ of $\mathscr{T}$ which associates with each space X the push-out of the cotriad

$$T' \times X \leftarrow T'_0 \times X \rightarrow T'_0,$$

and similarly with maps. In particular when $(T', T'_0) = (I, \{0\})$ the functor Φ_B thus defined is called the fibre-cone and denoted by Γ_B, while when $(T', T'_0) = (I, \{0, 1\})$ the functor is called the fibre-suspension and denoted by Σ_B.

Returning to the general case we observe that there is a "coarse" alternative to the "fine" topology of the topological push-out.

For consider, at the set-theoretic level, the push-out $\Phi_B X$ of the cotriad

$$T \times_B X \leftarrow T_0 \times_B X \rightarrow T_0.$$

This comes equipped with two "coordinate" functions, namely $t: \Phi_B(X) \to T$, given by the first projection, and $x: t^{-1}(T - T_0) \to X$, given by the second projection. Both t and x are functions over B, so that we can give $\Phi_B(X)$ the topology generated after supplementation by the family of subsets $t^{-1}U$ for all open $U \subset T$ and by the family of subsets $x^{-1}V$ for all open

$V \subset X$. In general this topology is coarser than the previous one but it has various advantages; for example Φ_B is a continuous endofunctor with the coarse topology.

As it happens the example which is most important in our work is that of a bifunctor, called the fibre-join and denoted by $*_B$. This too exists in a fine version and a coarse version. Thus let X and Y be spaces over B. In the fine version $X *_B Y$ is defined as the quotient space of

$$I \times X \times_B Y + X + Y$$

with respect to the relations

$$(0, x, y) \sim x, \qquad (1, x, y) \sim y.$$

With this topology the fibre-join operation is commutative, in the sense that $X *_B Y$ is homeomorphic to $Y *_B X$ under the transformation given by $(t, x, y) \mapsto (1 - t, y, x)$. However the fibre-join is not associative with the fine topology. Notice that the fibre-cone $\Gamma_B X$ can be identified with $X *_B (B \times \{0\})$ and the fibre-suspension $\Sigma_B X$ with $X *_B (B \times \{0, 1\})$ still using the fine topology of course.

At the set-theoretical level the set $X *_B Y$ comes equipped with three coordinate functions

$$t: X *_B Y \to I \times B, \qquad x: t^{-1}([0, 1) \times B)) \to X, \qquad y: t^{-1}((0, 1] \times B)) \to Y$$

defined in the obvious way. In the coarse version we give $X *_B Y$ the coarsest topology which makes these three functions over B continuous. Then, by using the same formulae as in the case of (2.49) we obtain

Proposition (3.17). *For all spaces X and Y over B there is a natural homeomorphism*

$$(\Gamma_B(X *_B Y), X *_B Y) \to (\Gamma_B X \times_B \Gamma_B Y, \Gamma_B X \times_B Y \cup X \times_B \Gamma_B Y),$$

in the coarse topology.

Proposition (3.18). *The fibre-join operation, with the coarse topology, is associative. Specifically there is a natural equivalence*

$$X *_B (Y *_B Z) \equiv (X *_B Y) *_B Z$$

over B for all spaces X, Y, Z over B.

For what we are going to do next the following notation is convenient. Let X be a space over B and let H be a subset of X, with closure $\mathscr{Cl}\, H$. For each point $b \in B$ we denote by $\mathscr{Cl}_b\, H$ the fibre $X_b \cap \mathscr{Cl}\, H$ of $\mathscr{Cl}\, H$ over b. Thus $\mathscr{Cl}_b\, H$ consists of those points x of X_b such that each neighbourhood of x in X meets H. Note that if $\phi: X \to Y$ is a closed map over B then $\phi\, \mathscr{Cl}_b = \mathscr{Cl}_b\, \phi$, for each point $b \in B$.

One may use this notation in making a "Kuratowski" approach to the topologizing of a set X over B. In this approach one assumes given a family of operators $\mathscr{Cl}_b$ ($b \in B$), where $\mathscr{Cl}_b$ assigns to each subset H of X a subset $\mathscr{Cl}_b H$ of X_b such that $H_b \subset \mathscr{Cl}_b H$. Suppose that for each point $b \in B$ the following three conditions are satisfied:

(i) $\mathscr{Cl}_b \varnothing = \varnothing$;
(ii) $\mathscr{Cl}_b(H_1 \cup H_2) = \mathscr{Cl}_b H_1 \cup \mathscr{Cl} H_2$, for each pair of subsets H_1, H_2;
(iii) $\mathscr{Cl}_b(\mathscr{Cl}_b H) = \mathscr{Cl}_b H$, for each subset H.

Then X can be topologized so that a subset H of X is closed if and only if $\mathscr{Cl}_b H = H_b$ for each point $b \in B$. Moreover $\mathscr{Cl}_b H = X_b \cap \mathscr{Cl} H$ in this topology.

Various kinds of spaces have been discussed in the previous chapter, for example discrete spaces, indiscrete spaces, compact spaces, Hausdorff spaces and regular spaces. In each case a topological property P of spaces is concerned. One of the aims of the present chapter is to extend such properties to spaces over a base, in a natural fashion. Specifically we aim to define, for each space B, a property P_B of spaces over B such that the following three conditions are satisfied.

Condition (3.19). *If X, Y are equivalent spaces over B and if X has property P_B then so does Y.*

Condition (3.20). *The space X has property P if and only if the space X over the point $*$ has property P_*.*

Condition (3.21). *If the space X over B has property P_B then the space $\xi^* X$ over B' has property $P_{B'}$ for each space B' and map $\xi: B' \to B$.*

It follows at once from these three conditions that $B \times T$ has property P_B over B if and only if T has property P. It is convenient to refer to P_B either as "P over B" or as "fibrewise P" according to circumstances.

I do not suggest that there is any canonical extension of P to P_B; in fact this is not possible. For example, take P to be the universal property, which is satisfied by every space. Then we can take P_B to be the universal property, which is satisfied by every space over B. Or we can take P_B to be the following property.

Definition (3.22). The space X over B is open over B if the projection $X \to B$ is an open map.

Proposition (3.23). *Let $\phi: X \to Y$ be a function over B, where Y is a space over B and X is an open space over B. If*

$$\phi \times \mathrm{id}: X \times_B X \to Y \times_B X$$

is open then so is ϕ.

For let U be a neighborhood of a given point x of X. Then $U \times_B U$ is open in $X \times_B X$ and so $\phi U \times_B U$ is open in $Y \times_B X$. Hence $\phi U \cap q^{-1}pU$ is a neighbourhood of ϕx in Y.

Let us say that the space X over B is compact over B if the projection $p: X \to B$ is compact, and similarly with locally compact in place of compact. Thus X is compact over B if and only if each fibre X_b is compact and the projection $X \to B$ is closed. In particular $B \times T$ is compact over B if and only if the space T is compact. If X is compact over B then each closed subspace of X is also compact over B. Similar results hold with locally compact in place of compact.

Recall that if $p: X \to B$ is compact then so is

$$p \times \mathrm{id}: X \times T \to B \times T,$$

for each space T. If, further, T is a space over B then it follows from (2.55) that

$$p \times \mathrm{id}: X \times_B T \to B \times_B T = T$$

is compact. We can reformulate this in two different ways.

Proposition (3.24). *If X is compact over B then ξ^*X is compact over B' for each space B' and map $\xi: B' \to B$.*

Proposition (3.25). *Let X be a space over B. Let B' be a space and let $\xi: B' \to B$ be a compact map. Then the canonical map $\xi^*X \to X$ is compact.*

Similar results hold with locally compact in place of compact. The following are also straightforward adaptations of results in the previous chapter.

Proposition (3.26). (i) *Let X be compact over B. If B is compact, or locally compact, then so is X.*
(ii) *Let X be locally compact over B. If B is locally compact then so is X.*

Proposition (3.27). *Let $\phi: X \to Y$ be a map over B.*
(i) *If X is compact over B and ϕ is surjective then Y is compact over B.*
(ii) *If Y is compact over B and ϕ is a closed embedding then X is compact over B.*

Proposition (3.28). *Let T be compact over B and let T_0 be a closed subspace of T. Then for each compact space X over B the push-out of the cotriad*

$$T \times_B X \leftarrow T_0 \times_B X \to T_0$$

is also compact over B.

This is true in the fine topology and hence is true in any coarser topology. It shows, in particular, that the fibre cone and fibre suspension of a compact space over B are also compact over B.

Bearing in mind the way in which the notion of compactness was generalised from ordinary spaces to spaces over a base we now seek to generalize the separation axioms in a similar fashion, as follows.

Definition (3.29). The space X over B is Hausdorff over B if for each point b of B and each distinct pair $x, x' \in X_b$ there exist neighbourhoods U of x, U' of x' in X, which are disjoint.

Thus B is Hausdorff over itself. More generally $B \times T$ is Hausdorff over B if, and only if, T is a Hausdorff space. Clearly each subspace of a Hausdorff space over B is also Hausdorff over B. Also the sum and fibre product of two Hausdorff spaces over B are again Hausdorff over B. Again, if X is Hausdorff over B then ξ^*X is Hausdorff over B' for each space B' and map $\xi: B' \to B$.

An alternative characterization is provided by

Proposition (3.30). *The space X over B is Hausdorff over B if and only if the diagonal embedding*

$$\Delta: X \to X \times_B X$$

is closed.

For if Δ is closed each point (x, x') of $X \times_B X - \Delta X$ has a neighbourhood W which does not meet ΔX and hence has a neighbourhood $U \times_B U' \subset W$ which does not meet ΔX; then U does not meet U'. The argument in the reverse direction is similar.

Corollary (3.31). *Let Y be a Hausdorff space over B. Then for each space X over B and each pair of maps $\phi, \phi': X \to Y$ over B the points $x \in X$ such that $\phi x = \phi' x$ form a closed subset of X.*

For consider

$$X \xrightarrow{\Delta} X \times_B X \xrightarrow{\phi \times \phi'} Y \times_B Y \xleftarrow{\Delta} Y.$$

The subset of X we are concerned with is precisely $\Delta^{-1}(\phi \times \phi')^{-1}\Delta Y$. Since ΔY is closed, by (3.30), the conclusion follows immediately.

Corollary (3.32). *Let Y be a Hausdorff space over B. Then for each space X over B and each map $\phi: X \to Y$ over B the graph*

$$\Gamma_\phi: X \to X \times_B Y$$

is a closed embedding.

This follows similarly from consideration of the commutative diagram shown below.

$$\begin{array}{ccc} X & \xrightarrow{\Gamma_\phi} & X \times_B Y \\ \downarrow{\scriptstyle \phi} & & \downarrow{\scriptstyle \phi \times \mathrm{id}} \\ Y & \xrightarrow[\Delta]{} & Y \times_B Y \end{array}$$

For $\Gamma_\phi X = (\phi \times \mathrm{id})^{-1} \Delta Y$, and ΔY is closed.

Proposition (3.33). *Let $\phi\colon X \to Y$ be a map over B, where X is compact over B and Y is Hausdorff over B. Then ϕ is compact.*

For consider the diagram shown below, where r is the standard equivalence.

$$\begin{array}{ccc} X & \xrightarrow{\Gamma_\phi} & X \times_B Y \\ \downarrow{\scriptstyle \phi} & & \downarrow{\scriptstyle p \times \mathrm{id}} \\ Y & \xrightarrow[r]{} & B \times_B Y \end{array}$$

Now Γ_ϕ is a closed embedding, by (3.32) and so is compact. Also p is compact, by hypothesis, and so $p \times \mathrm{id}$ is compact. Therefore $(p \times \mathrm{id}) \circ \Gamma_\phi = r \circ \phi$ is compact, and so ϕ is compact, since r is an equivalence.

Corollary (3.34). *Let $\theta\colon X \times B \to Y$ be a map, where X is compact and Y is Hausdorff. Then the map*

$$\phi\colon X \times B \to Y \times B$$

is compact, where

$$\phi(x, b) = (\theta(x, b), b) \qquad (x \in X, b \in B).$$

This follows at once from (3.33) with X replaced by $X \times B$ and Y by $Y \times B$.

Proposition (3.35). *Let X be a Hausdorff space over B. Let C, D be compact subsets of a fibre X_b of X. Then there exist neighbourhoods U of C and V of D in X such that $U \cap V$ does not meet X_b.*

This follows at once from (2.60) with $X = Y$ and $W = X \times_B X - \Delta X$.

Proposition (3.36). *Let $\phi\colon X \to Y$ be a compact surjection over B. If X is Hausdorff over B then so is Y.*

For since ϕ is a compact surjection so is $\phi \times \phi$, in the diagram shown below.

$$\begin{array}{ccc} X & \xrightarrow{\Delta} & X \times_B X \\ \phi \downarrow & & \downarrow \phi \times \phi \\ Y & \xrightarrow{\Delta} & Y \times_B Y \end{array}$$

By (3.30) the diagonal ΔX is closed, since X is Hausdorff over B, hence $(\phi \times \phi)\Delta X = \Delta\phi X$ is closed. But $\Delta\phi X = \Delta Y$, since ϕ is surjective, and so Y is Hausdorff over B, by (3.30) again.

The most familiar examples of compact spaces are closed and bounded subspaces of the euclidean spaces $\mathbb{R}^n$ ($n = 0, 1, \ldots$). For spaces over a base we have

Proposition (3.37). *Let X be a closed subspace of $B \times \mathbb{R}^n$. Then X is compact over B if there exists a map $\alpha\colon B \to \mathbb{R}$ such that X_b is bounded by $\alpha(b)$ for each point $b \in B$.*

For suppose that the condition is satisfied. The norm $D\colon \mathbb{R}^n \to \mathbb{R}$ is compact, by (2.62), and so the product

$$\mathrm{id} \times D\colon B \times \mathbb{R}^n \to B \times \mathbb{R}$$

is compact. Consider the graph Γ of α in $B \times \mathbb{R}$. The projection $\Gamma \to B$ is a homeomorphism and so compact. Since $\mathrm{id} \times D$ is compact so is the map $(\mathrm{id} \times D)^{-1}\Gamma \to \Gamma$ which $\mathrm{id} \times D$ determines, by (2.55), and therefore $(\mathrm{id} \times D)^{-1}\Gamma$ is compact over B. But X is closed in $B \times \mathbb{R}^n$ and so closed in $(\mathrm{id} \times D)^{-1}\Gamma$. Therefore X is compact over B, as asserted.

Proposition (3.38). *Let T be Hausdorff over B and let T_0 be a closed subspace of T. Then for each Hausdorff space X over B the push-out of the cotriad*

$$T \times_B X \leftarrow T_0 \times_B X \to T_0$$

is also Hausdorff over B.

That this is true with the coarse topology follows at once from consideration of the coordinate functions t and x. Hence the result is true with any finer topology. In particular the fibre-cone and fibre-suspension of a Hausdorff space over B are also Hausdorff over B. A similar argument shows that the fibre join $X *_B Y$ of Hausdorff spaces X, Y over B is Hausdorff over B in the coarse topology and hence in any finer topology.

Taken together (3.28) and (3.38) imply

Proposition (3.39). *Let T be compact Hausdorff over B and let T_0 be a closed subspace of T. Then for each compact Hausdorff space X over B the push-out of the cotriad*

$$T \times_B X \leftarrow T_0 \times_B X \rightarrow T_0$$

is also compact Hausdorff over B.

Hence it follows, using (3.33), that the coarse topology and the fine topology on the push-out coincide, under the hypotheses of (3.39). In particular this is true for the fibre-cone and the fibre-suspension of a compact Hausdorff space over B. Similarly if X, Y are compact Hausdorff over B then so is $X *_B Y$, moreover the coarse and fine topologies coincide.

We turn now to the regularity axiom, where the right extension turns out to be as in

Definition (3.40). The space X over B is regular over B if for each point x of X and each neighbourhood V of x there exists a neighbourhood U of x such that $\mathscr{Cl}_b\, U \subset V$, where $b = px$.

It is important to notice that if the condition holds for all V in a subbasis for the topology of X, after supplementation, then it holds for all V and so X is regular over B.

Another way to formulate the definition is as follows. Let X be a space over B with projection p. Let E be a subset of a fibre X_b, with E closed in X, and let x be a point of X_b which is not contained in E. Then X is regular over B if, in this situation, there exist neighbourhoods U of x and V of E in X which are disjoint.

Thus B is regular over itself. More generally $B \times T$ is regular over B if and only if T is regular. It follows at once from the definition that a closed subspace of a regular space over B is also regular over B. Moreover the fibre sum and fibre product of two regular spaces over B are again regular over B. Finally if X is regular over B then ξ^*X is regular over B' for each space B' and map $\xi\colon B' \rightarrow B$.

If X is indiscrete over B then obviously X is regular over B. If X is compact Hausdorff over B then X is regular over B, from (3.35).

Next we extend the shrinking theorems to regular spaces over a base, beginning with

Proposition (3.41). *Let X be regular over B. Let C be a compact subset of some fibre X_b of X and let V be a neighbourhood of C in X. Then there exists a neighbourhood U of C in X such that $\mathscr{Cl}_b\, U \subset V$.*

Since V is a neighbourhood of each point x of C there exist, by fibrewise regularity, disjoint neighbourhoods U_x of x and W_x of $X_b - V_b$. The open sets $\{U_x\}$ $(x \in C)$ cover the compact C; extract a finite subcovering indexed by $x_1, \ldots, x_m$, say. Then the union

$$U = U_{x_1} \cup \cdots \cup U_{x_m}$$

is a neighbourhood of C and the intersection

$$W = W_{x_1} \cap \cdots \cap W_{x_m}$$

is a neighbourhood of $X_b - V_b$. These neighbourhoods do not intersect and so we have $\mathscr{Cl}_b\, U \subset V$, as required. We at once generalize this to

Proposition (3.42). *Let X be regular over B. Let C be a compact subset of some fibre X_b of X, and let $\{V_j\}$ $(j = 1, \ldots, n)$ be a covering of C by open sets of X. Then there exists a covering $\{U_j\}$ $(j = 1, \ldots, n)$ of C by open sets of X such that $\mathscr{Cl}_b\, U_j \subset V_j$ for each j.*

Write $V = V_2 \cup \cdots \cup V_n$; then $X - V$ is closed in X. Hence $C \cap (X - V)$ is closed in C and so is compact. Applying the previous proposition to the neighbourhood V_1 of $C \cap (X - V)$, we obtain a neighbourhood U of $C \cap (X - V)$ such that $\mathscr{Cl}_b\, U \subset V_1$. Now $C \cap V$ and $C \cap (X - V)$ cover C, hence V and U cover C. Thus $U_1 = U$ is a shrinking of V_1, in the present sense. We continue by repeating the argument for $\{U_1, V_2, \ldots, V_n\}$, so as to shrink V_2, and so on. Thus the result is obtained.

Proposition (3.43). *Let $\phi: X \to Y$ be a compact surjection over B. If X is regular over B then so is Y.*

For let X be regular over B. Let V be a neighbourhood of the point y of Y. Then $\phi^{-1}V$ is a neighbourhood of the compact $\phi^{-1}y \subset X$ and so, by (3.41), there exists a neighbourhood U of $\phi^{-1}y$ such that $\mathscr{Cl}_b\, U \subset \phi^{-1}V$; here $b = qy$. Now by (2.27) there exists a neighbourhood W of y such that $\phi^{-1}W \subset U$, and then $\mathscr{Cl}_b\, W \subset V$ since

$$\mathscr{Cl}_b\, W = \mathscr{Cl}_b\, \phi\phi^{-1}W = \phi\, \mathscr{Cl}_b\, \phi^{-1}W \subset \phi\, \mathscr{Cl}_b\, U \subset \phi\phi^{-1}V = V.$$

Thus Y is regular over B, as asserted.

Corollary (3.44). *Let T_0 be a closed subset of the compact regular space T over B. Then for each compact regular X over B the push-out of the cotriad*

$$T \times_B X \leftarrow T_0 \times_B X \rightarrow T_0$$

is also compact regular over B.

In particular the fibre-suspension $\Sigma_B X$ and the fibre-cone $\Gamma_B X$ of a compact regular X over B are again compact regular over B. Similarly one can show that the fibre-join of two compact regular spaces over B is again compact regular over B.

Proposition (3.45). *Let X be locally compact regular over B. Let C be a compact subset of some fibre X_b and let V be a neighbourhood of C in X. Then there exists a neighbourhood U of C in X such that $\mathscr{Cl}_b\, U \subset V$ and such that $\mathscr{Cl}\, U$ is compact over B.*

First take the special case where C is a point, say the point x of X_b, and V is a neighbourhood of x. Since X is locally compact over B there exists a neighbourhood W of x such that $\mathscr{C}\ell\ W$ is compact over B. Since X is regular over B so is $\mathscr{C}\ell\ W$. Now $V \cap \mathscr{C}\ell\ W$ is a neighbourhood of x in $\mathscr{C}\ell\ W$ and so there exists a neighbourhood N of x in $\mathscr{C}\ell\ W$ such that, in $\mathscr{C}\ell\ W$, $\mathscr{C}\ell_b\ N \subset V \subset \mathscr{C}\ell\ W$. So $N \cap W$ is a neighbourhood of x in X such that $\mathscr{C}\ell_b\ U \subset V$ and such that $\mathscr{C}\ell\ U$ is compact over B, as required.

In the general case, therefore, each point x of C admits a neighbourhood U_x such that $\mathscr{C}\ell\ U_x$ is compact over B and such that $\mathscr{C}\ell_b\ U_x \subset V$. The sets $\{U_x\}$ $(x \in C)$ constitute an open covering of the compact C; extract a finite subcovering indexed by $x_1, \ldots, x_n$, say. Then the union

$$U = U_{x_1} \cup \cdots \cup U_{x_n}$$

is a neighbourhood of C such that $\mathscr{C}\ell_b\ U \subset V$ and $\mathscr{C}\ell\ U$ is compact over B.

Definition (3.46). Let X be a space over B. Suppose that for each point b of B and each point x of X_b there exists a neighbourhood V of b and a neighbourhood U of x such that the projection maps U homeomorphically onto V. Then X is discrete over B.

In other words the condition is that the projection $X \to B$ is locally a homeomorphism, in particular is an open map. Note that $B \times T$ is discrete over B if and only if T is discrete. Also the fibre sum and fibre product of two discrete spaces over B are again discrete over B. Moreover if X is discrete over B then ξ^*X is discrete over B' for each space B' and map $\xi\colon B' \to B$.

If X is discrete over B then X is both Hausdorff and regular over B. Further each section of X over B is an open embedding. In general a subspace of a discrete space over B is not discrete over B. However if X is discrete over B then each open subspace X' of X is also discrete over B. In particular if X is discrete over B and X is locally connected then each component of X is discrete over B.

Proposition (3.47). *Let $\phi\colon X \to Y$ be a function over B, where X is discrete over B and Y is open over B. Then ϕ is continuous.*

For let $W \subset Y$ be open, and let x be a point of $\phi^{-1}W$. Then there exists a neighbourhood U of x and a neighbourhood V of px related as in (3.46). Therefore $U \cap p^{-1}qW$ is a neighbourhood of x contained in $\phi^{-1}W$. Thus ϕ is continuous.

Proposition (3.48). *Let $\theta, \phi\colon X \to Y$ be maps over B, where Y is discrete over B. If θ and ϕ coincide at some point x of X then they coincide on some neighbourhood of x. If in addition X is connected then $\theta = \phi$ throughout X.*

Let $p: X \to B$, $q: Y \to B$ be the projections, so that $q\theta = p = q\phi$. Since Y is discrete over B the projection q maps a neighbourhood U of $\theta x = \phi x$ in Y homeomorphically onto a neighbourhood V of px in B. Then θ and ϕ coincide with $q^{-1}p$ on the neighbourhood $\theta^{-1}U \cap \phi^{-1}U \cap p^{-1}V$ of x. Thus the set M of points of X where θ and ϕ coincide is open in X. However M is also closed in X, by (3.31), since Y is Hausdorff over B. Therefore $M = X$ if X is connected.

In discussing spaces over a base there are several other conditions which may be satisfied in particular cases. Each of these has a local form as well as a global form.

Definition (3.49). The space X over B is sectionable if there exists a section of X over B, locally sectionable if there exists an open covering of B such that X_U is sectionable over U for each member U of the covering.

If X is locally sectionable over B then the sectional category of X is defined to be the least number $\operatorname{secat}_B X$ of open sets required to cover B such that X is sectionable over each of the open sets. Note that the fibre product of two locally sectionable spaces over B is again locally sectionable over B. Also that if X is locally sectionable over B then ξ^*X is locally sectionable over B' for each space B' and map $\xi: B' \to B$; moreover

$$\operatorname{secat}_{B'} \xi^*X \leq \operatorname{secat}_B X.$$

Definition (3.50). The space X over B is sliceable if for each point b of B and each point x of X_b there exists a section $s: B \to X$ such that $sb = x$, locally sliceable if there exists an open covering of B such that X_U is sliceable over U for each member U of the covering.

If X is locally sliceable over B then the slicing category $\operatorname{slicat}_B X$ of X is defined to be the least number of open sets required to cover B such that X_U is sliceable over U for each of the open sets U.

Note that the fibre product of two locally sliceable spaces over B is again locally sliceable over B. Also that if X is locally sliceable over B then ξ^*X is locally sliceable over B' for each space B' and map $\xi: B' \to B$; moreover

$$\operatorname{slicat}_{B'} \xi^*X \leq \operatorname{slicat}_B X.$$

Definition (3.51). The space X over B is trivial if there exists a space T such that X is equivalent to $B \times T$, as a space over B, locally trivial if there exists an open covering of B such that X_U is trivial over U for each member U of the covering.

An equivalence $X \to B \times T$ or $B \times T \to X$ over B is called a trivialization of X with fibre T. Note that if X is trivial over B the fibre T is unique up to equivalence.

For an example of a space X over B which is locally trivial but not trivial consider the circle S. Represent points of S, in the usual way, by complex numbers z of unit modulus. Regard S as a space over itself with projection given by $z \mapsto z^2$. Then S is trivial over any proper subset of S. However S is not trivial over the whole of S, since the fibre is the point pair $\dot{I}$ and S is not homeomorphic to $S \times \dot{I}$, for reasons of connectivity.

Let X be a trivial space over B with discrete fibre T. A trivialization $\phi: B \times T \to X$ determines a T-indexed family of open sections $\phi_t: B \to X$, where $\phi_t(b) = \phi(b, t)$. Moreover for each point b of B and each point x of X_b there exists precisely one index t such that $\phi_t(b) = x$. Conversely if X is a space over B and if $\{\phi_t\}$ is a T-indexed family of open sections of X satisfying the uniqueness condition then X is trivial over B with discrete fibre T.

If X is locally trivial over B then the triviality category $\operatorname{trivcat}_B X$ of X is defined to be the least number of open sets required to cover B such that X_U is trivial over U for each of the open sets U. Note that the sum and fibre product of locally trivial spaces over B are again locally trivial over B. Also that if X is locally trivial over B then ξ^*X is locally trivial over B' for each space B' and map $\xi: B' \to B$; moreover

$$\operatorname{trivcat}_{B'} \xi^*X \leq \operatorname{tricat}_B X.$$

To understand the difference between the three conditions better, take $B = I$ and take X to be one of the four subsets of $I \times I$ shown below

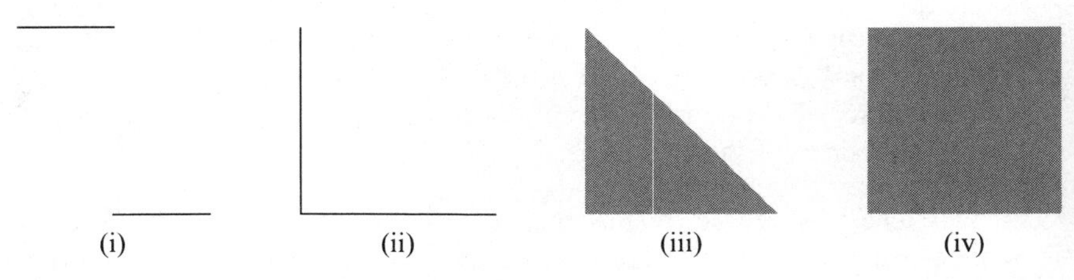

In each the projection is downwards onto the interval I. Then (i) is not sectionable, (ii) is sectionable but not sliceable, (iii) is sliceable but not trivial, and (iv) is trivial.

The following result is basic.

Proposition (3.52). *Let A be a space. Let X be a space over the cylinder $I \times A$. Suppose that X is trivial over $[0, \frac{1}{2}] \times A$ and over $[\frac{1}{2}, 1] \times A$. Then X is trivial over $I \times A$.*

Since $[0, \frac{1}{2}]$ and $[\frac{1}{2}, 1]$ intersect we may assume, without real loss of generality, that the fibre T is the same for both trivializations. We denote these by

$$f: [0, \tfrac{1}{2}] \times A \times T \to X|([0, \tfrac{1}{2}] \times A),$$

$$g: [\tfrac{1}{2}, 1] \times A \times T \to X|([\tfrac{1}{2}, 1] \times A).$$

Restrict f and g to the intersection $\{\frac{1}{2}\} \times A \times T$ of the domains. The composition of the restriction of g^{-1} with the restriction of f is an equivalence

$$h: \{\tfrac{1}{2}\} \times A \times T \to \{\tfrac{1}{2}\} \times A \times T$$

over $\{\frac{1}{2}\} \times A$. Disregard the point $\{\frac{1}{2}\}$ and take the product of the identity on $[\frac{1}{2}, 1]$ with h:

$$k: [\tfrac{1}{2}, 1] \times A \times T \to [\tfrac{1}{2}, 1] \times A \times T.$$

By mapping $[0, \frac{1}{2}] \times A \times T$ according to f and $[\frac{1}{2}, 1] \times A \times T$ according to gk^{-1} we obtain a trivialization of X over the whole of $I \times A$, as required.

The main discussion of locally trivial spaces, or fibre bundles as they are generally called, is reserved for a later chapter. However, I would like to give one or two more results here which are really consequences of (3.52). These results involve the following well-known property of compact metric spaces.

Proposition (3.53). *Let Γ be an open covering of the compact metric space X. Then there exists a positive number ε such that each open ball $B_\varepsilon(x)$ $(x \in X)$ is contained in some member of Γ.*

Such an ε is called a Lebesgue number of the covering. To show that it exists, first choose a positive number $\varepsilon(x)$, for each point x of X, such that $B_{\varepsilon(x)}(x)$ is contained in some member of Γ. From the open covering $\{B_{(1/2)\varepsilon(x)}(x)\}$ $(x \in X)$ of the compact X extract a finite subcovering indexed by $x_1, \ldots, x_n$, say. Then

$$\varepsilon = \tfrac{1}{2}\min(\varepsilon(x_{1'}), \ldots, \varepsilon(x_n))$$

is a Lebesgue number. For given any open ball $B_\varepsilon(x)$ we have that $x \in B_{(1/2)\varepsilon(x_i)}(x_i)$, for some $i = 1, \ldots, n$, and so for any point y of $B_\varepsilon(x)$ we have

$$\begin{aligned} d(y, x_i) &\leq d(y, x) + d(x, x_i) \\ &< \varepsilon + \tfrac{1}{2}\varepsilon(x_i) \leq \varepsilon(x_i), \end{aligned}$$

where d is the metric. Thus $B_\varepsilon(x) \subset B_{\varepsilon(x_i)}(x_i)$, and since the latter open ball is contained in some member of Γ, so is the former. Notice, incidentally, that since X is compact a finite subfamily of the $B_\varepsilon(x)$ will cover X.

Proposition (3.54). *Let X be locally trivial over I. Then X is trivial over I.*

For by (3.53) there exists a positive integer n such that X is trivial over each of the intervals

$$\left[\frac{i}{n}, \frac{i+1}{n}\right] \qquad (i = 0, \ldots, n-1).$$

Applying (3.52) over and over again, with $A = *$, the result follows.

Proposition (3.55). *Let X be locally trivial over $I \times I$. Then X is trivial over $I \times I$.*

Again by (3.53) there exists a positive integer n such that X is trivial over each of the subsets.

$$\left[\frac{i}{n}, \frac{i+1}{n}\right] \times \left[\frac{j}{n}, \frac{j+1}{n}\right] \qquad (i, j = 0, \dots, n-1).$$

For each j it follows from (3.52), with $A = [j/n, (j+1)/n]$, that X is trivial over $I \times [j/n, (j+1)/n]$, and then it follows from (3.52), with $A = I$, that X is trivial over $I \times I$, as asserted. Of course the argument can be extended to show that locally trivial spaces over I^n are trivial for any positive n.

Let us now turn to the problem of constructing an adjoint to the fibre product. This means, in effect, assigning a satisfactory topology to the set

$$\operatorname{map}_B(X, Y) = \coprod_{b \in B} \operatorname{map}(X_b, Y_b),$$

where X, Y are spaces over B. In other words we seek a fibrewise generalization of the compact-open topology.

For each pair of subsets $K \subset X$ and $V \subset Y$ let (K, V) denote the subset of $\operatorname{map}_B(X, Y)$ consisting, for each point $b \in B$, of maps $\phi: X_b \to Y_b$ such that $\phi K_b \subset V_b$. Here, as usual, we write $K_b = K \cap X_b$, $V_b = V \cap Y_b$. Note that if K_b is empty, for some b, then (K, V) contains all maps $X_b \to Y_b$. We have the relation

(3.56)
$$\left(\bigcup_{j \in J} K_j, V\right) = \bigcap_{j \in J} (K_j, V),$$

where $\{K_j\}$ $(j \in J)$ is a family of subsets of X, and the relation

(3.57)
$$\left(K, \bigcap_{j \in J} V_j\right) = \bigcap_{j \in J} (K, V_j),$$

where $\{V_j\}$ $(j \in J)$ is a family of subsets of Y.

Definition (3.58). The fibrewise compact-open topology for the set $\operatorname{map}_B(X, Y)$ over B is the topology generated, after supplementation, by the family of subsets (K, V), where $K \subset X$ is compact over B, and $V \subset Y$ is open.

We refer to the subbasic sets (K, V) as the fibrewise compact-open sets. For example $\operatorname{map}_B(X, Y) = (\varnothing, Y)$ itself is a fibrewise compact-open set. When the domain X is itself compact over B the fibrewise compact-open sets constitute a subbasis for the topology of $\operatorname{map}_B(X, Y)$ even before supplementation, since for each open subset $W \subset B$ we have

$$\operatorname{map}_W(X_W, Y_W) = (X - X_W, Y_W).$$

Moreover, when $B = *$ the fibrewise compact-open topology on $\mathrm{map}_*(X, Y)$ reduces to the compact-open topology on $\mathrm{map}(X, Y)$ as in the previous chapter.

If Y is indiscrete over B then so is $\mathrm{map}_B(X, Y)$ for all spaces X over B. In particular $\mathrm{map}_B(X, B) \equiv B$.

In case the domain space is B itself the compact subsets over B are precisely the closed subsets. We use this in the proof of

Proposition (3.59). *Let X be a space over B. Then $\mathrm{map}_B(B, X)$ is equivalent to X, as a space over B.*

For consider the function $\xi: \mathrm{map}_B(B, X) \to X$ which assigns to each map $\phi: \{b\} \to X_b$ the point $\phi(b) \in X_b$ $(b \in B)$. Clearly ξ is bijective. If $K \subset B$ is closed and $U \subset X$ is open then the direct image of the fibrewise compact-open set (K, U) is the open set $U \cap X_{B-K}$, so that ξ is open, by (3.12). Also ξ is continuous, since the inverse image of an open set U of X is precisely the fibrewise compact-open set (B, U) of $\mathrm{map}_B(B, X)$. This completes the proof.

Proposition (3.60). *Let X, Y be spaces over B. Let B' be a closed subspace of B and let X', Y' be the restrictions of X, Y to B'. Then the fibrewise compact-open topology for $\mathrm{map}_{B'}(X', Y')$ coincides with the induced topology which $\mathrm{map}_{B'}(X', Y')$ obtains as a subset of $\mathrm{map}_B(X, Y)$.*

The proof is straightforward and may be left to serve as an exercise.

Let X, Y, Z be spaces over the given space B. Precomposition with a map $\theta: X \to Y$ over B determines a map

$$\theta^*: \mathrm{map}_B(Y, Z) \to \mathrm{map}_B(X, Z)$$

over B, while postcomposition with a map $\phi: Y \to Z$ over B determines a map

$$\phi_*: \mathrm{map}_B(X, Y) \to \mathrm{map}_B(X, Z)$$

over B.

Proposition (3.61). *If θ is a compact surjection then θ^* is an embedding. If ϕ is an embedding then ϕ_* is an embedding.*

Let θ be a compact surjection. If $K \subset Y$ is compact over B then so is $\theta^{-1}K \subset X$. Since θ^* is injective we have $(K, V) = \theta^{*-1}(\theta^{-1}K, V)$, for each open set V of Z. Since $(\theta^{-1}K, V)$ is open this proves that θ^* is an embedding.

For the second part let ϕ be an embedding. Then ϕ_* is injective. Also for each fibrewise compact-open subset (K, U) of $\mathrm{map}_B(X, Y)$ there exists an open set V of Z such that $U = \phi^{-1}V$ and so (K, V) is a fibrewise compact-open subset of $\mathrm{map}_B(X, Z)$. Since $(K, U) = \phi_*^{-1}(K, V)$ this shows that ϕ_* is an embedding.

Proposition (3.62). *Let X_1, X_2 be spaces over B and let*

$$X_1 \underset{\sigma_1}{\rightarrow} X_1 + X_2 \underset{\sigma_2}{\leftarrow} X_2$$

be the standard insertions. Then the map

$$\mathrm{map}_B(X_1 + X_2, Y) \rightarrow \mathrm{map}_B(X_1, Y) \times_B \mathrm{map}_B(X_2, Y)$$

over B given by (σ_1^*, σ_2^*) is an equivalence of spaces over B, for all spaces Y over B.

This follows at once from the definition of the fibrewise compact-open topology since the subsets of $X_1 + X_2$ which are compact over B are precisely the sums $K_1 + K_2$, where $K_1 \subset X_1$, $K_2 \subset X_2$ are compact over B.

Proposition (3.63). *If Y is regular over B then so is $\mathrm{map}_B(X, Y)$ for all spaces X over B.*

For let $\phi: X_b \rightarrow Y_b$ be a map, regarded as a point of $\mathrm{map}_B(X, Y)$, and let (K, V) be a fibrewise compact-open neighbourhood of ϕ. Then V is a neighbourhood of the compact ϕK_b in Y and so, by (3.41), there exists a neighbourhood U of ϕK_b in Y such that $\mathscr{Cl}_b\, U \subset V$. Then $\phi \in (K, U)$ and $\mathscr{Cl}_b(K, U) \subset (K, V)$. For the latter, let $\psi: X_b \rightarrow Y_b$ be a map such that ψ is not contained in (K, V). Then $(K \cap (X - W), Y - \mathscr{Cl}\, U)$ is a fibrewise compact-open neighbourhood of ψ which does not meet (K, U). Thus $\mathscr{Cl}_b(K, U) \subset (K, V)$ as asserted, and the proof of (3.63) is complete.

After these preliminaries we can begin to show that $\mathrm{map}_B(\ ,\ )$ has the properties of an adjoint functor to $\times_B$. The first step is

Proposition (3.64). *Let X, Y, Z be spaces over B. If the function $h: X \times_B Y \rightarrow Z$ over B is continuous then so is the function $\hat{h}: X \rightarrow \mathrm{map}_B(Y, Z)$ over B given by*

$$\hat{h}(x)(y) = h(x, y) \qquad (x \in X,\, y \in Y).$$

For let (K, V) be a fibrewise compact-open subset of $\mathrm{map}_B(Y, Z)$ and let $x \in \hat{h}^{-1}(K, V)$. Then $h^{-1}V$ is a neighbourhood in $X \times_B Y$ of the compact subset $\{x\} \times K_{px}$. Since the projection $X \times_B K \rightarrow X \times_B B = X$ is compact there exists, by (2.27), a neighbourhood U of x in X such that $U \times_B K \subset h^{-1}V$. Then $U \subset \hat{h}^{-1}(K, V)$ and so h is continuous.

Proposition (3.65). *Let X be regular over B. Let Y be a space over B with topology generated, after supplementation, by a family Γ of subsets. Then the topology of $\mathrm{map}_B(X, Y)$ is generated, after supplementation, by the family of* *fibrewise compact-open subsets (K, V), where $K \subset X$ is compact over B and V is a member of Γ.*

In view of (3.57) we may assume, without real loss of generality, that Γ satisfies the basis condition. So let (K, V) be an arbitrary fibrewise compact-open subset of $\mathrm{map}_B(X, Y)$, and let $\phi: X_b \to Y_b$ be a map such that $\phi \in (K, V)$. Then $\phi K_b \subset V_b$, so $K_b \subset \phi^{-1} V_b$. Now V is the union of a family $\{V_j\}$ $(j \in J)$ of basic open sets. For each index j there exists an open set W_j of X such that $\phi^{-1}(V_j \cap Y_b) = W_j \cap X_b$. The family $\{W_j\}$ $(j \in J)$ covers the compact K_b. Extract a finite subcovering indexed by $j_1, \ldots, j_n$ say. Since X is regular over B we can shrink this to an open covering $\{U_j\}$ $(j = j_1, \ldots, j_n)$ of K_b such that $\mathscr{Cl}_b\, U_j \subset \phi^{-1}(V_j \cap Y_b)$ for each index j. Each of the sets $K \cap \mathscr{Cl}\, U_j$ is closed in K and so compact over B. Hence $(K \cap \mathscr{Cl}\, U_j, V)$ is a fibrewise compact-open neighbourhood of ϕ. Also $K \cap (X - U)$, where $U = \bigcup_{i=1}^{n} U_{j_i}$, is closed in K and so compact over B. Hence $(K \cap (X - U), V')$ is also a fibrewise compact-open neighbourhood of ϕ, for any open set V' of Y, since $K \cap (X - U)$ is empty over b. Choose V' be any of the members V_j of Γ. Then

$$(K \cap (X - U), V') \cap \bigcap_{i=1}^{n} (K \cap \mathscr{Cl}\, U_{j_i}, V_{j_i})$$

$$\subset (K \cap (X - U), V') \cap (K \cap U, V) \subset (K, V).$$

This proves (3.65) and we at once deduce

Proposition (3.66). *Let Y_1, Y_2 be spaces over B, and let*

$$Y_1 \underset{\pi_1}{\leftarrow} Y_1 \times_B Y_2 \underset{\pi_2}{\rightarrow} Y_2$$

be the standard projections. Then the map

$$\mathrm{map}_B(X, Y_1 \times_B Y_2) \to \mathrm{map}_B(X, Y_1) \times_B \mathrm{map}_B(X, Y_2)$$

over B given by π_{1*}, π_{2*} is an equivalence of spaces over B for all regular X over B.

Proposition (3.67). *Let X, Y be regular over B. Then for all spaces Z over B the fibrewise compact-open topology on $\mathrm{map}_B(X \times_B Y, Z)$ is generated, after supplementation, by the family of subsets $(E \times_B F, V)$, where $E \subset X$ and $F \subset Y$ are compact over B and where $V \subset Z$ is open.*

For let $K \subset X \times_B Y$ be compact over B and let N be a neighbourhood of K_b, for some point b of B. Since X and Y are regular over B, so is $X \times_B Y$. Hence there exists, for each point c of K_b, a neighbourhood W_c such that $\mathscr{Cl}_b\, W_c \subset N$. Hence, by the product topology, there exists a neighbourhood of the form $U_c \times_B V_c$, where $U_c \subset X$ and $V_c \subset Y$ are open, such that $\mathscr{Cl}_b\, U_c \times \mathscr{Cl}_b\, V_c$ is contained in N. Since K_b is compact we can extract from the open covering $\{U_c \times V_c\}$ $(c \in K_b)$ of K_b a finite subcovering indexed by $c_1, \ldots, c_n$, say.

Now let K_X, K_Y be the projections of K in X, Y respectively, and write

$$E_i = K_X \cap \mathscr{Cl}\, U_{c_i}, \qquad F_i = K_Y \cap \mathscr{Cl}\, V_{c_i} \qquad (i = 1, \ldots, n).$$

Then E_i and F_i are compact over B, since K_X and K_Y are compact over B, and we have

$$K_b \subset (X_b \times Y_b) \cap \bigcup_{i=1}^{n} (E_i \times_B F_i) \subset N.$$

So let (K, W) be a fibrewise compact-open subset of $\mathrm{map}_B(X \times_B Y, Z)$. If $\phi \in (K, W)$ where $\phi\colon X_b \to Y_b$, then $\phi K_b \subset W_b$ and so there exists a neighbourhood N of K_b in $X \times_B Y$ such that $N_b \subset \phi^{-1} W_b$. Hence, by the above, there exist subsets $E_i \subset X$, $F_i \subset Y$ $(i = 1, \ldots, n)$, which are compact over B, such that each fibrewise compact-open set $(E_i \times_B F_i, W)$ is a neighbourhood of ϕ, and such that

$$\bigcap_{i=1}^{n} (E_i \times_B F_i, W) = \left(\bigcup_{i=1}^{n} (E_i \times_B F_i), W \right) \subset (K, W).$$

This proves (3.67) and we at once deduce

Proposition (3.68). *Let X_i $(i = 1, 2)$ be regular over B and let Y_i $(i = 1, 2)$ be spaces over B. Then the function*

$$\mathrm{map}_B(X_1, Y_1) \times_B \mathrm{map}_B(X_2, Y_2) \to \mathrm{map}_B(X_1 \times_B X_2, Y_1 \times_B Y_2)$$

given by the fibre product functor is an embedding.

Proposition (3.69). *Let T be compact regular over B and let T_0 be a closed subspace of T. For each space X over B let $\Phi_B(X)$ be the push-out of the cotriad*

$$T \times_B X \leftarrow T_0 \times_B X \to T_0$$

and similarly for maps over B. Then the function

$$\Phi_{B\#}\colon \mathrm{map}_B(X, Y) \to \mathrm{map}_B(\Phi_B X, \Phi_B Y)$$

over B is continuous for all regular X over B and all Y over B.

To see this consider the diagram shown below, where T_B stands for the functor $T \times_B$ and where $p\colon T_B \to \Phi_B$ is the obvious natural transformation.

$$\begin{array}{ccc} \mathrm{map}_B(X, Y) & \xrightarrow{T_{B\#}} & \mathrm{map}_B(T_B X, T_B Y) \\ \downarrow{\scriptstyle \Phi_{B\#}} & & \downarrow{\scriptstyle p_*} \\ \mathrm{map}_B(\Phi_B X, \Phi_B Y) & \xrightarrow[p^*]{} & \mathrm{map}_B(T_B X, \Phi_B Y) \end{array}$$

Since T, and hence T_0, is compact regular over B the natural projection $p\colon T_B X \to \Phi_B X$ is a compact surjection, hence p^* is an embedding, by (3.61) above. Also $T_{B\#}$ and p_* are continuous, from first principles. Therefore $p^*\Phi_{B\#} = p_* T_{B\#}$ is continuous and so $\Phi_{B\#}$ is continuous, as asserted.

Composition of maps determines a function

$$\operatorname{map}_B(Y, Z) \times_B \operatorname{map}_B(X, Y) \to \operatorname{map}_B(X, Z)$$

over B. In particular (taking $X = B$ and replacing Y, Z by X, Y respectively) evaluation determines a function

$$\operatorname{map}_B(X, Y) \times_B X \to Y$$

over B. Although these functions are not continuous in general we have

Proposition (3.70). *If Y is locally compact regular over B the function*

$$\operatorname{map}_B(Y, Z) \times_B \operatorname{map}_B(X, Y) \to \operatorname{map}_B(X, Z)$$

over B is continuous for all spaces X, Z over B.

Corollary (3.71). *If X is locally compact regular over B the function*

$$\operatorname{map}_B(X, Y) \times_B X \to Y$$

over B is continuous for all spaces Y over B.

To prove (3.70), let (K, V) be a fibrewise compact-open subset of $\operatorname{map}_B(X, Z)$. Let $\theta\colon X_b \to Y_b$, $\phi\colon Y_b \to Z_b$, for some $b \in B$, be maps such that $\phi\theta \in (K, V)$. Then $\phi^{-1}V_b$ is a neighbourhood in Y_b of the compact θK_b. Since Y is locally compact regular over B there exists, by (3.45), a neighbourhood U of θK_b in Y such that $\mathscr{Cl}\, U$ is compact over B and such that $\mathscr{Cl}_b\, U \subset \phi^{-1}V_b$. The composition function sends $(K, U) \times_B (\mathscr{Cl}\, U, V)$ into (K, V). Since $\theta \in (K, U)$ and $\phi \in (\mathscr{Cl}\, U, V)$ this proves (3.70) and hence (3.71).

Corollary (3.72). *Let $\hat{h}\colon X \to \operatorname{map}_B(Y, Z)$ be a map over B, where Y is locally compact regular over B. Then $h\colon X \times_B Y \to Z$ is a map over B, where*

$$h(x, y) = \hat{h}(x)(y) \qquad (x \in X,\ y \in Y).$$

This constitutes a converse of (3.64), subject to the restriction on Y. It is usual to refer to h as the adjoint of $\hat{h}$, when the functions are related as above. For the proof of (3.72) it is only necessary to observe that h may be expressed as the composition

$$X \times_B Y \to \operatorname{map}_B(Y, Z) \times_B Y \to Z$$

of $\hat{h} \times_B \mathrm{id}_Y$ and the evaluation map.

Another application of these results is to prove

Proposition (3.73). *If $\phi: X \to Y$ is a quotient map over B then so is*

$$\phi \times \mathrm{id}_T: X \times_B T \to Y \times_B T,$$

where T is locally compact regular over B.

For let Z be a space over B and let f, g be functions over B making the diagram on the left commutative.

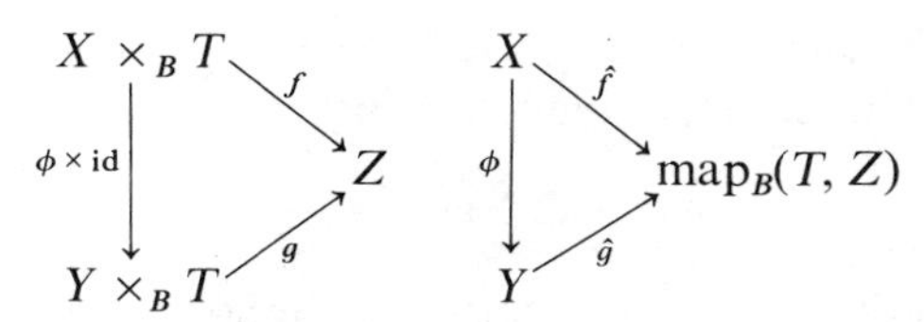

Suppose that f is continuous. Then the adjoints $\hat{f}$, $\hat{g}$ are defined, as shown on the right, and $\hat{f}$ is continuous. Therefore $\hat{g}$ is continuous, since ϕ is a quotient map, and so g is continuous, by (3.72). Therefore $\phi \times \mathrm{id}$ is a quotient map as asserted.

Of course this result can perfectly well be proved directly but the proof given is somewhat easier. As an application the reader may wish to demonstrate

Exercise (3.74). Show that if either
(i) X and Z are locally compact regular over B, or
(ii) X and Y (or Y and Z) are compact regular over B then

$$(X *_B Y) *_B Z \equiv X *_B (Y *_B Z),$$

in the fine topology.

We are now ready to establish the exponential law, which justifies the use of the term adjoint in this context.

Proposition (3.75). *Let X, Y, Z be spaces over B and let*

$$\xi: \mathrm{map}_B(X \times_B Y, Z) \to \mathrm{map}_B(X, \mathrm{map}_B(Y, Z))$$

be the injection defined by taking adjoints as in (2.90). *If X is regular over B then ξ is continuous. If Y is also regular over B then ξ is an open embedding. If, further, Y is locally compact over B then ξ is a homeomorphism.*

First observe that if $E \subset X$, $F \subset Y$ and $V \subset Z$ then

$$\xi(E \times_B F, V) = (E, (F, V)), \qquad \xi^{-1}(E, (F, V)) = (E \times_B F, V).$$

When X is regular over B the topology of $\mathrm{map}_B(X, \mathrm{map}_B(Y, Z))$ is generated after supplementation by fibrewise compact-open sets of the form $(E, (F, V))$, where E, F are compact over B and V is open. The inverse image of this subset is $(E \times_B F, V)$, which is also fibrewise compact-open, and so ξ is continuous.

When Y is also regular over B the topology of $\mathrm{map}_B(X \times_B Y, Z)$ is generated after supplementation by fibrewise compact-open sets of the form $(E \times_B F, V)$, where E and F are compact over B and V is open. The direct image of $(E \times_B F, V)$ is $(E, (F, V))$, which is also fibrewise compact-open, and so ξ is an open embedding. The final assertion follows at once from (2.100).

Let X, Y be spaces over B. For each space B' and map $\xi: B' \to B$ we have a continuous bijection

$$\xi_{\#}: \mathrm{map}_{B'}(\xi^* X, \xi^* Y) \to \xi^* \mathrm{map}_B(X, Y)$$

over B'. We prove

Proposition (3.76). *Let X be locally compact regular over B, and let Y be a space over B. Then $\xi_{\#}$ is an equivalence of spaces over B'.*

Since the evaluation function

$$\mathrm{map}_B(X, Y) \times_B X \to Y$$

is continuous so is the pull-back

$$\xi^*(\mathrm{map}_B(X, Y) \times_B X) \to \xi^* Y.$$

Rewriting the domain here according to (3.9) we regard the pull-back as a map

$$\xi^* \mathrm{map}_B(X, Y) \times_{B'} \xi^* X \to \xi^* Y$$

over B'. Taking the adjoint we obtain a map

$$\xi^* \mathrm{map}_B(X, Y) \to \mathrm{map}_{B'}(\xi^* X, \xi^* Y)$$

over B' which is obviously the inverse of $\xi_{\#}$. This proves (3.76) and, taking $B = *$ and replacing B, X, Y by B, X_0, Y_0, respectively we obtain

Corollary (3.77). *For any space B there is a natural equivalence*

$$\mathrm{map}_B(B \times X_0, B \times Y_0) \equiv B \times \mathrm{map}(X_0, Y_0)$$

where X_0 is locally compact regular and Y_0 is any space.

The space $\mathrm{map}_B(X, Y)$ over B should not be confused with the space $\mathrm{MAP}_B(X, Y)$ of maps $X \to Y$ over B. The former has the fibrewise compact-open topology while the latter has the topology induced by the ordinary compact-open topology on the space $\mathrm{map}(X, Y)$.

However, for any space Z over B we may consider the space $\mathrm{sec}_B\, Z$ of sections $B \to Z$, with the ordinary compact-open topology. For any space X over B there is a map $B \to \mathrm{map}_B(X, X)$ given by $b \mapsto \mathrm{id}_{X_b}$; this may be regarded as the adjoint of the equivalence between X and $B \times_B X$. Hence, taking $Z = \mathrm{map}_B(X, Y)$ the function

$$\sigma: \mathrm{MAP}_B(X, Y) \to \mathrm{sec}_B\, \mathrm{map}_B(X, Y)$$

is defined, where σ transforms each map $\phi: X \to Y$ over B into the section $\sigma(\phi)$ which sends the point $b \in B$ into the map $\phi_b: X_b \to Y_b$. Moreover, σ is continuous for all Y. For consider the subbasic open set $(C, (K, V))$ of the codomain, where $C \subset B$ is compact, where $K \subset X$ is compact over B, and where $V \subset Y$ is open. Then $\sigma^{-1}(C, (K, V)) = (K_C, V)$, where K_C is compact, since C is compact and K is compact over B. This shows that σ is continuous and we go on to prove

Proposition (3.78). *Let X be compact regular over B. Then*

$$\sigma: \mathrm{MAP}_B(X, Y) \to \mathrm{sec}_B\, \mathrm{map}_B(X, Y)$$

is an equivalence of spaces over B, for all spaces Y over B.

First observe that σ is bijective. For an inverse function is defined by associating with each section $s: B \to \mathrm{map}_B(X, Y)$ the composition

$$X \equiv B \times_B X \to \mathrm{map}_B(X, Y) \times_B X \to Y$$

of $s \times_B \mathrm{id}_X$ with the evaluation function. To see that σ^{-1} is continuous, i.e. that σ is open, take a compact-open subset (C, V) of $\mathrm{MAP}_B(X, Y)$. Then $\sigma(C, V) = (pC, (C, V))$, which is open since pC is compact and since C is closed in X and so compact over B. This completes the proof.

We began this chapter with a discussion of spaces under a given space and then went on to consider spaces over a given space. Now, in the last part of this chapter, we combine these ideas and outline the theory of spaces over and under a given space. Of course a space over and under a given space may be regarded either as a cosectioned space under or as a sectioned space over the given space. On the whole the latter point of view seems to be the more satisfactory and so we will choose our terminology accordingly. What we are now going to describe is the natural generalization of the theory of pointed spaces.

The space we work over and under is called the base and usually denoted by B. A sectioned space over B is a triple consisting of a space X and maps

$$B \xrightarrow{s} X \xrightarrow{p} B$$

such that $ps = \mathrm{id}_B$. Usually X alone is sufficient notation. The map p is called the projection and the map s the section. We regard B as a sectioned space over itself with projection and section the identity.

Let X, Y be sectioned spaces over B with projections p, q and sections s, t respectively. By a map $\phi: X \to Y$ of sectioned spaces over B we mean a map in the ordinary sense such that $q\phi = p$ and such that $\phi s = t$. The term fibrewise pointed map may also be used. The pointed set of such maps is denoted by $\mathrm{MAP}_B^B(X, Y)$, the basepoint being the nul-map tp. With this definition of morphism the category $\mathscr{T}_B^B$ of sectioned spaces over B is defined. We know from the generalities in the first chapter that B constitutes a nul object for the category. We also know that the pull-back in $\mathscr{T}$ of a triad in $\mathscr{T}_B^B$ is automatically a pull-back in $\mathscr{T}_B^B$ and that the push-out in $\mathscr{T}$ of a cotriad in $\mathscr{T}_B^B$ is automatically a push-out in $\mathscr{T}_B^B$.

Returning to the category $\mathscr{T}_B$ for a moment, let X, X' be spaces over B and let $\phi: X' \to X$ be a continuous injection over B. The push-out of the cotriad

$$X \overset{\phi}{\leftarrow} X' \to B$$

is called the space over B obtained from X by fibre collapsing with respect to ϕ. Provided the projection $X' \to B$ is surjective the push-out may be regarded as a quotient space of X alone; when X' is empty the push-out is just the disjoint union $X + B$. Usually X' is a subspace of X and ϕ the inclusion. Then the push-out is called the space over B obtained from X by fibre collapsing X', and denoted by $X/_B X'$. Note that $X/_B X'$ has a natural section $X'/_B X'$ which is closed when X' is closed in X.

Exercise (3.79). Let X, Y be spaces over B and let X' be a closed subspace of X. Either suppose that X' is compact over B or that Y is locally compact regular over B. Then

$$\frac{X *_B Y}{X' *_B Y} \equiv \left(\frac{X}{X'}\right) *_B Y$$

in the fine topology.

If X, Y are sectioned spaces over B, with sections s, t respectively, then the push-out of the cotriad

$$X \overset{s}{\leftarrow} B \overset{t}{\rightarrow} Y$$

is denoted by $X \vee_B Y$ and called the fibre-wedge sum. The sections s, t also define a triad

$$X \underset{u}{\rightarrow} X \times_B Y \underset{v}{\leftarrow} Y,$$

where the components of u are (id_X, tp) and the components of v are (sq, id_Y). Since u, v are injections so is the push-out

$$X \vee_B Y \to X \times_B Y;$$

the fibre-collapse of the push-out is denoted by $X \wedge_B Y$ and called the fibre-smash product. Both the fibre-wedge and the fibre-smash constitute bifunctors

$$\mathscr{T}_B^B \times \mathscr{T}_B^B \to \mathscr{T}_B^B.$$

Sectioned spaces where the section is closed play a special role as in

Proposition (3.80). *Let X, Y, Z be sectioned spaces over B with closed sections. Then*

$$(X \wedge_B Z) \vee_B (Y \wedge_B Z) \equiv (X \vee_B Y) \wedge_B Z.$$

Proposition (3.81). *Let X, Y, Z be sectioned spaces over B with closed sections. Suppose that either*

(i) *X and Z are locally compact regular over B, or that*
(ii) *X and Y (or Y and Z) are compact regular over B. Then*

$$(X \wedge_B Y) \wedge_B Z \equiv X \wedge_B (Y \wedge_B Z).$$

The proofs are similar to those in case $B = *$ and will therefore be omitted.

Proposition (3.82). *Let X, Y be sectioned spaces over B with closed sections, and let X' be a closed subspace of X. If Y is locally compact regular over B or if X' is compact over B then*

$$(X \wedge_B Y)/_B(X' \wedge_B Y) \equiv (X/_B X') \wedge_B Y.$$

The proof is left to serve as an exercise.

Note that a functor $B \times \colon \mathscr{T}^* \to \mathscr{T}_B^B$ is given by $(B \times)T = B \times T$, and similarly for maps, with the section being determined by the basepoint of T. A sectioned space X over B is said to be trivial if there exists a pointed space T such that $B \times T$ is equivalent to X, as a sectioned space over B. Such a pointed space T, if it exists, is unique up to equivalence; it is therefore customary to refer to T as *the* fibre.

Proposition (3.83). *Let A be a space. Let X be a sectioned space over the cylinder $I \times A$. Suppose that X is trivial over $[0, \frac{1}{2}] \times A$ and over $[\frac{1}{2}, 1] \times A$. Then X is trivial over $I \times A$.*

The proof is essentially the same as that of the corresponding result (3.52) in the non-sectioned case, and so will be left as an exercise.

Suppose that we have an endofunctor Φ_B of $\mathscr{T}_B$. For each sectioned space X over B the section $B \to X$ has a left inverse, i.e. the projection, hence the transform $\Phi_B B \to \Phi_B X$ has a left inverse. We define $\Phi_B^B X$ to be the push-out of the cotriad

$$\Phi_B X \leftarrow \Phi_B B \to B,$$

and similarly for maps of sectioned spaces. The endofunctor Φ_B^B of $\mathscr{T}_B^B$ thus defined is called the reduction of the original endofunctor Φ_B.

For example, let T be a locally compact regular space over B, and let T_0 be a closed subspace of T. Define $\Phi_B X$, for each space X over B, to the push-out of the cotriad

$$T \times_B X \leftarrow T_0 \times_B X \to T_0,$$

and similarly for maps over B. Using (3.82) and transitivity of quotients it follows at once that in this case $\Phi_B^B X$, for each space X over B with closed section, is equivalent to the fibre-smash product $(T/_B T_0) \wedge_B X$, and similarly for the maps of sectioned spaces over B. In particular the reduced fibre-cone functor Γ_B^B is equivalent to $(B \times I)\wedge_B$ and the reduced fibre-suspension functor to $(B \times S)\wedge_B$, where I has basepoint 0 and S has basepoint $\dot{I}/\dot{I}$.

Another relationship involving the fibre-smash product may be deduced from (3.17) above. First observe that the fibre-suspension Σ_B may be regarded as a functor $\mathscr{T}_B \to \mathscr{T}_B^B$ instead of an endofunctor of $\mathscr{T}_B$. Specifically we embed X in the fibre-cone $\Gamma_B X$, through the correspondence $x \mapsto (1, x)$, and then take $\Sigma_B X$ to be the sectioned space $\Gamma_B X/_B X$; of course this is equivalent to the previous definition of fibre-suspension in the fine topology. Note that the section of $\Sigma_B X$ is always closed. Now (3.17) shows that for all spaces X, Y over B we have an equivalence

$$(\Gamma_B(X *_B Y), X *_B Y) \to (\Gamma_B X \times_B \Gamma_B Y, \Gamma_B X \times_B Y \cup X \times_B \Gamma_B Y)$$

of pairs of spaces over B, and hence an equivalence

$$\frac{\Gamma_B(X *_B Y)}{X *_B Y} \to \frac{\Gamma_B X \times_B \Gamma_B Y}{\Gamma_B X \times_B Y \cup X \times_B \Gamma_B Y}$$

of sectioned spaces over B. We therefore conclude that

(3.84) $$\Sigma_B(X *_B Y) \equiv (\Sigma_B X) \wedge_B (\Sigma_B Y),$$

for all spaces X, Y over B.

We now turn to the process of fibrewise compactification. This may be regarded as a functor from the category of locally compact Hausdorff spaces over B and compact maps over B to the category of compact Hausdorff sectioned spaces over B and maps of sectioned spaces over B. As we shall see this leads, inter alia, to a better understanding of the relation between the sum $+$ and product $\times_B$ in $\mathscr{T}_B$, on the one hand, and the wedge $\vee_B$ and smash $\wedge_B$ in $\mathscr{T}_B^B$ on the other.

Let X be locally compact Hausdorff over B. Then a compact Hausdorff space X_B^+ over B can be constructed as follows. Take X_B^+, as a set, to be the disjoint union $X + B$, and give X_B^+ the following topology. The generating open sets, before supplementation, are of two kinds. The first kind are the open sets of X, regarded as subsets of X_B^+. The second kind are the complements in X_B^+ of the subsets of X which are compact over B. Clearly the inclusion $X \to X_B^+$ is an open embedding. Moreover X_B^+ reduces to the topological sum $X + B$ when X is compact Hausdorff over B.

I assert that X_B^+ is Hausdorff over B. Certainly each pair of distinct points in a fibre of X can be separated in X and so can be separated in X_B^+. It remains to be shown that each point x of X can be separated from its projection $px = b$ in the other summand of X_B^+. Since X is locally compact over B there exists a neighbourhood U of x such that $\mathscr{Cl}\ U$ is compact over B. Then U and $X_B^+ - \mathscr{Cl}\ U$ are disjoint neighbourhoods of x and b, respectively. Thus X_B^+ is Hausdorff over B, as asserted.

I also assert that X_B^+ is compact over B. To see this we use (3.16). We have given X_B^+ the topology generated, after supplementation, by the two families of subsets described above. The first family, the open sets of X, is closed under finite intersection and unrestricted union. The second family, the complements of the fibrewise compact sets of X, is also closed under finite intersection and unrestricted union, since X is Hausdorff over B. Thus a neighbourhood U of a fibre X_b^+, in the topology before supplementation, is a union $V \cup (X_B^+ - K)$, where $V \subset X$ is open and $K \subset X$ is compact over B. Now the section $s: B \to X_B^+$ does not meet X and so does not meet V. Therefore $s(b) \in X_B^+ - K$, hence $b \notin pK$. Since pK is closed in B there exists a neighborhood N of b which does not meet pK. Then $(p^+)^{-1}N$ does not meet K and so is contained in $X_B^+ - K$ and hence in U. Thus p^+ is closed in the topology before supplementation. Also X_b^+ is the compactification of the locally compact Hausdorff space X_b and so compact. Thus X_B^+ is compact over B in the topology before supplementation and so compact over B after supplementation by (3.16).

Let X, Y be locally compact Hausdorff over B. Clearly each function $\phi: X \to Y$ over B determines a section-preserving function $\phi_B^+ : X_B^+ \to Y_B^+$ over B, and vice versa. I assert that ϕ_B^+ is continuous if and only if ϕ is compact. For suppose that ϕ is compact. If $K \subset Y$ is compact over B then $\phi^{-1}K \subset X$ is compact over B. Hence if $Y_B^+ - K$ is a neighbourhood of some point of the section of Y_B^+ then $X_B^+ - \phi^{-1}K$ is a neighbourhood of the corresponding point of the section of X_B^+. This proves continuity at points of the section; continuity away from such points is obvious.

Conversely suppose that ϕ_B^+ is continuous. Then $\phi^{-1}(Y_B^+ - K) = X_B^+ - \phi^{-1}K$ is open for each compact K over B, and hence $\phi^{-1}K$ is compact over B. So let $\{U_j\}$ $(j \in J)$ be an open covering of Y such that $\mathscr{Cl}\, U_j$ is compact over B for each index j. Then each of the maps $\phi^{-1}\,\mathscr{Cl}\, U_j \to \mathscr{Cl}\, U_j$ determined by ϕ is closed and so compact. Therefore ϕ is compact, by (2.55).

Fibrewise compactification has satisfactory naturality properties. For let X be a locally compact Hausdorff space over B. Then ξ^*X is a locally compact Hausdorff space over B' for each space B' and map $\xi: B' \to B$. Hence the fibrewise compactifications of X and ξ^*X are defined, as sectioned spaces over B and B', respectively. When ξ is compact there is a continuous bijection

$$\theta: (\xi^*X)_{B'}^+ \to \xi^*(X_B^+)$$

of sectioned spaces over B', given by the canonical function. Now X_B^+ is Hausdorff over B, hence $\xi^*(X_B^+)$ is Hausdorff over B', and $(\xi^*X)_{B'}^+$ is compact over B'. Therefore θ is an equivalence of sectioned spaces over B'.

Taking $B = *$ and replacing B' by B we obtain

Corollary (3.85). *Let B be compact. Then*

$$(B \times T)_B^+ \equiv B \times T^+,$$

as sectioned spaces over B, for all locally compact Hausdorff spaces T.

Exercise (3.86). Let X be locally compact Hausdorff over B. Let $\phi: X' \to X$ be a closed embedding over B. Then X' is locally compact Hausdorff over B. Also $\phi^+: X_B'^+ \to X_B^+$ is a closed embedding. Finally $X/_B X'$ is equivalent to $X_B^+/_B X_B'^+$ as a sectioned space over B.

Proposition (3.87). *Let X, Y be locally compact Hausdorff over B. Then*

$$X_B^+ \vee_B Y_B^+ \equiv (X + Y)_B^+, \qquad X_B^+ \wedge_B Y_B^+ \equiv (X \times_B Y)_B^+,$$

as sectioned spaces over B.

Corollary (3.88). *Let X be locally compact Hausdorff over B. Then*

$$(X \times \mathbb{R}^n)_B^+ \equiv X \wedge_B (B \times S^n)$$

as sectioned spaces over B.

To prove the first part of (3.87) consider the standard insertions

$$X \to X + Y \leftarrow Y.$$

Since these are both compact they induce maps

$$X_B^+ \to (X + Y)_B^+ \leftarrow Y_B^+$$

of sectioned spaces over B. The fibre-wedge

$$X_B^+ \vee_B Y_B^+ \to (X + Y)_B^+$$

of these maps over B is a continuous bijection and hence an equivalence of sectioned spaces, since $X_B^+ \vee_B Y_B^+$ is compact over B and $(X + Y)_B^+$ is Hausdorff over B.

To prove the second part of (3.87), consider the bijection

$$\xi: (X \times_B Y)_B^+ \to X_B^+ \wedge_B Y_B^+$$

which is given by the identity on $X \times_B Y$ and preserves sections. Clearly ξ sends $U \times_B V$, where $U \subset X$ and $V \subset Y$ are open, into $U \times_B V$, which is open in the codomain. Also $\xi | (X \times_B Y)$ is continuous and so sends the fibrewise compact subset K of $X \times_B Y$ into the fibrewise compact subset $\xi K \subset X_B^+ \wedge_B Y_B^+$. Moreover $X_B^+ \wedge_B Y_B^+$ is Hausdorff over B, by (3.38), since X_B^+ and Y_B^+ are Hausdorff over B and the projection $X_B^+ \times_B Y_B^+ \to X_B^+ \wedge_B Y_B^+$ is compact. Therefore ξK is closed, and hence the image of the complement of K in $(X \times_B Y)_B^+$ is open. Thus ξ is continuous. However $(X \times_B Y)_B^+$ is compact over B and $X_B^+ \wedge_B Y_B^+$ is Hausdorff over B, as we have seen, so that ξ is an equivalence, as asserted.

Finally let us turn our attention to the problem of constructing an adjoint functor for the fibre-smash product. If X and Y are sectioned spaces over B we denote by $\mathrm{map}_B^B(X, Y)$ the subspace of $\mathrm{map}_B(X, Y)$ consisting of pointed maps, where the basepoints in the fibres are determined by the sections in the usual way. Here $\text{map}_B^B(X, Y)$ itself is regarded as a sectioned space over B, the section being that which sends each point $b \in B$ into the nul-map $X_b \to Y_b$.

Proposition (3.89). *Let X be a space over B. Then*

$$\text{map}_B^B(X/_B \varnothing, Y) \equiv \text{map}_B(X, Y)$$

for any sectioned space Y over B.

Since $X/_B \varnothing = X + B$ we have the natural equivalence

$$\text{map}_B(X/_B \varnothing, Y) \equiv \text{map}_B(X, Y) \times_B \text{map}_B(B, Y)$$

by (3.62). By inspection this transforms $\text{map}_B^B(X/_B \varnothing, Y)$ bijectively, and therefore homeomorphically, into $\text{map}_B(X, Y) \times_B \text{map}_B^B(B, Y)$. Since

$$\text{map}_B^B(B, Y) \equiv B$$

the result follows at once.

Corollary (3.90). *Regard $B \times \dot{I}$ as a sectioned space over B with section given by 0. Then evaluation at points $(b, 1)$, for each $b \in B$, determines a natural equivalence*

$$\text{map}_B^B(B \times \dot{I}, X) \equiv X$$

for each sectioned space X over B.

Proposition (3.91). *Let X_1, X_2 be sectioned spaces over B with closed sections and let*

$$X_1 \underset{\sigma_1}{\rightarrow} X_1 \vee_B X_2 \underset{\sigma_2}{\leftarrow} X_2$$

be the standard insertions. Then the map

$$\text{map}_B^B(X_1 \vee_B X_2, Y) \to \text{map}_B^B(X_1, Y) \times_B \text{map}_B^B(X_2, Y)$$

given by σ_1^*, σ_2^* is an equivalence of sectioned spaces for all sectioned spaces Y over B.

Obviously the map in question is a continuous bijection. Moreover the quotient map

$$X_1 + X_2 \to X_1 +^B X_2 \equiv X_1 \vee_B X_2$$

is closed, by (2.45), and hence compact, since the fibres are finite and so compact. Hence and from (3.61) the result follows.

Proposition (3.92). *Let X, Y, Z be sectioned spaces over B. If the fibrewise pointed function $h: X \wedge_B Y \to Z$ is continuous then so is the fibrewise pointed function $\hat{h}: X \to \text{map}_B^B(Y, Z)$, where*

$$\hat{h}(x)(y) = h(x, y) \qquad (x \in X, y \in Y).$$

This follows at once from (3.64). Similarly (3.66) implies

Proposition (3.93). *Let Y_1, Y_2 be sectioned spaces over B and let*

$$Y_1 \overset{\pi_1}{\leftarrow} Y_1 \times_B Y_2 \overset{\pi_2}{\rightarrow} Y_2$$

be the standard projections. Then the map

$$\mathrm{map}_B^B(X, Y_1 \times_B Y_2) \to \mathrm{map}_B^B(X, Y_1) \times_B \mathrm{map}_B^B(X, Y_2)$$

given by π_{1*}, π_{2*} is an equivalence of sectioned spaces for all regular sectioned spaces X over B.

Proposition (3.94). *Let X_1, X_2 be compact regular sectioned spaces over B and let Y_1, Y_2 be sectioned spaces over B. Then the injection*

$$\mathrm{map}_B^B(X_1, Y_1) \wedge_B \mathrm{map}_B^B(X_2, Y_2) \to \mathrm{map}_B^B(X_1 \wedge_B X_2, Y_1 \wedge_B Y_2)$$

given by the fibre-smash product is continuous.

This can be established by consideration of the diagram shown below, where ρ is the generic quotient map, where ξ is given by the ordinary fibre-product, and where η is given by the fibre-smash product.

$$\begin{array}{ccc}
\mathrm{map}_B^B(X_1, Y_1) \times_B \mathrm{map}_B^B(X_2, Y_2) & \xrightarrow{\xi} & \mathrm{map}_B^B(X_1 \times_B X_2, Y_1 \times_B Y_2) \\
 & & \downarrow \rho_* \\
\downarrow \rho & & \mathrm{map}_B^B(X_1 \times_B X_2, Y_1 \wedge_B Y_2) \\
 & & \uparrow \rho^* \\
\mathrm{map}_B^B(X_1, Y_1) \wedge_B \mathrm{map}_B^B(X_2, Y_2) & \xrightarrow[\eta]{} & \mathrm{map}_B^B(X_1 \wedge_B X_2, Y_1 \wedge_B Y_2)
\end{array}$$

The top horizontal ξ is an embedding, by (3.68); in particular ξ is continuous. Therefore $\rho_*\xi$ is continuous and so $\rho^*\eta$ is continuous since ρ is a quotient map. However ρ^* is an embedding, by (3.61) since $\rho: X_1 \times_B X_2 \to X_1 \wedge_B X_2$ is compact. Therefore η is continuous, as asserted. I do not know whether η is an embedding in general.

In the fibrewise pointed theory composition of pointed maps determines a fibrewise pointed function

$$\mathrm{map}_B^B(Y, Z) \wedge_B \mathrm{map}_B^B(X, Y) \to \mathrm{map}_B^B(X, Z)$$

for all sectioned spaces X, Y, Z. In particular (take $X = B \times \dot{I}$ and replace Y, Z by X, Y respectively) evaluation determines a fibrewise pointed function

$$\mathrm{map}_B^B(X, Y) \wedge_B X \to Y.$$

Proposition (3.95). *Suppose that the sectioned space Y is locally compact regular over B. Then the composition function*

$$\mathrm{map}_B^B(Y, Z) \wedge_B \mathrm{map}_B^B(X, Y) \to \mathrm{map}_B^B(X, Z)$$

is continuous for all sectioned spaces X, Z over B.

Corollary (3.96). *Suppose that the sectioned space X is locally compact regular over B. Then the evaluation function*

$$\mathrm{map}_B^B(X, Y) \wedge_B X \to Y$$

is continuous for all sectioned spaces Y over B.

Corollary (3.97). *Let $\hat{h}: X \to \mathrm{map}_B^B(Y, Z)$ be a map of sectioned spaces, where X, Y, Z are sectioned spaces with Y locally compact regular over B. Then $h: X \wedge_B Y \to Z$ is continuous, where*

$$h(x, y) = \hat{h}(x)(y) \qquad (x \in X, y \in Y).$$

All these results follow at once from the corresponding results of the non-sectioned theory. It is usual to refer to h as the adjoint of $\hat{h}$, and to $\hat{h}$ as the adjoint of h, when they are related as above.

Proposition (3.98). *Let X, Y, Z be sectioned spaces, with X, Y compact regular over B. Then the exponential function*

$$\mathrm{map}_B^B(X \wedge_B Y, Z) \to \mathrm{map}_B^B(X, \mathrm{map}_B^B(Y, Z)),$$

given by taking adjoints, is an equivalence of sectioned spaces over B.

This follows from (2.63), (3.61) and (3.75).

Recall that if X, Y are sectioned spaces over B then $\mathrm{MAP}_B^B(X, Y)$ is defined, as a pointed space, the basepoint being the nul-map. Also that if Z is a sectioned space over B then the pointed space $\sec_B Z$ is defined; the basepoint being the section of Z. Taking Z to be $\mathrm{map}_B^B(X, Y)$ in particular we have the pointed map

$$\sigma: \mathrm{MAP}_B^B(X, Y) \to \sec_B \mathrm{map}_B^B(X, Y),$$

where σ transforms each section-preserving map $\phi: X \to Y$ over B into the section which assigns to each point $b \in B$ the pointed map $\phi_b: X_b \to Y_b$. Using (3.78) we obtain

Proposition (3.99). *Let X be a compact regular sectioned space over B. Then*

$$\sigma: \mathrm{MAP}_B^B(X, Y) \to \sec_B \mathrm{map}_B^B(X, Y)$$

is an equivalence of pointed spaces, for each sectioned space Y over B.

References

P. I. Booth. The exponential law of maps, I. *Proc. London Math. Soc.* (3) **20**, (1970), 179–192.

P. I. Booth and R. Brown. On the application of fibred mapping spaces to exponential laws for bundles, ex-spaces and other categories of maps. *Gen. Top. and Its Appl.* **8** (1978), 165–179.

P. I. Booth and R. Brown. Spaces of partial maps, fibred mapping spaces and the compact-open topology. *Gen. Top. and Its Appl.* **8** (1978), 181–195.

J. Dugundji. *Topology*. Allyn and Bacon, Boston, 1966.

D. Husemoller. *Fibre Bundles*. McGraw-Hill, New York, 1966.

I. M. James. General topology over a base. Aspects of Topology, London Math. Soc. Lecture Notes, 1984.

S. B. Niefield. Cartesianness: topological spaces, uniform spaces and affine schemes. *J. Pure Applied Alg.* **23** (1982), 147–168.

CHAPTER 4

Topological Transformation Groups

This chapter contains an outline of the basic theory of topological groups, particularly topological transformation groups. The theory is not only of great interest and importance in itself but also contains striking illustrations of the ideas we have discussed earlier.

Definition (4.1). A topological group G is a space equipped with group structure such that the multiplication $G \times G \to G$ and the inversion $G \to G$ are continuous.

In fact the two continuity conditions can be combined into the single condition that the division function $G \times G \to G$, given by $(g, h) \mapsto gh^{-1}$, is continuous. However it is not sufficient for the multiplication alone to be continuous; such topological binary systems are called paratopological groups.

For example, the group aut(X) of self-equivalences of a space X is paratopological when X is locally compact regular, by (2.98), and topological if X is compact Hausdorff.

Of course any group at the algebraic level may be regarded as a topological group by using the discrete topology. In this sense the group theory of the algebraists may be regarded as a special case of the theory of topological groups. Alternatively the indiscrete topology may be used.

The euclidean space $\mathbb{R}^n$ ($n = 0, 1, \ldots$) forms a topological group with vector addition as the binary operation. The group $Gl(n, \mathbb{R})$ of automorphisms of $\mathbb{R}^n$ also forms a topological group, known as the general linear group. To see this, take a basis in $\mathbb{R}^n$ and so represent automorphisms by non-singular $n \times n$ matrices. The set of such matrices may be regarded as a subset of euclidean $n \times n$ space. In this way $Gl(n, \mathbb{R})$ obtains its topology. Matrix multiplication is given by bilinear functions of the elements and is obviously continuous. Matrix inversion is given by dividing the adjoint matrix by the determinant; both these operations are continuous and so inversion is continuous.

For any topological group G the binary operation is usually written $(g, h) \mapsto gh$. Moreover if A, B are subsets of G then AB denotes the subset consisting of the products gh, where $g \in A$, $h \in B$. The inversion in G is usually written $g \mapsto g^{-1}$ and if A is a subset of G then A^{-1} denotes the subset consisting of the inverses g^{-1}, where $g \in A$. Note that for each element $g \in G$ the (left) translation $g_{\#}: G \to G$ is continuous, where $g_{\#}(h) = gh$. Moreover $(g^{-1})_{\#}$ is inverse to $g_{\#}$, and so $g_{\#}$ is a homeomorphism. It follows at once that the underlying space of a topological group must be homogeneous. Since $g_{\#}$ is a homeomorphism the direct image gU of each open set U is again an open set. Hence $A.U$, for each subset A of G, is the union of the open sets gU, for $g \in A$, and so is open. Similar observations may be made in the case of right translation.

Neighbourhoods of the neutral element e play a particular role in investigations of the topological group G. Note that each neighbourhood U of e contains a neighbourhood V of e such that $V = V^{-1}$, namely $U \cap U^{-1}$; such neighbourhoods V are said to be symmetric. Also that U contains a neighbourhood W of e such that $WW^{-1} \subset U$. To see this consider the division map $G \times G \to G$. By continuity there exist neighbourhoods W_1, W_2 of e such that $W_1 W_2^{-1} \subset U$. Then $W = W_1 \cap W_2$ meets the requirement.

Definition (4.2). A homomorphism $\phi: G \to G'$ of topological groups is a homomorphism at the algebraic level which is also a continuous function.

We denote the category of topological groups and (continuous) homomorphisms by $\mathscr{T}\mathscr{G}$. The equivalences in the category are called isomorphisms of topological groups. They have to be not only isomorphisms at the algebraic level but also homeomorphisms at the topological level. An isomorphism at the algebraic level may be continuous and yet fail to have a continuous inverse. For example, take the identity function on the set of real numbers where the domain has the discrete topology and the codomain the metric topology.

Definition (4.3). A subgroup of a topological group G is a subgroup at the algebraic level with the induced topology.

For example, consider the general linear group $Gl(n, \mathbb{R})$ of automorphisms of $\mathbb{R}^n$. The automorphisms which are also isometries, i.e. the orthogonal transformations, form a subgroup $O(n, \mathbb{R})$, called the orthogonal group.

Proposition (4.4). *Let H be a subgroup of the topological group G. Then the closure $\mathscr{Cl}\ H$ is also a subgroup. Moreover $\mathscr{Cl}\ H$ is normal if H is normal.*

Since G is a topological group the shearing map

$$k: G \times G \to G \times G$$

is a homeomorphism, where $k(g, h) = (g, gh^{-1})$. Since H is a subgroup we have that $k(H \times H) = H \times H$. Now $\mathscr{C}\ell(H \times H) = \mathscr{C}\ell\, H \times \mathscr{C}\ell\, H$ and so, by (2.22), k determines a homeomorphism of $\mathscr{C}\ell\, H \times \mathscr{C}\ell\, H$ with itself. It follows at once that $\mathscr{C}\ell\, H$ is a subgroup. When H is normal, consider any inner automorphism α of G. Then α is a homeomorphism and $\alpha H = H$ since H is normal. Hence $\alpha(\mathscr{C}\ell\, H) = \mathscr{C}\ell\, H$, by (2.22) again, and so $\mathscr{C}\ell\, H$ is normal.

Proposition (4.5). *Let H be a subgroup of the topological group G. If H is open in G then H is closed in G. If H is closed in G and of finite index in G then H is open in G.*

To see this, observe that $G - H$ is the union of cosets of H. If H is open then each of the cosets is also open, hence $G - H$ is open and so H is closed. If H is closed then each of the cosets is also closed. If, further, the index of H is finite, the number of cosets is finite, hence $G - H$ is closed and so H is open.

Proposition (4.6). *Let H be a discrete normal subgroup of the connected group G. Then H is a central subgroup.*

For let h be any element of H, and let $\alpha: G \to G$ be given by $\alpha(g) = ghg^{-1}$. Since G is connected, so is $\alpha(G)$. But $\alpha(G) \subset H$, since H is normal, and H is discrete. Therefore $\alpha(G)$ reduces to h itself and so H is central.

For reasons given in the first chapter the category $\mathscr{T}\mathscr{G}$ admits a product, called the direct product and denoted by $\times$. Also triads in $\mathscr{T}\mathscr{G}$ admit pullbacks. The category does not admit a coproduct.

Definition (4.7). Let H be a subgroup of the topological group G. The (left) factor set G/H, with the quotient topology determined by the natural projection, is the (left) factor space of G by H.

The right factor space is defined similarly. Of course the decomposition of G by left cosets of H is different, in general, from the decomposition by right cosets. However the operation of inversion induces a homeomorphism of the left factor space with the right factor space. The two decompositions coincide if, and only if, H is normal in G, in which case there is no point in distinguishing between the left factor space and the right.

Note that the quotient map $\pi: G \to G/H$ is open, since if $U \subset G$ is open then so is its saturation $U.H$. It follows at once that not only π itself but also

$\mathrm{id}_G \times \pi$ is a quotient map, in the diagram shown below.

$$\begin{array}{ccc} G \times G & \xrightarrow{\;m\;} & G \\ {\scriptstyle \mathrm{id}\times\pi}\downarrow & & \downarrow{\scriptstyle \pi} \\ G \times G/H & \longrightarrow & G/H \end{array}$$

Moreover, suppose that H is normal in G. Then $\pi \times \pi$ is an open surjection, in the next diagram, and hence a quotient map.

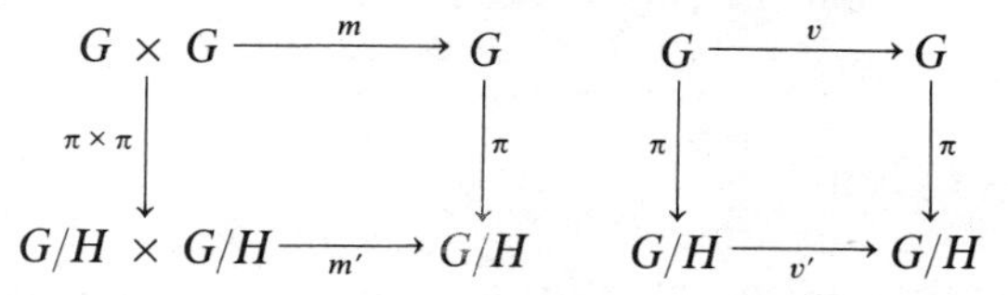

Therefore a continuous multiplication m', and a continuous inversion v', are defined on G/H. Thus G/H obtains the structure of a topological group in such a way that π is a homomorphism with kernel H. In this situation we refer to G/H as the factor group of G by H.

Example (4.8). Let H be a subgroup of the topological group G. Then G/H is discrete if and only if H is open in G.

For if G/H is discrete then (H) is open in G/H, hence H is open in G. Conversely if H is open in G then each coset of H is open in G, hence each point of G/H is open and so G/H is discrete.

Exercise (4.9). Let G be a topological group and let Δ be the diagonal subgroup of $G \times G$. Show that the factor space $(G \times G)/\Delta$ is homeomorphic to G.

Exercise (4.10). Show that for irrational values of α the factor group of the real plane $\mathbb{R} \times \mathbb{R}$ by the line of gradient α is isomorphic to the real line $\mathbb{R}$, with the indiscrete topology.

As we have noted earlier, the induced topology and the quotient topology do not "commute" with each other in general. However, let G be a topological group and let H, K be subgroups with $K \subset H$. Instead of topologizing H/K as a factor space of H, we may topologize H/K as a subset of G/K. In fact these two topologies coincide. For let $\pi: G \to G/K$ be the natural projection and let ρ be the induced map $H \to H/K$. The assertion will follow if we can show that ρ is an open map. So let U be an open set of H. Then $U = V \cap H$, where V is open in G. Then πV is open in G/K, since π is open, and therefore $\pi V \cap \pi H$ is open in πH. But since H is a union of cosets of K we have $\pi V \cap \pi H = \pi U$, so that πU is open in $\pi H = H/K$, as asserted.

Exercise (4.11). Show that if K is a normal subgroup of the topological group G then every subgroup of the factor group G/K is isomorphic to a factor group H/K for some subgroup H of G containing K.

Proposition (4.12). *Let $\phi: G \to G'$ be a surjective homomorphism of topological groups, with kernel K. The continuous bijection $\psi: G/K \to G'$ induced by ϕ is an isomorphism if ϕ is either open or closed.*

For if ϕ is either open or closed then so is ψ, since G/K has the quotient topology.

Example (4.13). Let $Sl(n, \mathbb{R})$ denote the normal subgroup of $Gl(n, \mathbb{R})$ consisting of unimodular transformations. Then the factor group $Gl(n, \mathbb{R})/Sl(n, \mathbb{R})$ is isomorphic to the multiplicative group $\mathbb{R} - \{0\}$ of non-zero real numbers.

For let $D: Gl(n, \mathbb{R}) \to \mathbb{R} - \{0\}$ be the determinant function, which is an epimorphism of topological groups. Let A be an element of $Gl(n, \mathbb{R})$, and let U be a neighbourhood of A. Since U is open in $Gl(n, \mathbb{R})$ there is an open interval J containing unity such that $tA \in U$ for all $t \in J$. As t ranges over J, t^n takes every value in some open interval J' containing unity, and so $t^n D(A) = D(tA)$ takes every value in some open interval J'' containing $D(A)$. Thus $D(U) \supset J''$ and $D(U)$ is open in $\mathbb{R} - \{0\}$. Since $D^{-1}(1) = Sl(n, \mathbb{R})$ the result now follows.

We now turn to the subject of topological transformation groups. Here the basic definition is

Definition (4.14). A (continuous) action of the topological group G on the space X is a map $G \times X \to X$ which constitutes an action at the algebraic level.

Given an action of the topological group G on the space X we say that X is a G-space. Of course the trivial action $\pi_2: G \times X \to X$ is always continuous; thus any space X can be regarded as a trivial G-space for any topological group G.

Let X be a G-space. The action is usually written $(g, x) \mapsto gx$. Moreover if A is a subset of G and B is a subset of X then AB denotes the subset consisting of the products gx, where $g \in A$ and $x \in B$. For each element $g \in G$ the translation $g_\#: X \to X$ is continuous, where $g_\# x = gx$. Moreover $(g^{-1})_\#$ is inverse to $g_\#$, and so $g_\#$ is a homeomorphism. It follows at once that if G acts transitively on X then X must be homogeneous. Since $g_\#$ is a homeomorphism the direct image gU of each open set U of X is again an open set. Hence $A.U$, for each subset A of G, is the union of the open sets gU, for $g \in A$, and so is open. It follows that the action $G \times X \to X$ is an open map.

Example (4.15). For any subgroup H of the topological group G, an action of G on the factor space G/H is defined using left translation. When H is normal another action is defined using conjugation.

Let G be a topological group. For each G-space X the orbit space X/G is defined as the orbit set with the quotient topology determined by the natural projection $\pi: X \to X/G$. For example $X/G = X$ when G acts trivially. Note that when G is a subgroup of the topological group Γ and G acts on Γ by translation the present use of the notation Γ/G is consistent with the former use.

Proposition (4.16). *Let G be a topological group. For any G-space X the projection $\pi: X \to X/G$ is an open map.*

For if $U \subset X$ is open then so is $G.U = \pi^{-1}\pi U$. This generalizes the result proved earlier for the case when G is a subgroup of a topological group Γ.

Before going any further let us introduce the category of G-spaces as follows.

Definition (4.17). Let X and Y be G-spaces, where G is a topological group. A G-map $\phi: X \to Y$ is a G-function, at the algebraic level, which is also continuous.

With this notion of morphism the category G–$\mathscr{T}$ of G-spaces is defined. The equivalences of the category are called G-equivalences (some authors prefer the term isomorphism). Note that a G-map $\phi: X \to Y$ determines, by restriction, a map

$$\phi^G: X^G \to Y^G$$

of the fixed-point spaces, and induces a map

$$\phi/G: X/G \to Y/G$$

of the orbit spaces. In this way functors are defined from G–$\mathscr{T}$ to $\mathscr{T}$, namely the fixed-point space functor and the orbit space functor.

Notice that if $\xi: H \to G$ is a homomorphism of topological groups then ξ transforms each G-space X into an H-space X through precomposition of the action with

$$\xi \times \mathrm{id}: H \times X \to G \times X.$$

Thus ξ determines a functor $\xi^\#$ from G–$\mathscr{T}$ to H–$\mathscr{T}$.

Sums and products of G-spaces are defined according to the programme outlined in the first chapter. Thus if X_i $(i = 1, 2)$ is a G-space then $X_1 + X_2$ is the G-space with action given by the sum of the action on X_1 and the action on X_2, as follows:

$$G \times (X_1 + X_2) \equiv (G \times X_1) + (G \times X_2) \to X_1 + X_2.$$

Also $X_1 \times X_2$ is the G-space with the diagonal action given by

$$g.(x_1, x_2) = (g.x_1, g.x_2).$$

Pull-backs in the category G–$\mathscr{T}$ are defined just as in $\mathscr{T}$ itself, but for push-outs there is a problem, due to the difficulty of mixing initial and final topologies. I will return to this point later.

It is a simple matter, as we have seen, to extend an endofunctor Φ of the category $\mathscr{S}$ of sets to an endofunctor Φ of the category of G-sets, for any group G. Provided Φ is a continuous functor the same procedure works in the topological case. If X is a G-space we can regard the projection $\pi: X \to X/G$ as a G-map by making G act trivially on the orbit space. Then the transform $\Phi(\pi): \Phi(X) \to \Phi(X/G)$ is also a G-map, with G acting trivially on $\Phi(X/G)$. Thus $\Phi(\pi)$ induces a G-map

$$\frac{\Phi(X)}{G} \to \Phi\left(\frac{X}{G}\right).$$

In general, however, this G-map is not an equivalence, even at the algebraic level.

Let T_0 be a subspace of the space T. Let Φ be the endofunctor of G–$\mathscr{T}$ which assigns to each G-space X the push-out of the cotriad

$$T \times X \leftarrow T_0 \times X \to T_0,$$

in the coarse topology. Then the orbit space $(\Phi X)/G$ is equivalent to $\Phi(X/G)$. For example

$$(\Gamma X)/G \equiv \Gamma(X/G), \qquad (\Sigma X)/G \equiv \Sigma(X/G).$$

In general the corresponding results with the fine topology are false.

For example, consider the join $G * G$, with the coarse topology and the action given by

$$g(t, g_1, g_2) = (t, gg_1, gg_2).$$

The orbit space $(G * G)/G$ is equivalent to the suspension ΣG. More generally, consider the join

$$E_G^{(n)} = G * \cdots * G \quad (n \text{ factors})$$

with the coarse topology and the action defined similarly. Regard $E_G^{(n)}$ as a space over the orbit space $B_G^{(n)} = E_G^{(n)}/G$. Clearly $E_G^{(n)}$ is trivial over U_i $(i = 1, \ldots, n)$ where U_i denotes the subspace of $B_G^{(n)}$ consisting of orbits of points of $E_G^{(n)}$ for which the ith coordinate function is non-zero. Thus we obtain

Proposition (4.18). *For any topological group G, the triviality category of the n-fold join $E_G^{(n)}$ over its orbit space $B_G^{(n)}$ does not exceed n.*

Recall that a subset A of the G-space X is invariant if $G.A = A$. In that case we can regard A itself as a G-space with the action which makes the inclusion $A \to X$ a G-map. In general, however, the injection $A/G \to X/G$ induced by the inclusion is not an embedding. We shall return to this point later.

For each point $x \in X$ the saturation $G.x$ is called the orbit of x under the action of G. The orbit is invariant and so may be regarded as a G-space. The G-map $\eta: G \to G.x$, given by $g \mapsto g.x$, induces a continuous bijection $\xi: G/G_x \to G.x$. In general ξ is not an equivalence. However there are two conditions we can give now, and another we shall give later, which ensure that ξ is an equivalence.

Proposition (4.19). *Either suppose that η is an open map or that η admits a right inverse. Then the map ξ induced by η is a homeomorphism.*

With the first alternative the conclusion is obvious. With the second the conclusion follows from the remark that if $s: G.x \to G$ is a right inverse of η then $\pi s: G.x \to G/G_x$ is a right inverse of ξ.

Example (4.20). The modular group $Sl(2, \mathbb{R})$ acts transitively on the upper half-plane $\mathbb{H}$. The stabilizer of the complex number i is the special orthogonal group $SO(2, \mathbb{R})$. Show that the factor space

$$Sl(2, \mathbb{R})/SO(2, \mathbb{R})$$

is equivalent to $\mathbb{H}$, under the map ξ given by evaluation at i.

Here $Sl(2, \mathbb{R})$ denotes the group of unimodular 2×2 matrices of the form

$$\begin{pmatrix} a & b \\ c & d \end{pmatrix} \qquad \text{where} \qquad ad - bc = 1,$$

acting on the upper half-plane $\mathbb{H}$ through $z \mapsto (az + b)/(cz + d)$. Obviously the action is transitive with stabilizer as stated, and a section $s: \mathbb{H} \to Sl(2, \mathbb{R})$ is given by the formula

$$s(a^2 i + b) = \begin{pmatrix} a & b \\ 0 & a^{-1} \end{pmatrix}$$

Thus $Sl(2, \mathbb{R})/SO(2, \mathbb{R})$ is equivalent to $\mathbb{H}$, by the second alternative of (4.19).

So far we have only been concerned with continuous actions on the left. Continuous actions on the right may also be used, as at the algebraic level. The terms left G-space and right G-space, etc., are used to distinguish the two situations. Similarly the notations $G \backslash X$ and X/G are used to distinguish between the left and right factor spaces.

Proposition (4.21). *Suppose that the topological group G acts on the left of the space X while the topological group H acts on the right of X, and that the* *actions commute as in* (1.47). *Then there are induced actions of G on the left of X/H and of H on the right of $G\backslash X$ such that*

$$G\backslash(X/H) \equiv (G\backslash X)/H.$$

We have already seen this, in Chapter 1, at the algebraic level, and so what remains to be done is to establish the continuity of the functions concerned. Consider first the diagram shown below, where h is the action of G on X and k is the induced action of G on X/H.

$$\begin{array}{ccc} G \times X & \xrightarrow{h} & X \\ \downarrow{\scriptstyle \mathrm{id} \times \pi} & & \downarrow{\scriptstyle \pi} \\ G \times X/H & \xrightarrow[k]{} & X/H \end{array}$$

Now π is open, by (4.16), and so id $\times$ π is open. Also π, and hence id $\times$ π, is surjective. Hence id $\times$ π is a quotient map, since π is a quotient map, and so k is continuous, since πk is continuous. Thus X/H is a left G-space and π is a G-map. Consider the induced map

$$G\backslash\pi : G\backslash X \to G\backslash(X/H).$$

This is invariant under the action of H on $G\backslash X$ and so induces a map

$$\xi : (G\backslash X)/H \to G\backslash(X/H).$$

Here both domain and codomain are ultimately quotient spaces of X, and ξ is induced by the identity on X. The same is true of the analogous map

$$\eta : G\backslash(X/H) \to (G\backslash X)/H.$$

Since ξ and η are mutually inverse this proves the result.

Mixed products have already been discussed at the algebraic level. At the topological level the mixed product of a right G-space X and a left G-space Y is the set $X \times_G Y$ with the quotient topology obtained from the natural projection

$$\pi : X \times Y \to X \times_G Y.$$

Proposition (4.22). *Let X, Y, Z be spaces. Suppose that the topological group G acts on the right of X and the left of Y, while the topological group H acts on the right of Y and the left of Z. Also suppose that the action of G on the left of Y commutes with the action of H on the right of Y. Then G acts on the left of $Y \times_H Z$, and H acts on the right of $X \times_G Y$, so that*

$$(X \times_G Y) \times_H Z \equiv X \times_G (Y \times_H Z).$$

The proof is straightforward and is therefore left as an exercise. The last two results justify the omission of brackets in the situations concerned. Our last result of this type is also left as an exercise.

Proposition (4.23). *Suppose that the topological group G acts on the right of the space X and on the left of the space Y. Also suppose that the topological group H acts on the right of Y and that the actions of G and H on Y commute with each other. Then there is an induced action of H on the right of* $X \times_G Y$ *such that*

$$(X \times_G Y)/H \equiv X \times_G (Y/H).$$

This shows, for example, that if X is any right G-space then

$$X \times_G (G/H) \equiv (X \times_G G)/H \equiv X/H,$$

a result which is frequently useful.

Proposition (4.24). *Let H be a subgroup of the topological group G. Then for each H-space X and G-space Y the H-maps of X into* $\alpha^\# Y$ *correspond precisely to the G-maps of* $G/H \times_H X$ *into Y.*

Here $\alpha: H \to G$ denotes the inclusion. To see this, let $\theta: X \to \alpha^\# Y$ be an H-map. Then a G-map $\phi: G/H \times_H X \to Y$ is given by $\phi[g, x] = g.\theta x$. Conversely, let $\phi: G/H \times_H X \to Y$ be a G-map. Then an H-map $\theta: X \to \alpha^\# Y$ is given by $\theta(x) = \phi[e, x]$.

We turn next to a set of results where the topological group in question is required to be compact. This enables much more to be said and for many purposes the restriction to compact transformation groups is acceptable. Later, however, we shall consider the effect of dropping the compactness restriction on the group and requiring that the action satisfy a compactness condition instead. This is sufficient to cover all the applications we have in mind.

Let X be a G-space, where G is a compact topological group. Then each of the orbits of G in X is compact, being the continuous image of the compact G. We prove

Proposition (4.25). *Let X be a G-space, where G is compact. Let A be a subspace of X and let g be an element of G such that* $g.A \subset A$. *Then* $g.A = A$.

For consider the subset $S \subset G$ consisting of nonnegative powers of g. Then

$$H = S \cup S^{-1} \cup \{e\}$$

is the subgroup generated by g. The closure $\mathscr{Cl}\, H$ is compact and a subgroup by (4.4). If $\mathscr{Cl}\, H$ is discrete then $\mathscr{Cl}\, H$ is finite, hence H is finite, hence $S = H$ and the result follows at once.

Suppose then that $\mathscr{Cl}\, H$ is not discrete. Then some element of $\mathscr{Cl}\, H$ is not isolated and so e is not isolated, by homogeneity. Thus for each neighbourhood U of e, and for each symmetric neighbourhood $V \subset U$ of e, there exists a positive integer n such that $g^n \in V$ and hence $g^{n-1} \in g^{-1}V \subset g^{-1}U$. Therefore S meets each neighbourhood of g^{-1} in G and so $g^{-1} \in \mathscr{Cl}\, S$, hence $\mathscr{Cl}\, S = \mathscr{Cl}\, H$.

Now $S.A \subset A$, since $g.A \subset A$, and so $\mathscr{Cl}\, S.A \subset A$ by continuity. But $g^{-1} \in \mathscr{Cl}\, S$, as we have just seen, hence $g^{-1}.A \subset A$. Thus $g.A = A$ and (4.25) is proved.

Corollary (4.26). *Let G' be a subgroup of the compact group G. Suppose that $g^{-1}G'g \subset G'$ for some element $g \in G$. Then $g^{-1}G'g = G'$.*

In the topological group G we say that a subgroup K is subconjugate to a subgroup H if K is conjugate to a subgroup of H. When G is compact we see, by (4.26), that if K is subconjugate to H and H is subconjugate to K then H and K are conjugate subgroups. Let us put this in another way. The conjugacy class of a subgroup G' of G is denoted by $[G']$. A partial ordering $\leq$ on the conjugacy classes is defined so that $[K] \leq [H]$ if and only if K is subconjugate to H. When G is compact the partial ordering has the property that if $[K] \leq [H]$ and $[H] \leq [K]$ then $[H] = [K]$.

Proposition (4.27). *Let X be a G-space, where G is compact. Then the action*

$$h: G \times X \to X$$

and the natural projection

$$\pi: X \to X/G$$

are compact maps.

To prove the first assertion, observe that $h = \pi_2 \phi$, where $\phi: G \times X \to G \times X$ is given by

$$\phi(g, x) = (g, g.x) \qquad (g \in G, x \in X).$$

Since π_2 is compact, and since ϕ is always a closed embedding, it follows at once that h is compact. In particular h is closed and so, if $A \subset X$ is closed then $G.A$ is also closed. Since $G.A = \pi^{-1}\pi A$, the saturation of A, this shows that π is closed. Finally the orbits of G, i.e. the fibres of π, are compact, as we have observed, and so π is compact, by (2.61). This proves (4.27) and we have at once

Corollary (4.28). *Let X and Y be G-spaces, where G is compact, and let $\phi: X \to Y$ be a G-map. If ϕ is a closed embedding then so is*

$$\phi/G: X/G \to Y/G.$$

In particular if A is a closed invariant subspace of X then, for G compact, we may regard A/G as a (closed) subspace of X/G.

Proposition (4.29). *Let Z be the push-out of the cotriad*

$$X \xleftarrow{u} A \xrightarrow{\phi} Y$$

of G-spaces and G-maps, where G is compact and u is injective. If u and ϕ are closed then Z/G is the push-out of the cotriad

$$X/G \xleftarrow{u/G} A/G \xrightarrow{\phi/G} Y.$$

To prove (4.29), consider the commutative diagram shown below, where the maps are the obvious ones.

$$\begin{array}{ccc} X + Y & \longrightarrow & X/G + Y/G \\ \downarrow & & \downarrow \\ X +^A Y & \longrightarrow & X/G +^{A/G} Y/G \end{array}$$

By (4.27) the natural projections to the orbit spaces are closed, hence the upper horizontal map is closed. Moreover it follows from (4.28) and (2.46) that the right-hand vertical is also closed. Since $X +^A Y$ is a quotient space of $X + Y$ these conclusions imply that the bottom horizontal is closed, and hence that the continuous bijection

$$(X +^A Y)/G \to X/G +^{A/G} Y/G$$

is closed and therefore a homeomorphism as asserted.

Proposition (4.30). *Let X be a G-space, where G is compact. Let A be a closed subset of X which meets each orbit of X under G in precisely one point. Then a section $s\colon X/G \to X$ is given by*

$$s\pi(x) = A \cap (G.x) \qquad (x \in X).$$

For if $V \subset X$ is closed then so is $s^{-1}V = \pi(V \cap A)$, by (4.27).

Proposition (4.31). *Let X and Y be G-spaces, where G is compact. Let A be a closed subset of X and let $\phi\colon A \to Y$ be a map such that $g.\phi(x) = \phi(g.x)$ whenever both x and $g.x$ lie in A. Then ϕ can be extended uniquely to a G-map $\psi\colon G.A \to Y$.*

The only problem here is to show that ϕ, extended in the obvious way, satisfies the continuity condition. So consider the diagram shown below, where k and l are given by the respective actions

$$\begin{array}{ccc} G \times A & \xrightarrow{k} & G.A \\ {\scriptstyle \mathrm{id} \times \phi}\downarrow & & \downarrow{\scriptstyle \psi} \\ G \times Y & \xrightarrow[l]{} & Y \end{array}$$

Since $\psi k = l(\mathrm{id} \times \phi)$ and since k is closed the assertion follows at once.

Proposition (4.32). *Let X be a G-space, with G compact, and let $\xi: B \to X/G$ be a map, for some space B. Then the pull-back ξ^*X of X, regarded as a G-space in the obvious way, has orbit space $(\xi^*X)/G$ equivalent to B.*

Here we regard B as a trivial G-space. Now $\pi: X \to X/G$ is compact, by (4.27), hence $\pi \times \mathrm{id}_B : X \times B \to X/G \times B$ is compact, hence the restriction

$$\xi^*X = X \times_{X/G} B \to X/G \times_{X/G} B = B$$

is compact. Since the restriction is invariant, with respect to the diagonal action of G, it induces a map $\xi^*X/G \to B$. This map is bijective and closed, therefore an equivalence, as asserted.

The results obtained so far do not involve any separation assumptions. Let us see now what more can be said if we make use of the Hausdorff or the regularity axiom. It turns out that the regularity condition is largely irrelevant. In fact factor spaces are automatically regular, as shown in

Proposition (4.33). *Let H be a subgroup of the topological group G. Then the factor space G/H is regular.*

By homogeneity it is sufficient to consider a closed set F of G/H which does not contain the neutral coset πe. Since $E = \pi^{-1}F$ is closed and does not contain e there exists a neighbourhood of e which does not meet E. For reasons explained above we can choose this neighbourhood to be of the form $V^{-1}V$, where V is a neighbourhood of e. Then V does not meet VE, and so VH does not meet VE, since E is saturated. Now VH and VE are open, hence $\pi(VH)$ and $\pi(VE)$ are open. Since $\pi(VH)$ does not meet $\pi(VE)$ this establishes regularity.

This result may be contrasted with

Proposition (4.34). *Let H be a subgroup of the topological group G. The factor space G/H is Hausdorff if and only if H is closed in G.*

Consider the natural projection $\pi: G \to G/H$. If G/H is Hausdorff then $\pi(e) = (H)$ is closed, hence $\mathrm{sat}(e) = H$ is closed. Conversely let H be closed. Then $d^{-1}(G - H)$ is open, where $d: G \times G \to G$ is the division map. However π is open and so

$$\pi \times \pi: G \times G \to G/H \times G/H$$

is open. Now $(\pi \times \pi)d^{-1}(G - H)$ is the complement of the diagonal. Hence the diagonal is closed and so G/H is Hausdorff.

Proposition (4.35). *Let X be a Hausdorff G-space. Then the fixed-point set X^G is closed in X.*

For suppose that $x \in X - X^G$. Then $g.x \neq x$ for some element $g \in G$. By (2.69) the set of points $x' \in X$ such that $g.x' \neq x'$ is open. Hence there exists a neighbourhood N of x such that $g.x' \neq x'$ for all $x' \in N$; thus N does not meet X^G. This proves (4.35). It follows at once that if $\phi: X \to Y$ is a closed G-map, where X is Hausdorff, then $\phi^G: X^G \to Y^G$ is a closed map.

Proposition (4.36). *Let X be a Hausdorff G-space, where G is compact. Then for each point x of X evaluation at x determines an equivalence between the factor space G/G_x and the orbit $G.x$, as G-spaces.*

Clearly the G-map $G/G_x \to G.x$, given by evaluation at x, is a continuous bijection. Also G/G_x is compact, since G is compact, and $G.x$ is Hausdorff, since X is Hausdorff. Hence and from (2.73) the result follows.

In view of (4.36) we see that the partial ordering of orbits of a G-space, determined by the existence of G-maps, corresponds inversely to the partial ordering of the conjugacy classes of subgroups of G, under the stated conditions. Specifically, let X be a Hausdorff G-space with G compact. Then there exists a G-map from $G.x$ to $G.x'$, where $x, x' \in X$, if and only if G_x is subconjugate to $G_{x'}$. It follows that if there exist G-maps from $G.x$ to $G.x'$ and from $G.x'$ to $G.x$ then $G.x$ and $G.x'$ are equivalent as G-spaces.

Exercise (4.37). The orthogonal group $O(n) = O(n, \mathbb{R})$ acts transitively on the spheres S^{n-1}. The stabilizer is equivalent to $O(n - 1)$, and the factor space $O(n)/O(n - 1)$ is equivalent, as an $O(n)$-space, to S^{n-1}.

Exercise (4.38). Let X be the space of symmetric real 3×3 matrices of trace zero (thus X is a real vector space of dimension 5). Regard X as an $SO(3)$-space under matrix conjugation. Then the orbit space $X/SO(3)$ is a real vector space of dimension 3.

The following device (due to G. Segal) which I call the colon construction is often useful in the study of G-spaces where G is compact. For G-spaces X and Y let us denote by $(X : Y)$ the quotient space W/G, where $W \subset X \times Y$ is the invariant subspace consisting of pairs (x, y) such that $G_x \subset G_y$. It is easy to see that the diagram shown below is cartesian at the set-theoretic level.

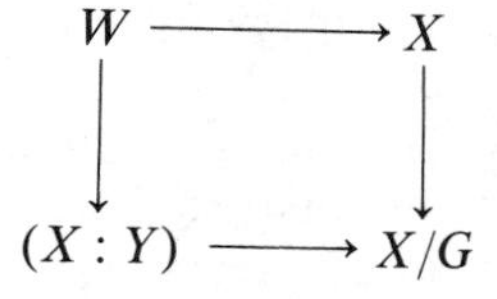

Proposition (4.39). *If G is compact and X is Hausdorff over X/G then the above diagram is cartesian at the topological level.*

For since X is Hausdorff over X/G the graph function

$$W \to X \times_{X/G} W$$

of the projection $W \to X$ is a closed embedding. Also the projection $W \to W/G$ is compact, by (4.27), and so the product $X \times_{X/G} W \to X \times_{X/G} W/G$ is compact. Hence the composition

$$W \to X \times_{X/G} W \to X \times_{X/G} W/G$$

is compact. However the composition is a continuous bijection and so is a homeomorphism as asserted. Moreover, if we regard the orbit spaces as trivial G-spaces then the above diagram is cartesian for the category of G-spaces. Hence we obtain

Proposition (4.40). *Let X and Y be G-spaces, where G is compact. If X is Hausdorff over X/G then the G-maps $X \to Y$ correspond precisely to the sections of $(X : Y)$ over X/G.*

The correspondence is given by assigning to each G-map $\phi\colon X \to Y$ the section Γ_ϕ/G, where $\Gamma_\phi\colon X \to X \times Y$ is the graph function. In fact the spaces $\mathrm{map}_G(X, Y)$ and $\mathrm{sec}_{X/G}(X : Y)$ are equivalent under this correspondence.

When X is a free G-space we have that $W = X \times Y$, hence $(X : Y) = X \times_G Y$, the mixed product: here we have converted X from a left G-space to a right G-space in the standard way. In this case the previous diagram reduces to

$$\begin{array}{ccc} X \times Y & \longrightarrow & X \\ \downarrow & & \downarrow \\ X \times_G Y & \longrightarrow & X/G \end{array}$$

As an application we prove

Proposition (4.41). *Let X and Y be free G-spaces, where G is compact. Also suppose that X is Hausdorff over X/G and that Y is Hausdorff over Y/G. Let $\phi\colon X \to Y$ be a G-map such that $\phi/G\colon X/G \to Y/G$ is an equivalence of spaces. Then ϕ is an equivalence of G-spaces.*

For the G-map ϕ determines a section $s\colon X/G \to X \times_G Y$, as before. Precompose s with the inverse of ϕ/G, and then postcompose with the switching map, to obtain a section $t\colon Y/G \to Y \times_G X$. Now apply (4.40) and obtain a G-map $\psi\colon Y \to X$. It follows by uniqueness that $\psi\phi = \mathrm{id}_X$ and $\phi\psi = \mathrm{id}_Y$, hence the result. In this proof we convert X and Y from left G-spaces to right G-spaces, and vice versa, in the standard way.

Proposition (4.42). *Let H be a closed subgroup of the compact topological group G. Then the group $\mathrm{aut}_G(G/H)$ of G-equivalences of G/H with itself is* *isomorphic to the Weyl group N/H, where $N = N(H)$ is the normalizer of H in G.*

We make N act on G/H by conjugation; then $H \subset N$ acts trivially and so there is an induced action of N/H. The map

$$\lambda: N/H \to \mathrm{aut}_G(G/H)$$

thus determined is an isomorphism at the set-theoretical level. Also N is closed in the compact G, hence compact, and so N/H is compact. Moreover G/H is Hausdorff, by (4.34), and so $\mathrm{aut}_G(G/H)$ is Hausdorff, by (2.89). Therefore λ is an isomorphism at the topological level which proves (4.42).

There are a number of important results about G-spaces in which all the orbits are of the same type, such as

Proposition (4.43). *Let H be a closed subgroup of the compact group G, with normalizer $N = N(H)$ and Weyl group $K = N/H$. Let X be a Hausdorff G-space, each of whose orbits is of type $[H]$. Then X is equivalent, as a G-space, to the mixed product $G \times_N X^H$. Also X^H/N is equivalent to X/G. Finally $(X : Y)$, for each G-space Y, is equivalent to the mixed product $X^H \times_K Y^H$, as a space over X/G.*

For consider the G-map

$$\xi: G \times_N X^H \to X$$

defined by $\xi[g, x] = g.x$. Then ξ is closed, as well as continuous, since X^H is closed in X and the action $G \times X \to X$ is closed. I assert that ξ is bijective and therefore an equivalence of G-spaces.

Surjectivity is obvious: the problem is to show that ξ is injective. So suppose that $g.x = g'.x'$, where $g, g' \in G$ and $x, x' \in X^H$. Then $G_x = H = G_{x'}$, since G is compact. Write $n = g^{-1}.g'$. Then $n(x') = x$ and

$$H = G_x = G_{n(x')} = nG_{x'}n^{-1} = nHn^{-1},$$

so that $n \in N$. Thus $[g, x] = [gn, n^{-1}x] = [g', x']$, which shows that ξ is injective.

Turning to the second assertion, observe that X^H/N is equivalent to $(G \times_N X^H)/G$ under the transformation $N(a) \mapsto G[e, a]$. Since $(G \times_N X^H)/G$ is equivalent to X/G, by what we have just proved, we obtain that X^H/N is equivalent to X/G, as asserted.

For the final assertion, consider the obvious map

$$\eta: X^H \times_K Y^H \to (X : Y)$$

where Y is any G-space. If $\eta[x, y] = \eta[x', y']$, for $x, x' \in X^H$ and $y, y' \in Y^H$, then $gx = x'$ and $gy = y'$ for some $g \in G$. The latter implies that $H = G_{y'} = G_{gy} = gG_yg^{-1} = gHg^{-1}$, and so $g \in N$; therefore $[x, y] = [gx', gy']$ in $X^H \times_K Y^H$. Finally if $(x, y) \in X \times Y$ are such that $G_x \subset G_y$ then $G_x = gHg^{-1}$, hence $G_{gx} = H$; thus $(x, y) \sim (gx, gy)$, hence $G_{gy} \supset H$ and $gy \in X^H$. This completes the proof.

Next a few words about slices and sections.

Definition (4.44). Let X be a G-space with G compact. A slice through a point x of X is a subset S, containing x, such that

(i) GS is open in X,
(ii) Gx is a G-retract of GS, and
(iii) there exists a G-retraction $r: GS \to Gx$ for which $r^{-1}x = S$.

The definition is in line with that given in the previous chapter, where X is regarded as a space over X/G, but has been made more specific.

Note that if S is a slice through x then gS is a slice through gx for each element $g \in G$. Also note that if x is closed in Gx, as is the case when X is Hausdorff, then $S = r^{-1}x$ is closed in GS. Finally note that if $y \in GS$ then G_y is subconjugate to G_x. For if $y = gz$, where $z \in S$, then $G_y = gG_zg^{-1}$ and $G_z \subset G_{rz} = G_x$, so that $g^{-1}G_yg^{-1} \subset G_x$.

There are various results in the literature establishing the existence of slices in certain situations. For example, slices exist at each point of a completely regular G-space when G is a compact Lie group. To obtain such results requires the use of differential geometry. Since they are not directly relevant to our purposes, we will not go into details here.

For a simple example where slices exist consider the join $X * Y$, where X is a G-space and Y is a trivial G-space. Given a point x of X, regarded as a subspace of $X * Y$ in the standard way, take S to be the open cone consisting of points (x, y, t), where $y \in Y$ and $0 \leq t < 1$. Then S satisfies the conditions for a slice through x. In fact slices exist at each point of $X * Y$.

Notice that if S is a slice through the point x of the G-space X then G_x acts on S. For $gs \in S$, where $s \in S$, if and only if

$$x = r(gs) = gr(s) = gx,$$

i.e. if and only if $g \in G_x$.

Exercise (4.45). Let X be a Hausdorff G-space, with G compact. If S is a slice at the point x of X then the natural map

$$S/G_x \to X/G$$

is an open embedding.

Slices are not unrelated to local sections of the natural projection $X \to X/G$. A case of particular interest is when G is a subgroup of a topological group Γ and Γ is regarded as a G-space using left translation. By homogeneity Γ admits local sections over Γ/G if (and only if) there exists a neighbourhood of the neutral coset (G) over which Γ has a section. It can be shown that this is the case when Γ is a Lie group and G a closed subgroup. Indeed if we accept the existence of slices in this case, and suppose G compact, then the embedding of the slice in (4.45) provides just such a neighbourhood. Notice further that if Γ admits local sections over Γ/G then so does Γ/H for each subgroup H of G.

Recall that the compactification X^+ of a locally compact Hausdorff space X is compact Hausdorff. I assert that if X is a G-space, with G compact, then X^+ is also a G-space, in fact a pointed G-space; the action of G on X is extended to an action on X^+ so that the "point at infinity" $*$ is left fixed. The continuity of the extended action follows from the fact that, by (2.55), the inverse image of a compact subset of X under the original action is compact, since G is compact. Now X/G is locally compact Hausdorff, since X is locally compact Hausdorff and the projection $X \to X/G$ is compact, by (4.27). Hence the projection determines a pointed G-map $X^+ \to (X/G)^+$ where, as usual, we regard the orbit spaces as trivial G-spaces. The pointed G-map induces a continuous bijection

$$X^+/G \to (X/G)^+.$$

However X^+/G is compact, since X^+ is compact, and $(X/G)^+$ is Hausdorff. Therefore the continuous bijection is a homeomorphism and so X^+/G is equivalent to $(X/G)^+$, as a pointed space.

We have seen, in the last few pages, that compact transformation groups have many nice properties. Unfortunately the compactness restriction is unacceptably severe for some purposes. Fortunately however, there is a weaker condition which is satisfied in the most important applications and which implies many of the nice properties found in the case of a compact group.

Let X be a G-space, where G is a topological group. Consider the continuous surjection $\theta\colon G \times X \to X \times_{X/G} X$ given by

$$\theta(g, x) = (g.x, x) \qquad (g \in G, x \in X).$$

We may regard θ as a map over X, and hence over X/G, using the second projection in both domain and codomain. Note that θ is bijective if and only if the action is free.

Definition (4.46). The action of the topological group G on the G-space X is proper if the map $\theta\colon G \times X \to X \times_{X/G} X$ is compact.

A G-space where the action is proper is called a proper G-space. The condition for an action to be proper does not involve inversion and so makes sense for a paratopological group. However the action of a paratopological group on itself by translation is proper if and only if the inversion is continuous: in this sense topological groups are proper paratopological groups.

Note that for any G-space X the graph

$$\Gamma: G \times X \to G \times X \times_{X/G} X$$

of the action is an embedding. If X is Hausdorff over X/G the image of Γ is closed and so Γ is compact. If in addition G is compact then the projection

$$\pi: G \times X \times_{X/G} X \to X \times_{X/G} X$$

is compact, hence $\theta = \pi\psi$ is compact. Thus we have

Proposition (4.47). *Let X be a G-space, where G is compact. If X is Hausdorff over X/G then the action is proper.*

In the other direction we have

Proposition (4.48). *Let X be a proper G-space, where G is Hausdorff. Then X is Hausdorff over X/G.*

For the diagonal in the fibre product can be expressed as the composition

$$X \xrightarrow{\Gamma} G \times X \xrightarrow{\theta} X \times_{X/G} X,$$

where Γ is the graph of the nul-map $X \to G$, after switching factors. Now Γ is compact, by (2.70), since G is Hausdorff and so $\theta\Gamma$ is compact, and therefore closed.

Thus for G compact Hausdorff, the G-space X is proper if and only if X is Hausdorff over X/G.

Exercise (4.49). Let $\phi: X \to Y$ be a G-map, where X is a Hausdorff G-space and Y is a G-space. If Y is a proper G-space then X is a proper G-space.

Note that if the topological group G acts properly on the space X then G also acts properly on each invariant subspace of X. Also that if X is a proper G-space and $\xi: H \to G$ is a compact homomorphism, for some topological group H, then H acts properly on X through ξ.

If G acts properly on X then the stabilizer G_x of each point x of X is compact, by (2.61). This shows for example that the action of the general linear group $Gl(V)$ on the non-trivial real vector space V cannot be proper.

Proposition (4.50). *Let X be a proper G-space. Then for each point x of X the G-map*

$$G/G_x \to G.x,$$

given by evaluation at x, is an equivalence of G-spaces.

By (2.55) the map

$$G \times \{x\} \to G.x \times \{x\}$$

determined by θ is compact. Discarding the $\{x\}$ factor this means that the evaluation map $G \to G.x$ is compact. Now (4.50) follows at once.

Proposition (4.51). *Let H be a subgroup of the topological group G. The translation action of G on G/H is proper if and only if H is compact.*

The result in the "only if" direction is given at once by (4.50). To prove the result in the "if" direction, suppose that H is compact and consider the diagram shown below.

$$\begin{array}{ccc} G \times G & \xrightarrow{\theta} & G \times G \\ \downarrow{\scriptstyle \mathrm{id} \times \pi} & & \downarrow{\scriptstyle \pi \times \pi} \\ G \times G/H & \xrightarrow[\theta]{} & G/H \times G/H \end{array}$$

The upper θ is compact, since G acts properly on itself by translation. Also π is compact, by (4.27) above and so $\pi \times \pi$ is compact. Hence $\theta(\mathrm{id} \times \pi) = (\pi \times \pi)\theta$ is compact, and so the lower θ is compact, by (2.53). Hence the action on G/H is proper, as asserted.

Proposition (4.52). *Suppose that the topological group G acts freely on the space X. Consider the function $\tau: X \times_{X/G} X \to G$ defined by*

$$y = \tau(x, y).x \qquad (x, y \in X).$$

The action of G is proper if and only if τ is continuous.

Clearly θ is injective, since the action is free. If θ is compact then τ is given by the first projection of the inverse of θ, and so is continuous. The converse is obvious.

Now suppose that G is locally compact Hausdorff, so that the compactification G^+ is defined, as a compact Hausdorff space. Then a useful necessary and sufficient condition for a continuous G-action to be proper is given by

Proposition (4.53). *Let X be a G-space, where G is locally compact Hausdorff. Suppose that X is Hausdorff over X/G. Then the action of G on X is proper if and only if for each pair of points x, y in the same orbit of X there exist neighbourhoods V_x of x, V_y of y such that*

$$\{g \in G \mid gV_x \cap V_y \neq \varnothing\}$$

is compact.

Consider the graph Γ of the action $G \times X \to X$, as a map over X/G. This subset of $G \times X \times_{X/G} X$ is closed, since X is Hausdorff over X/G. I assert that Γ is closed in $G^+ \times X \times_{X/G} X$ if and only if the action of G on X is proper. Assuming this our result follows easily. For if Γ is closed in

$G^+ \times X \times_{X/G} X$ then each point of the form $(*, (x, y))$, with x, y in the same orbit, has a neighbourhood which does not meet Γ, hence has a product neighbourhood, say

$$(G - K) \times V_x \times_{X/G} V_y \qquad (K \text{ compact})$$

which does not meet Γ. This is equivalent to the condition given in (4.53). Conversely if that condition is satisfied then each point $(*, (x, y))$, with x, y in the same orbit, has a neighbourhood which does not meet Γ, by reversing the above argument. However, any point of $G \times X \times_{X/G} X$ which is not in Γ has a neighbourhood in $G \times X \times_{X/G} X$ which does not meet Γ, hence a neighbourhood (the same) in $G^+ \times X \times_{X/G} X$ with that property, since G is open in G^+. Thus Γ is closed in $G^+ \times X \times_{X/G} X$.

It remains for the assertion itself to be proved. First observe that the projection

$$G^+ \times X \times_{X/G} X \to X \times_{X/G} X$$

is compact, since G^+ is compact. Also that the map $G \times X \to G \times X \times_{X/G} X$ given by

$$(g, x) \mapsto (g, g.x, x) \qquad (g \in G, x \in X)$$

is compact, by (3.32), since X is Hausdorff over X/G. So if the image Γ of this map is closed in $G^+ \times X \times_{X/G} X$ then the corresponding map $G \times X \to G^+ \times X \times_{X/G} X$ is compact, and composing this with the above projection we obtain that θ is compact, as asserted

Conversely suppose that θ is compact. Then (identifying the domain with Γ) we see that the map $\psi: G^+ \times \Gamma \to G^+ \times X \times_{X/G} X$ is closed, where

$$\psi(g, (h, x, y)) = (g, x, y) \qquad (g \in G^+, h \in G, x \in X, y \in Y).$$

Also the graph Γ' of the inclusion $G \subset G^+$ is closed in $G^+ \times G$, hence the product $\Gamma' \times X \times_{X/G} X$ is closed in $G^+ \times G \times X \times X$, hence the intersection $(\Gamma' \times X \times_{X/G} X) \cap (G^+ \times \Gamma)$ is closed in $G^+ \times \Gamma$. But Γ is precisely the image of this intersection under the closed map ψ. This completes the proof of the assertion and hence of (4.53).

Corollary (4.54). *Let X be a G-space, where G is discrete. Suppose that X is Hausdorff over X/G. The action is proper if and only if for each pair of points x, y in the same orbit of X there exist neighbourhoods V_x of x, V_y of y such that the subset*

$$\{g \in G \mid gV_x \cap V_y \neq \emptyset\}$$

is finite.

Corollary (4.55). *Let X, Y be G-spaces, where G is discrete. Suppose that X is Hausdorff over X/G and Y is Hausdorff over Y/G. If G acts properly on X then G acts properly on $X \times Y$.*

The case when G acts freely is of special interest. In that case, under the assumptions of (4.54), one sees that the action is proper if and only if there exist neighbourhoods such that the subset contains at most one element. An action which is both free and proper is called a properly discontinuous action.

Proposition (4.56). *Let X be a G-space, where G is discrete. Suppose that the action is proper and that X is Hausdorff over X/G. Then for each point x of X there exists a neighbourhood U of x with the following four properties:*

(i) *U is invariant under the action of G_x,*
(ii) *the orbits of G_x in U are the intersections with U of the orbits of G in X,*
(iii) *for each element $g \in G - G_x$ the sets U and gU do not intersect,*
(iv) *the canonical map $U/G_x \to X/G$ maps U/G_x homeomorphically onto a neighbourhood of the image of x in X/G.*

To construct a neighbourhood U which satisfies the four conditions, first observe that there exists, by (4.54), a neighbourhood U_0 of x such that the set K of elements $g \in G$ for which gU_0 does not meet U_0 is finite. Obviously $G_x \subset K$; let $g_1, \ldots, g_n$ be the elements of $K - G_x$, and write $x_i = g_i x$ $(i = 1, \ldots, n)$. Now $x_i \neq x$ for each i and so, since X is Hausdorff over X/G, there exists for each i a neighbourhood V_i of x and a neighbourhood W_i of x_i which do not meet. The intersection $U_i = V_i \cap g_i^{-1} W_i$ is then a neighbourhood of x, and

$$U_i \cap g_i U_i \subset V_i \cap W_i = \varnothing.$$

Thus

$$U' = U_0 \cap U_1 \cap \cdots \cap U_n$$

is a neighbourhood of x such that U' does not meet $g'U'$ for any $g' \notin G_x$. Set

$$U = \bigcap_{g \in G_x} gU';$$

then U is a G_x-invariant neighbourhood of x such that U does not meet gU for any element g of $G - G_x$.

It remains to be shown that U satisfies the fourth condition. But since U is open and since the equivalence relation on U defined by G is open, the assertion that U/G_x is mapped homeomorphically onto a neighbourhood of x in X/G follows at once.

To round off the present chapter I will now give an outline of the theory of covering spaces. At the outset this may not appear to have much to do with what we have been discussing but it will gradually emerge that the connection with the theory of actions of discrete groups is very close.

Definition (4.57). Let X be a space over B. Suppose that for each point of B there exists a neighbourhood U such that X_U is trivial, with discrete fibre. Then X is a covering space over B.

The notion of covering space over B should not be confused with the notion of covering of B, which means something quite different. Instead of saying that X is a covering space over B it is often said that X covers B or that the projection $X \to B$ is a covering map. Note that a covering space of B is discrete over B, and hence regular and Hausdorff over B, in the sense of the previous chapter. The triviality condition may equally well be expressed in terms of open sections as in Chapter 3. Specifically the condition is that for each point b of B there exists a neighbourhood U and an X_b-indexed family of open sections $s_x: U \to X_U$ such that (i) $s_x b = x$ and (ii) $s_x U$ is disjoint from $s_{x'} U$ when $x \neq x'$. An open set U of B is said to be elementary, with respect to X, if U is connected and X is trivial over U. When B is locally connected each point b of B admits an elementary neighbourhood U, and then each component of X_U maps homeomorphically onto U. For X_U is the disjoint union of open sets, each of which projects homeomorphically onto U. Since U is connected, so is each of these open sets, and because they are open and disjoint each is a component of X_U.

Proposition (4.58). *Let G be a closed discrete subgroup of the topological group Γ. Then Γ is a covering space over Γ/G.*

For let V be a neighbourhood of e in Γ which does not meet $G - e$, and let W be a neighborhood such that $W^{-1}W \subset V$. Then the projection $\pi W = U$ is a neighbourhood of πe over which Γ is trivial, with fibre G. More generally $\pi(gW)$ is a neighbourhood of πg with the same property for each element g of Γ.

For example, take the subgroup $\mathbb{Z}$ of integers in the real line $\mathbb{R}$. We see that $\mathbb{R}$ is a covering space over the circle $\mathbb{R}/\mathbb{Z}$. If the circle is represented, in the usual way, as the group of complex numbers of unit modulus then the projection p is the covering map $\mathbb{R} \to S^1$. Notice that although $p|(-1, 1)$ is an open surjection and $(-1, 1)$ is discrete over S^1 the projection $p|(-1, 1)$ is not a covering, since the fibre over $p(0)$ has one point while the fibre over every other point has two.

Notice, incidentally, that if X is a connected covering space over B with non-trivial fibre then X cannot be trivial, as a space over B. For example, $\mathbb{R}$ is non-trivial as a space over S^1. We can deduce from this that S^1 is not a retract of the unit disc B^2. For suppose, to obtain a contradiction, that there exists a retraction $\xi: B^2 \to S^1$. The induced covering space $\xi^*\mathbb{R}$ over B^2, is trivial, by (3.55), and so the restriction of $\xi^*\mathbb{R}$ to S^1 is trivial. But since $\xi|S^1$ is the identity the restriction of $\xi^*\mathbb{R}$ is just $\mathbb{R}$ itself, whence the contradiction.

Proposition (4.59). *Let X be a covering space over B. Let G be a discrete group acting freely on X by fibre-preserving maps which are fibrewise-transitive. Then the action is proper.*

The projection $p: X \to B$ induces a map $p': X/G \to B$. Since p is an open surjection so is p'. Also p' is injective and so a homeomorphism. We may therefore identify B with X/G. Consider, therefore, our map

$$\theta: G \times X \to X \times_B X,$$

as a map over B. The map is bijective since the action is free. Since p is a covering map there exists an open covering of B such that X is trivial over each member U of the covering. Now θ is a homeomorphism on each fibre, since the fibres are discrete, and so is an equivalence over U. By (3.15), therefore, θ is an equivalence over B. Therefore the action is proper, as asserted.

Exercise (4.60). Let X be discrete over B. If all the fibres X_b have the same finite cardinality then X is a covering space over B.

Proposition (4.61). *Let X be a space over the locally connected space B. Then X is a covering space over B if and only if X_C is a covering space over C for each component C of B.*

Suppose that X is a covering space over B. Let C be a component of B, let b be a point of C, and let U be a neighbourhood of b over which X is trivial. Let V be a component of U containing b. Since B is locally connected, V is open in B and so is open in U. Evidently X_C is trivial over V. Thus X_C is a covering space over C.

Conversely suppose that X_C is a covering space over C for each component C of B. Let $b \in B$ and let U be a neighbourhood of b in C over which X_C is trivial. Since B is locally connected, C is open in B. Then U is also open in B and X is trivial over U. Thus X is a covering space over B.

Proposition (4.62). *Let X and Y be covering spaces over the locally connected space B. Then each continuous surjection $\phi: X \to Y$ over B is a covering map.*

For consider the commutative diagram shown below

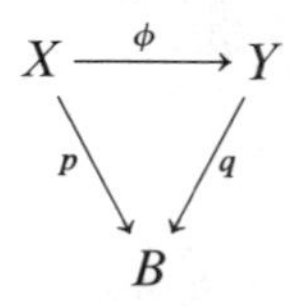

Given a point b of B, let U be an elementary neighbourhood of b with respect to X and let V be an elementary neighbourhood of b with respect to Y. Then the component W of b in $U \cap V$ is an elementary neighbourhood with respect to both X and Y. Hence each component of Y_W is an elementary open set of Y with respect to X. Since these components, for each $b \in B$, form a covering of Y, the result follows at once.

Proposition (4.63). *Let X be a covering space over the locally connected space B. Then each component Y of X covers some component of B. In particular Y covers B itself if B is connected.*

For let $p: X \to B$ be the projection, and let Y be a component of X. Then pY is connected. I assert that pY is both open and closed in B, and so a component of B. For let b be a point of $\mathscr{Cl}(pY)$ and let U be an elementary neighbourhood of b. Since U meets pY, X_U meets Y. So some component V of X_U meets Y. Since Y is a component of X, $Y \supset V$ and hence $pY \supset U$. So $\mathscr{Cl}(pY) \subset \mathscr{Int}(pY)$, and therefore pY is open and closed, as asserted. The same argument shows that if $b \in pY$ and if Y is an elementary neighbourhood of b, then $U \subset pY$ and Y_U is the disjoint union of those components of X_U which meet X. Hence Y is trivial over U and so Y covers pY. This completes the proof. To show that the converse of (4.63) is false we give

Example (4.64). Let X be a covering space over B, with projection p. Let $B_\infty = B \times B \times \ldots$ be a countable product of copies of B and, for $n \geq 1$, let $E_n = X^n \times B_\infty$. Define $p_n : E_n \to B_\infty$ by

$$p_n(x_1, \ldots, x_n, b_1, b_2, \ldots) = (px_1, \ldots, px_n, b_1, b_2, \ldots).$$

Let $E_\infty = \Pi_n E_n$, with $p_\infty : E_\infty \to B_\infty$ being given by $p_\infty | E_n = p_n$. Then each p_n is a covering projection but p_∞ in general is not.

Proposition (4.65). *Let X be a connected covering space over the locally connected space B. Then the group $\operatorname{aut}_B(X)$ of self-equivalences of X over B acts freely on X.*

For let $\theta, \phi: X \to X$ be equivalences over B such that $\theta(x) = \phi(x)$ for some point x of X. Then θ and ϕ are covering maps, by (4.62), and so are locally homeomorphisms. Hence $\theta = \phi$, since X is connected. The group $\operatorname{aut}_B(X)$ considered here is usually called the group of the covering and its elements are called covering transformations.

Proposition (4.66). *Let X be a G-space, where G is discrete. Suppose that the action is free and proper, also that X is Hausdorff over X/G. Then X is a covering space over X/G.*

For by (4.56) each point x of X admits a neighbourhood U satisfying certain conditions. Since G_x is trivial the last of these conditions is that the projection $\pi: X \to X/G$ maps U homeomorphically onto $\pi U = V$, say. Now the action $G \times U \to GU \subset X$ is a homeomorphism, and determines a trivialization $G \times V \to X_V$ as required. A similar argument proves the more general

Proposition (4.67). *Let X be a G-space, where G is discrete. Suppose that the action is free and proper, also that X is Hausdorff over X/G. Then X/H is a covering space over X/G for each subgroup H of G.*

Proposition (4.68). *Let X be a covering space over the locally connected space B. If the discrete group G acts on X by maps over B then the orbit space X/G is also a covering space over B. Moreover if G acts effectively and if X is both Hausdorff and connected then G acts freely.*

To prove the first assertion we need to show that each point b of B admits an elementary neighbourhood with respect to the projection $q: X/G \to B$. To see this, let U be an elementary neighbourhood of b with respect to the projection $p: X \to B$. Let $\{V_j\}$ $(j \in J)$ be the components of X_U. I assert that q maps the components $\{W_k\}$ of X_U/G homeomorphically onto U.

To establish this, first observe that if gV_j meets $V_{j'}$, for some element $g \in G$, then $gV_j \subset V_{j'}$, since $V_{j'}$ is connected. Also observe that if $gV_j \subset V_j$ then $g_{\#}$ maps V_j identically since $pg_{\#} = p$ and p maps V_j homeomorphically onto U. Suppose then that $\pi V_j \subset W_k$, where $\pi: X \to X/G$ is the natural projection. Then π maps V_j homeomorphically onto W_k, from the above observations. So, given W_k, choose a V_j such that $\pi V_j = W_k$ and then, since $q\pi = p$, the projection $W_k \to U$ is a homeomorphism as required.

This proves the first assertion. To prove the second, let g be an element of G with $g \neq e$. The fixed-point set X^g of g is closed, since X is Hausdorff. Since X^g is also open, from the second observation above, we conclude that $X^g = X$ or $X^g = \varnothing$, since X is connected. The first alternative is ruled out since the action is effective and so G acts freely, as asserted.

Proposition (4.69). *Let X be a connected covering space over the locally connected space B. Suppose that X is Hausdorff over B. Then the group $G =$ $\mathrm{aut}_B(X)$ of the covering acts properly and freely on X.*

We have already seen, in (4.65), that G acts freely; it remains to be shown that the action is proper. So let x, y be points in the same orbit such that $px = py = b$, say. Let g be the element such that $gx = y$. Then there exist neighbourhoods U_x of x and U_y of y such that g maps U_x homeomorphically onto U_y. Also there exists an elementary neighbourhood V of b and open sections s_x, s_y of X over V such that $s_x b = x$, $s_y b = y$. Take $V_x = U_x \cap s_x V$, $V_y = U_y \cap s_y V$. If $g'x' = y'$, where $x' \in V_x$, $y' \in V_y$, then $gx' = y'$ also, since p maps V_y homeomorphically, and so $g' = g$, since the action is free. Now (4.69) follows at once from (4.54).

In the previous chapter we have briefly discussed the theory of fibre bundles, i.e. locally trivial spaces, over a base. The theory can be greatly enriched by including a topological group G in the structure. Before outlining the theory of principal G-bundles let us discuss briefly the more general concept of principal G-spaces, over a base.

Let B be a space and let X be an open space over B. Suppose that the topological group G acts freely on X and acts trivially on B. Suppose that the projection $p: X \to B$ is surjective and equivariant, so that the action of G on X is fibre-preserving. Suppose that the action is fibrewise-transitive so that the induced map $X/G \to B$ is a homeomorphism. When all these conditions are satisfied we say that X, with the given action of G, is a principal G-space over B.

A morphism $\phi: X \to Y$, where X, Y are G-spaces over B, is an equivariant map over B. In this way a category is defined and the equivalences of the category are called isomorphisms, of G-spaces over B. An isomorphism with $B \times G$ is called a trivialization. Note that if X and Y are isomorphic G-spaces over B then G acts freely (resp. properly) on X if G acts freely (resp. properly) on Y. The following result is basic.

Proposition (4.70). *Let X, Y be principal G-spaces over B and let $\phi: X \to Y$ be a morphism. If the action of G on Y is proper then ϕ is an isomorphism.*

By (3.23) it is sufficient to show that

$$\phi \times \mathrm{id}: X \times_B X \to Y \times_B X$$

is an equivalence. Now the shearing map

$$G \times Y \to Y \times_B Y$$

is a homeomorphism, since the action is proper; let

$$\psi: Y \times_B Y \to G$$

be the first projection of the inverse homeomorphism. Then the composition

$$Y \times_B X \xrightarrow[\mathrm{id} \times \Delta]{} Y \times_B X \times_B X \xrightarrow[\mathrm{id} \times \phi \times \mathrm{id}]{} Y \times_B Y \times_B X \xrightarrow[\psi \times \mathrm{id}]{} G \times X \xrightarrow[\theta]{} X \times_B X$$

is an inverse of $\phi \times \mathrm{id}$, and so the result follows.

Corollary (4.71). *Let X be a principal G-space over B. Suppose that the action of G on X is proper. If X is sectionable (resp. locally sectionable) over B, then X is trivial (resp. locally trivial) as a G-space over B.*

For if $s: B \to X$ is a section then a morphism $\phi: B \times G \to X$ is given by

$$\phi(b, g) = s(b).g \qquad (b \in B,\ g \in G).$$

Similarly in the local case.

Definition (4.72). A principal G-space over B is a principal G-bundle over B if the action is proper and the projection is locally trivial.

Example (4.73). Let G be a closed subgroup of the topological group Γ. Suppose that Γ is locally sectionable over Γ/G. Then Γ is a principal G-bundle over Γ/G.

For each principal G-bundle E over B the mixed product $E \times_G X$ is locally trivial over B, for each G-space X. We refer to $E \times_G X$ as the G-bundle with fibre X associated with the principal G-bundle E. When $X = G$ with G acting by left translation, the associated G-bundle is E itself. When $X = G$ with G acting by conjugation, the associated G-bundle is not in general principal. For conjugation leaves the neutral element of G fixed and so $E \times_G X$ with that action has a section.

If E is a principal G-bundle over B then for each closed subgroup H of G the orbit space E/H may, by (4.23), be regarded as the associated G-bundle with fibre G/H. Thus if G is itself a closed subgroup of the topological group Γ then Γ/H may be regarded as the G-bundle with fibre G/H over Γ/G associated with the principal bundle Γ, assuming the existence of local sections.

Note that if E is a principal G-bundle over B then ξ^*E is a principal G-bundle over B' for each space B' and map $\xi: B' \to B$. Moreover if $E \times_G X$ is the G-bundle over B with fibre the G-space X associated with E then $\xi^*(E \times_G X) = \xi^*E \times_G X$ is the G-bundle over B' with fibre X associated with ξ^*E.

Now suppose that we have a homomorphism $\alpha: G' \to G$ of topological groups. Let E' be a principal G'-bundle over the space B. Regarding G as a G'-space via α we obtain the principal G-bundle $E' \times_{G'} G = E$ say, also over B, together with a fibre-preserving G'-map $h: E' \to E$ given by $x \to [x, e]$. We say that E is the principal G-bundle over B obtained from E' via α.

Conversely let E be a principal G-bundle over B and let $h: E' \to E$ be a G'-map over B, where G' acts on E via α. Then an equivalence $E' \times_{G'} G \to E$ of G-bundles is given by $[x', g] \mapsto x'g$.

In case G' is a subgroup of G and α the inclusion we describe E as the principal G-bundle over B obtained from E' by extending the structural group from G' to G, and we describe E' as the principal G'-bundle over B obtained from E by reducing the structural group from G to G'. Although the structural group can always be extended to a larger group, as we have seen, reduction of the structural group to a given subgroup is not always possible (for example, reduction to the trivial subgroup is equivalent to trivialization). Moreover two reductions to a given subgroup G' of the same G-bundle are not necessarily equivalent as G'-bundles.

Proposition (4.74). *Suppose that G admits local sections over G/G'. Then the principal G-bundle E over B admits a reduction of structural group to G' if and only if the associated bundle E'/G with fibre G/G' admits a section.*

In one direction this is easy: if E' is a G'-bundle over B and $\phi: E' \to E$ is a G-map over B then

$$\phi/G': E'/G' \to E/G'$$

constitutes a section of the associated bundle. Conversely let $s: E/G \to E/G'$ be a section. Regarding E as a principal G'-bundle over E/G', consider the induced bundle s^*E over E/G. This comes equipped with a G'-map $s^*E \to E$ over s, and hence over $E/G = B$ as required.

Example (4.75). Let G be a compact Lie group and let H be a closed subgroup of G with normalizer $N = N(H)$. Let X be a G-space all of whose orbits are of type $[H]$. Then X is a fibre bundle over G/N with fibre X^H and group N.

For we have seen in (4.43) above that X is equivalent, as a G-space, to $G \times_N X^H$, which is an associated bundle of the principal N-bundle $G \to G/N$. Curiously enough the projection

$$X = G \times_N X^H \to X^H/N = X/G$$

is also a fibre bundle with fibre N/H, but we shall not attempt to prove this.

It is interesting to combine the ideas of the present chapter with those of the previous one. Thus let B be a G-space. By a G-space over B I mean a G-space X together with a G-map $p: X \to B$, called the projection. If X, Y are G-spaces over B then a G-map $X \to Y$ over B, or fibre-preserving G-map, is a map in the ordinary sense which is both fibre-preserving and equivariant. In this way the category $G\text{–}\mathscr{T}_B$ of G-spaces and G-maps over B is defined, and the reader may be interested to investigate its properties.

References

N. Bourbaki. *Topologie Générale*. Hermann, Paris, 1965.

G. Bredon. *Introduction to Compact Topological Transformation Groups*. Academic Press, New York, 1972.

G. Godbillon. *Eléments de Topologie Algébrique*. Hermann, Paris, 1971.

R. S. Palais. *The Classification of G-spaces*. Memoir 36, Amer. Math. Soc. Providence, RI, 1960.

R. S. Palais. On the existence of slices for actions of non-compact Lie Groups. *Ann. of Math.* **73** (1961), 295–323.

CHAPTER 5

The Notion of Homotopy

In this chapter we study the classification of maps by homotopy or continuous deformation. The subject is best approached by first considering a special case. Recall that a path in a space X is simply a map $\lambda: I \to X$. The parameter $t \in I$ is usually thought of as a measure of time, so that as t runs from 0 to 1 the image point $\lambda(t)$ runs from $\lambda(0)$ to $\lambda(1)$. When $\lambda(0) = \lambda(1) = x$, say, the path λ is called a loop, specifically a loop based at x.

The standard domain $I = [0, 1]$ may be replaced by any other closed interval $[a, b]$, where $a < b$, the conversion being accomplished by precomposition with the linear self-map of $\mathbb{R}$ which sends 0 into a and 1 into b. Such adjustments to the domain will be made without comment in what follows. Fundamentally the important thing is the replicating property of the domain: each such interval is the union of a pair of subintervals with one common point.

To each point x of X there corresponds the stationary path (or loop) at x, given by the constant map $\lambda_x: I \to X$ to x. To each path $\lambda: I \to X$ there corresponds the reverse path $\bar{\lambda}: I \to X$, given by $\bar{\lambda}(t) = \lambda(1 - t)$. To each pair of paths $\lambda, \mu: I \to X$ such that $\lambda(1) = \mu(0)$ there corresponds the path $\nu: I \to X$ given by

$$\nu(t) = \begin{cases} \lambda(2t) & (0 \leq t \leq \frac{1}{2}) \\ \mu(2t - 1) & (\frac{1}{2} \leq t \leq 1). \end{cases}$$

We refer to ν as the juxtaposition of λ and μ, although the term composition is often used in the literature. Sometimes we shall write ν as $\lambda + \mu$, and $\bar{\lambda}$ as $-\lambda$, but this notation should be treated with caution. Notice that $-(\lambda + \mu) = (-\mu) + (-\lambda)$. Also notice that $\lambda + \mu$ is a loop if λ and μ are loops.

Postcomposition with a map $\phi: X \to Y$ transforms the path λ in X into the path $\phi \circ \lambda$ in Y. If λ is stationary at x then $\phi \circ \lambda$ is stationary at ϕx. We have that $-(\phi \circ \lambda) = \phi \circ (-\lambda)$, and that $\phi \circ (\lambda + \mu) = \phi \circ \lambda + \phi \circ \mu$.

Let us say that a pair of points of a space X are related if there exists a path in X from one to the other. From the above this is an equivalence relation. The equivalence classes are called path-components and the set of equivalence classes is denoted by $\pi(X)$. If there exists precisely one path-component, then X is said to be path-connected.

If X is discrete the set $\pi(X)$ of path-components is equivalent to X itself. In particular the discrete point-pair is not path-connected. However the point-pair becomes path-connected if we coarsen the topology by allowing only one of the points to be an open set.

Note that if $\phi: X \to Y$ is a continuous surjection then Y is path-connected if X is path-connected.

From the formal point of view π constitutes a functor from $\mathscr{T}$ to $\mathscr{S}$. There is a natural equivalence

$$\pi(X_1 \times X_2) \equiv \pi(X_1) \times \pi(X_2)$$

for all spaces X_1, X_2. It follows that $\pi(G)$ is a group when G is a topological group and that $\pi(X)$ is a $\pi(G)$-set when X is a G-space.

Example (5.1). The join $X * Y$ is path-connected for all non-empty spaces X and Y.

We prove this for the join with the fine topology since that implies the result for any coarser topology. Recall that X, Y are embedded in $X * Y$ as the subspaces where the parameter has value 0, 1, respectively. The join of $x \in X$ to $y \in Y$ constitutes a path from x to y. So we can join (t, x, y) to (t', x', y') in three stages: first join (t, x, y) to x, then join x to y', and finally join y' to (t', x', y'). This proves the result. Taking $Y = \dot{I}$, in particular, we see that the suspension ΣX of a non-empty space X is path-connected, hence the sphere S^n $(n = 1, 2, \ldots)$ is path-connected.

Proposition (5.2). *If X is path-connected then X is connected.*

This follows at once from the connectivity of the unit interval, established in Chapter 2. In general the converse of (5.2) is false. Some local condition is required, such as

Definition (5.3). The space X is locally path-connected if the path-components of X are open sets.

For example the euclidean space $\mathbb{R}^n$ $(n = 0, 1, \ldots)$ is locally path-connected. Evidently an open subset of a locally path-connected space is itself locally path-connected. Also a discrete space is locally path-connected.

Proposition (5.4). *Let X be connected and locally path-connected. Then X is path-connected.*

For let x be any point of X. The path-component H of x is open. Also the complement $K = X - H$ is open, since it is the union of the path-components of X which do not contain x. Thus H is both open and closed, also H is non-empty, so $H = X$. This proves (5.4).

Exercise (5.5). Show that the path-component G_e of the neutral element e in a topological group G is a normal subgroup.

We now come to the main subject of this chapter, the notion of homotopy. The object is to define a classification of maps based on the idea of continuous deformation. The definition can be formulated in two equivalent ways. In the first way we consider the cylinder $I \times X$ of the domain space X, say. For $0 \leq t \leq 1$ we have the embedding $\sigma_t \colon X \to I \times X$ given by

$$\sigma_t(x) = (t, x) \qquad (x \in X).$$

Definition (5.6). Let $\theta, \phi \colon X \to Y$ be maps. A homotopy of θ into ϕ is a map $f \colon I \times X \to Y$ such that $f\sigma_0 = \theta$ and $f\sigma_1 = \phi$.

If there exists a homotopy of θ into ϕ we say that θ is homotopic to ϕ and write $\theta \simeq \phi$. As t runs from 0 to 1 the family $f\sigma_t = f_t \colon X \to Y$ of maps runs from θ to ϕ. It is quite usual to refer to the family f_t as a homotopy but of course one must make sure that the map f so determined is continuous, and not simply the individual maps f_t. We say that f_t is a nulhomotopy of θ if $f_0 = \theta$ and f_1 is a nul-map or constant.

When $X = *$ the maps $* \to Y$ correspond to points of Y and the homotopies $I \times * \to Y$ to paths between the points. Guided by this special case let us now show that homotopy is an equivalence relation.

To each map $\theta \colon X \to Y$ there corresponds the stationary homotopy at θ, give by $(t, x) \mapsto \theta x$ for all t. To each homotopy $f \colon I \times X \to Y$ there corresponds the reverse homotopy $\bar{f} \colon I \times X \to Y$ given by $\bar{f}(t, x) = f(1 - t, x)$. To each pair of homotopies $f, g \colon I \times X \to Y$ such that $f(1, x) = g(0, x)$ there corresponds the homotopy $h \colon I \times X \to Y$ given by

$$h(t, x) = \begin{cases} f(2t, x) & (0 \leq t \leq \frac{1}{2}) \\ g(2t - 1, x) & (\frac{1}{2} \leq t \leq 1). \end{cases}$$

We refer to h as the juxtaposition of f and g, although the term composition is often used in the literature. These constructions make it clear that homotopy is an equivalence relation.

The set of homotopy classes of maps $X \to Y$ is denoted by $\pi(X, Y)$ (some authors prefer $[X, Y]$). Note that a homotopy between such maps determines a path between the corresponding points of the space map(X, Y) of maps. Thus a surjection

$$\pi(X, Y) \to \pi(\text{map}(X, Y))$$

is defined. When X is locally compact regular this function is bijective, by the results of Chapter 2, but in general these two ways of classifying maps must be distinguished.

For each map $\phi: Y \to Z$, postcomposition with ϕ transforms a homotopy $f: I \times X \to Y$ into a homotopy $\phi \circ f: I \times X \to Z$. For each map $\theta: X \to Y$, precomposition with $\mathrm{id} \times \theta$ transforms a homotopy $g: I \times Y \to Z$ into a homotopy $g \circ (\mathrm{id} \times \theta): I \times X \to Z$. We see in this way that the operation of composition for maps induces an operation of composition for homotopy classes:

$$\pi(Y, Z) \times \pi(X, Y) \to \pi(X, Z).$$

Thus we may, when convenient, pass from the category $\mathscr{T}$ of spaces and maps to the category $\mathscr{T}/\simeq$ of spaces and homotopy classes of maps. Maps which are equivalences in the latter category are called homotopy equivalences, and spaces which are equivalent in that sense are said to be of the same homotopy type. In particular, spaces which are of the same homotopy type as a point are said to be contractible. Thus a space X is contractible if and only if the identity map id_X is nulhomotopic. More generally we say that a subspace A is contractible in X if the inclusion $A \to X$ is nulhomotopic.

Exercise (5.7). Let $X = \{a, b\}$ be the two-point space in which $\{a\}$ is open but $\{b\}$ is not. Show that X is contractible.

Clearly the space X is contractible if and only if X is a retract of the cone ΓX. In particular the sphere S^n is contractible if and only if S^n is a retract of the ball B^{n+1}. In the previous chapter we have seen that S^1 is not a retract of B^2, and so S^1 is not contractible. In fact S^n is not contractible for any value of n, but we shall not attempt to prove this here.

Definition (5.8). Let $\theta: X \to Y$ and $\phi: Y \to X$ be maps such that $\phi\theta \simeq \mathrm{id}_X$. Then ϕ is a homotopy left inverse of θ and θ is a homotopy right inverse of ϕ.

Note that if $\theta: X \to Y$ admits a homotopy left inverse ϕ and a homotopy right inverse ϕ' then $\phi \simeq \phi'$ and θ is a homotopy equivalence.

Definition (5.9). The space X is dominated by the space Y if there exists a map $X \to Y$ which admits a homotopy left inverse.

Recall that a subset X of $\mathbb{R}^n$ (or, more generally, of any topological vector space) is said to be convex if for each pair of points $x_0, x_1 \in X$ the line segment

$$(1 - t)x_0 + tx_1 \qquad (0 \le t \le 1)$$

is entirely contained in X. Also that X is said to be starlike from the point x_0 if for each point $x \in X$ the line segment

$$(1 - t)x_0 + tx \qquad (0 \le t \le 1)$$

is entirely contained in X. Thus X is convex if, and only if, X is starlike from each of its own points. If X is starlike from x_0, say, then X is contractible since a nul-homotopy $f: I \times X \to X$ of the identity is given by

$$f(t, x) = (1 - t)x_0 + tx \qquad (0 \leq t \leq 1).$$

In particular $\mathbb{R}^n$ is contractible and so is the n-ball $B^n \subset \mathbb{R}^n$.

Most of the specific homotopies which appear in our work can be traced back to convexity in some form. However, homotopies can arise naturally in other ways. For example, let X be a G-space, where G is a topological group. For each element g of G the translation $g_{\#}: X \to X$ is defined, as in the previous chapter. A path λ in G from g to g', say, determines a homotopy

$$I \times X \xrightarrow{\lambda \times \mathrm{id}} G \times X \xrightarrow{\text{action}} X$$

of $g_{\#}$ into $g'_{\#}$. Thus the translations defined by the elements of a given path-component of G all belong to the same homotopy class. When $X = S^n$, for example, this shows that all the rotations are homotopic and all the improper isometries are homotopic.

From the formal point of view π constitutes a binary functor from $\mathscr{T}$ to $\mathscr{S}$, contravariant in the first entry and covariant in the second. There is a natural equivalence

$$\pi(X, Y_1 \times Y_2) \equiv \pi(X, Y_1) \times \pi(X, Y_2)$$

for all spaces X, Y_1, Y_2 and a natural equivalence

$$\pi(X_1 + X_2, Y) \equiv \pi(X_1, Y) \times \pi(X_2, Y)$$

for all spaces X_1, X_2, Y. Moreover if G is a topological group then $\pi(A, G)$ is a group, for all spaces A, and if X is a G-space then $\pi(A, X)$ is a $\pi(A, G)$-set, for all spaces A.

Let Ψ be a continuous endofunctor of $\mathscr{T}$. As explained in Chapter 2, a homotopy $f: I \times X \to Y$ determines a homotopy $\bar{f}: I \times \Psi X \to \Psi Y$, according to the rule

$$\bar{f}_t = \Psi(f_t) \qquad (0 \leq t \leq 1).$$

Thus if $\theta, \phi: X \to Y$ are homotopic then so are $\Psi(\theta), \Psi(\phi): \Psi X \to \Psi Y$. Consequently Ψ determines a function

$$\Psi_{\#}: \pi(X, Y) \to \pi(\Psi X, \Psi Y),$$

and thereby induces an endofunctor of the quotient category $\mathscr{T}/\simeq$. Similarly in the case of multiple functors.

Our next result concerns the relation between homotopy and mapping spaces.

Proposition (5.10). *Let X, Y, Z be spaces. Then for any map $\theta: X \to Y$ the homotopy class of the precomposition*

$$\theta^*: \operatorname{map}(Y, Z) \to \operatorname{map}(X, Z)$$

depends only on the homotopy class of θ. *Also for any map* $\phi: Y \to Z$ *the homotopy class of the postcomposition*

$$\phi^*: \text{map}(X, Y) \to \text{map}(X, Z)$$

depends only on the homotopy class of ϕ.

To prove the first assertion consider the function

$$\lambda: I \times \text{map}(I \times X, Z) \to \text{map}(X, Z)$$

which is given by $\lambda(t, h) = h_t$. For each compact-open set (C, V) of map (X, Z) and point $(t, h) \in \lambda^{-1}(C, V)$ we have that $h(\{t\} \times C) \subset V$, i.e. $\{t\} \times C \subset h^{-1}V$. Since C is compact there exists a neighbourhood N of t such that $N \times C \subset h^{-1}V$. Since I is compact regular we can shrink N to a neighbourhood U of t such that $\mathscr{Cl}\, U$ is compact. Then $U \times (\mathscr{Cl}\, U \times C, V)$ is a neighbourhood of (t, h) which is contained in $\lambda^{-1}(C, V)$. Thus λ is continuous and the first part of (5.10) follows by precomposing with a homotopy $I \times X \to Y$.

To prove the second assertion, consider the function

$$\mu: I \times \text{map}(X, Y) \to \text{map}(X, I \times Y)$$

which is given by $\mu(t, f)(x) = (t, f(x))$. For each compact-open set (C, U) of map$(X, I \times Y)$ and point $(t, f) \in \mu^{-1}(C, U)$ we have that U is a neighbourhood of $\{t\} \times fC$. Since fC is compact there exist, by (2.60), neighbourhoods V of t and W of fC such that $V \times W$ is contained in U. Then $V \times (C, W)$ is a neighbourhood of (t, f) which is contained in $\mu^{-1}(C, U)$. Thus μ is continuous and the second part of (5.10) follows by postcomposing μ with a homotopy $I \times Y \to Z$.

We see from this that the homotopy type of map(X, Y) depends only on the homotopy types of X and Y. In particular map(X, Y) has the same homotopy type as Y when X is contractible, while map(X, Y) is contractible when Y is contractible.

There is a point which needs to be made about the quotient topology. Recall from (2.101) that when $p: X \to X'$ is a quotient map then so is id $\times$ $p: I \times X \to I \times X'$, since I is compact regular. For any space Y, therefore, a function $f': I \times X' \to Y$ is continuous if and only if the corresponding function $f'(\text{id} \times p): I \times X \to Y$ is continuous. In other words the family of functions $f'_t: X' \to Y$ $(0 \leq t \leq 1)$ constitutes a homotopy if and only if the family of functions $f'_t p: X \to Y$ constitutes a homotopy.

In particular, suppose that $X' = X/\sim$, where $\sim$ is an equivalence relation on X. We have seen earlier that maps of X' into Y correspond precisely to invariant maps of X into Y; now we see that homotopies of maps of X' into Y correspond precisely to invariant homotopies of invariant maps of X into Y.

For example, consider the cone ΓX on the space X, with the fine topology.

The homotopy $f_t: I \times X \to I \times X$ $(0 \leq t \leq 1)$ given by

$$f_t(s, x) = (st, x) \qquad (s \in I, x \in X)$$

induces a nulhomotopy of the identity on ΓX. Thus ΓX is contractible.

For another example, consider the join $X * Y$ of spaces X and Y, again with the fine topology. We have already seen that $*$ can be regarded as a binary functor $\mathscr{T} \times \mathscr{T} \to \mathscr{T}$. We see now that $*$ induces a binary functor at the $\mathscr{T}/\simeq$ level. In particular the homotopy type of $X * Y$ depends only on the homotopy types of X and Y. The corresponding results for the suspension and cone are special cases.

In fact, as Rutter has observed, the homotopy type of the join is the same whether the fine topology or the coarse topology is used. For consider the map

$$X * Y \text{ (fine)} \to X * Y \text{ (coarse)}$$

given by the identity function. A homotopy inverse of this continuous bijection is given by the formula

$$(t, x, y) \mapsto \begin{cases} x & (0 \leq t \leq \frac{1}{3}) \\ (3t - 1, x, y) & (\frac{1}{3} \leq t \leq \frac{2}{3}) \\ y & (\frac{2}{3} \leq t \leq 1). \end{cases}$$

Since the coarse join is associative, as we have seen in Chapter 2, it follows from this that the fine join is associative up to homotopy equivalence.

Infinite joins, with the coarse topology, have some interesting properties which we shall be using later. Thus let E_X, for any space X, denote the coarse join of a countable number of copies of X. We represent points of E_X in the form

$$\langle t, x \rangle = \langle t_1 x_1, t_2 x_2, \ldots, t_i x_i, \ldots \rangle,$$

where $x = (x_1, x_2, \ldots, x_i, \ldots)$, $t = (t_1, t_2, \ldots, t_i, \ldots)$; here $x_i \in X$, $t_i \geq 0$ and $\Sigma t_i = 1$, with all but a finite number of the $t_i = 0$. Consider the subspace $E'_X \subset E_X$ consisting of points $\langle t, x \rangle$ such that $t_i = 0$ whenever i is odd, and the subspace $E''_X \subset E_X$ consisting of points $\langle t, x \rangle$ such that $t_i = 0$ whenever i is even. I assert that E'_X and E''_X are weak deformation retracts of E_X, i.e. that the inclusions $E'_X \subset E_X$ and $E''_X \subset E_X$ are homotopy equivalences.

The two cases are similar: I will therefore give details only for the case of E'_X. Let $u: E'_X \to E_X$ be the inclusion and let $r: E_X \to E'_X$ be given by $r\langle t, x \rangle = \langle t', x' \rangle$, where

$$t'_{2i-1} = 0, \qquad x'_{2i-1} = x_i; \qquad t'_{2i} = t_i, \quad x'_{2i} = x_i \qquad (i = 1, 2, \ldots).$$

A deformation $h: I \times E_X \to E_X$ of ur into the identity may be constructed as follows. For $n = 0, 1, \ldots$ let

$$\mu_n: [1 - (\tfrac{1}{2})^n, 1 - (\tfrac{1}{2})^{n+1}] \to [0, 1]$$

be the order-preserving homeomorphism given by

$$\mu_n(s) = 2^{n+1}s - 2^{n+1} + 2.$$

For $1 - (\frac{1}{2})^n \leq s \leq 1 - (\frac{1}{2})^{n+1}$ the deformation is given by $h(s, \langle t, x\rangle) = \langle t^*, x^*\rangle$, where

$$t_i^* = t_i, \qquad x_i^* = x_i \qquad (i = 1, \ldots, n);$$

$$t_{n+2j-1}^* = (1 - s)t_{n+j}, \qquad x_{n+2j-1}^* = x_{n+j};$$

$$t_{n+2j}^* = st_{n+j}, \qquad x_{n+2j}^* = x_{n+j} \qquad (j = 1, 2, \ldots).$$

For instance the first stage in the homotopy is given by

$$\langle t_1 x_1, st_2 x_2, (1 - s)t_2 x_2, st_3 x_3, (1 - s)t_3 x_3, \ldots\rangle$$

which deforms $\langle t, x\rangle$ into

$$\langle t_1 x_1, t_2 x_2, 0, t_3 x_3, 0, \ldots\rangle.$$

Thus E'_X is a weak deformation retract of $E_{X'}$ and similarly so is E''_X. Note that E_X may be identified with the (coarse) join of E'_X and E''_X. A specific equivalence

$$\xi\colon E'_X * E''_X \to E_X$$

is given by

$$\xi(s, \langle t', x'\rangle, \langle t'', x''\rangle) = \langle t, x\rangle,$$

where

$$\begin{aligned} t_i x_i &= (1 - s)t'_i x'_i \qquad & i \text{ even},\\ &= st''_i x''_i & i \text{ odd}. \end{aligned}$$

These observations will be put to use in Chapter 7 below.

The definition of homotopy we have been using so far is based on the cylinder functor $I \times$ applied to the domain. The definition can be reformulated, however, in terms of the path-space functor map$(I, \;)$ applied to the codomain. In certain situations this is the more convenient form of the definition, and so I will describe it briefly.

Observe that the space $PY = \text{map}(I, Y)$ of paths in a space Y comes equipped with a family $\rho_t\colon PY \to Y$ $(0 \leq t \leq 1)$ of projections, where $\rho_t(\lambda) = \lambda(t)$.

Definition (5.11). Let $\theta, \phi\colon X \to Y$ be maps. A homotopy of θ into ϕ is a map $\hat{f}\colon X \to PY$ such that $\rho_0 \hat{f} = \theta$ and $\rho_1 \hat{f} = \phi$.

Thus if $f\colon I \times X \to Y$ is a homotopy of θ into ϕ in the previous sense then the adjoint $\hat{f}\colon X \to PY$ is a homotopy of θ into ϕ in the present sense, and vice versa.

In Chapter 3 we have already discussed the notion of sectional category, for a space over a given space. The classical notion of category, for a space, is as follows. Let us say that an open subspace U of a space X is categorical if U is contractible in X. Let us also say that X is locally categorical if X can be covered by categorical open sets. For example, locally contractible spaces (such as topological manifolds) are locally categorical. Products of locally categorical spaces are locally categorical.

Definition (5.12). Let X be a locally categorical space. The category cat X of X is the least number of categorical open sets required to cover X.

In general cat X is infinite; however cat X is clearly finite when X is compact. Evidently cat $X = 1$ if and only if X is contractible. Also cat $X \leq 2$ if X a suspension.

Proposition (5.13). *Suppose that X is dominated by Y. Then X is locally categorical if Y is locally categorical, and* cat $X \leq$ cat Y.

For let $\theta: X \to Y$ and $\phi: Y \to X$ be maps such that $\phi\theta \simeq \mathrm{id}_X$. If V is an open set of Y such that the inclusion $v: V \to Y$ is nul-homotopic then $U = \theta^{-1}V$ is an open set of X such that the inclusion $u: U \to X$ is nul-homotopic, since $\theta \circ u: U \to Y$ is nul-homotopic and $u \simeq \phi\theta u$. Thus each categorical open set of Y pulls back under θ to a categorical open set of X. Now (5.13) follows immediately.

From (5.13) we at once deduce that cat X depends only on the homotopy type of X. Further results about this numerical homotopy invariant will be obtained in Chapter 7.

We turn now to the category $\mathscr{T}^A$ of spaces under a given space A.

Definition (5.14). Let $\theta, \phi: X \to Y$ be maps under A. A homotopy under A of θ into ϕ is a homotopy in the ordinary sense which is a map under A at each stage of the deformation.

If there exists a homotopy under A of θ into ϕ we say that θ is homotopic to ϕ under A and write $\theta \simeq^A \phi$. A homotopy under A into a nul-map is called a nul-homotopy under A. The proof that homotopy under A is an equivalence relation between maps under A is the same as in the ordinary case. The set of equivalence classes is denoted by $\pi^A(X, Y)$. The operation of composition for maps under A induces an operation of composition for homotopy classes under A:

$$\pi^A(Y, Z) \times \pi^A(X, Y) \to \pi^A(X, Z).$$

Thus we may, when convenient, pass from the category $\mathscr{T}^A$ of spaces and maps under A to the category $\mathscr{T}^A/\simeq$ of spaces and homotopy classes under A. Maps under A which are equivalences in the latter category are called homotopy equivalences under A, and spaces under A which are equivalent in that sense are said to be of the same homotopy type under A. In particular spaces under A which are of the same homotopy type under A as A itself are said to be contractible under A. Clearly a space X under A is contractible under A if and only if the identity id_X is nul-homotopic under A. For example, if X is a space under A then the pull-back $P^A X$ of the triad

$$PX \xrightarrow{\rho_1} X \xleftarrow{u} A$$

is contractible under A. We call $P^A X$ the space of paths in X based in A.

Definition (5.15). Let $\theta\colon X \to Y$ and $\phi\colon Y \to X$ be maps under A such that $\phi\theta \simeq^A \mathrm{id}_X$. Then ϕ is a left inverse of θ, up to homotopy under A, and θ is a right inverse of ϕ, up to homotopy under A.

Note that if θ admits a left inverse ϕ, up to homotopy under A, and a right inverse ϕ', up to homotopy under A, then $\phi \simeq^A \phi'$ and θ is a homotopy equivalence under A. It is important to appreciate that a map under A may admit a homotopy inverse as an ordinary map but not as a map under A. However under certain conditions the existence of the former implies the existence of the latter, as we shall see in the next chapter.

From the formal point of view π^A constitutes a binary functor from $\mathscr{T}^A$ to $\mathscr{S}$, contravariant in the first entry and covariant in the second. There is a natural equivalence

$$\pi^A(X, Y_1 \times Y_2) \equiv \pi^A(X, Y_1) \times \pi^A(X, Y_2)$$

for all spaces X, Y_1, Y_2 under A, and a natural equivalence

$$\pi^A(X_1 \vee^A X_2, Y) \equiv \pi^A(X_1, Y) \times \pi^A(X_2, Y)$$

for all spaces X_1, X_2, Y under A. In fact the product $\times$ and wedge $\vee^A$ induce binary functors at the $\mathscr{T}^A/\simeq$ level. Hence the homotopy types under A of $X \times Y$ and $X \vee^A Y$ depend only on the homotopy types under A of X, Y.

A homotopy $f_t\colon X \to Y$ under A determines a homotopy $\xi_* f_t\colon \xi_* X \to \xi_* Y$ under A' for each space A' and continuous injection $\xi\colon A \to A'$. Thus ξ induces a function

$$\xi_\#\colon \pi^A(X, Y) \to \pi^{A'}(\xi_* X, \xi_* Y)$$

and so determines a natural transformation of π^A into $\pi^{A'}$.

Our next result is really just a corollary of (5.10) above.

Proposition (5.16). *Let X, Y, Z be spaces under A. Then for any map $\theta\colon X \to Y$ under A the homotopy class under A of the precomposition*

$$\theta^*\colon \mathrm{map}^A(Y, Z) \to \mathrm{map}^A(X, Z)$$

depends only on the homotopy class under A of θ. Also for any map $\phi\colon Y \to Z$ under A the homotopy class under A of the postcomposition

$$\phi_*\colon \mathrm{map}^A(X, Y) \to \mathrm{map}^A(X, Z)$$

depends only on the homotopy class under A of ϕ.

Let X be a space under A and let Y be a subspace of $\mathbb{R}^n$ or, more generally, of a topological vector space. Choose a map $A \to Y$ and so regard Y as a space under A. If Y is convex then any two maps $X \to Y$ under A are homotopic under A, by the same construction as was used in the ordinary case.

In dealing with situations where the insertion is an embedding (as is always the case when $A = *$) a different terminology is customary. It is sufficient to consider the case when A is a subspace of the space X and the insertion is the inclusion. We have already remarked in Chapter 2 that the term retraction is generally used instead of cosection, in this situation. In line with this a nul-homotopy under A of id_X is called a deformation retraction, and when X is contractible under A we say that A is a deformation retract of X.

Again let Y be a space and let $v\colon A \to Y$ be a map. By an extension of v to X we mean a map $\phi\colon X \to Y$ such that $\phi | A = v$. Evidently such extensions are simply maps under A. The classification of extensions by homotopy under A is called classification by homotopy relative to A, or rel A. We shall be using this terminology rather sparingly.

Another special case which has its own terminology is when the codomain is A itself. Then maps under A of a space X under A are cosections of X, and their classification by homotopy under A is known as classification by horizontal homotopy.

One of the applications of these ideas is in the classification of paths in a space which begin at one given point and end at another given point. Let us say that paths $\lambda, \mu\colon I \to X$ such that $\lambda | \dot{I} = \mu | \dot{I}$ are equivalent, and write $\lambda \sim \mu$, if $\lambda \simeq \mu$ rel $\dot{I}$. Then a category $\pi_1(X)$ can be defined as follows. The objects of the category are the points of X. Given objects $x_0, x_1 \in X$ the morphisms of the category, from x_0 to x_1, are the equivalence classes of paths in X, in the above sense. I assert that the two conditions (1.1) and (1.2) for a category are satisfied and that each morphism is an equivalence; in other words that $\pi_1(X)$ is a groupoid.

The proof of these three assertions depends ultimately on the convexity of the unit interval I. Let us begin by showing that juxtaposition of paths, although not itself an associative operation, is nevertheless associative up to equivalence. First consider the special case when the co-domain is the interval $I_3 = [0, 3]$. Let $e_i\colon I \to I_3$ $(i = 0, 1, 2)$ be given by

$$e_i(t) = t + i \qquad (0 \leq t \leq 1).$$

By convexity of I_3 we have at once that

(5.17) $$(e_0 + e_1) + e_2 \sim e_0 + (e_1 + e_2).$$

We are now ready to prove

Proposition (5.18). *Let λ, μ, v be paths in X such that $\lambda(1) = \mu(0)$ and $\mu(1) = v(0)$. Then*

$$(\lambda + \mu) + v \sim \lambda + (\mu + v).$$

For we have $\lambda = ke_0$, $\mu = ke_1$, $\nu = ke_2$, where $k\colon I_3 \to X$ is given by

$$k(t) = \begin{cases} \lambda(t) & (0 \le t \le 1) \\ \mu(t-1) & (1 \le t \le 2) \\ \nu(t-2) & (2 \le t \le 3). \end{cases}$$

It follows at once that

$$(\lambda + \mu) + \nu = k \circ ((e_0 + e_1) + e_2),$$

$$\lambda + (\mu + \nu) = k \circ (e_0 + (e_1 + e_2)).$$

Since $(e_0 + e_1) + e_2 \sim e_0 + (e_1 + e_2)$, as we have just seen, the conclusion follows immediately. A similar argument proves

Proposition (5.19). *Let λ be a path in X from x_0 to x_1. Then*

$$\lambda_0 + \lambda \sim \lambda \sim \lambda + \lambda_1,$$

where λ_i $(i = 0, 1)$ denotes the constant path at x_i.

Proposition (5.20). *Let λ be a path in X from x_0 to x_1. Then $\lambda + (-\lambda) \sim \lambda_0$, the constant path at x_0.*

The first two results show that $\pi_1(X)$ is a category and the last result that every morphism is an equivalence. We refer to $\pi_1(X)$ as the fundamental groupoid of X. Its functorial behavior is sufficiently obvious not to require comment. When the groupoid is trivial, i.e. when there is precisely one equivalence class of path between any two points in the same path-component, we say that the space is simply-connected. For example, contractible spaces are simply-connected, as well as path-connected.

For each point x_0 of X, the equivalence classes of loops based at x_0 form a group. This is called the fundamental group of X and denoted by $\pi_1(X, x_0)$. We may regard $\pi_1(X, x_0)$ as the group of automorphisms of x_0, as an object of the fundamental groupoid $\pi_1(X)$. From this point of view it is clear that the isomorphism class of the fundamental group is independent of the choice of base-point within a given path-component of X. Specifically, an equivalence class $[\mu]$ of paths from x_0 to x_1, say, determines an isomorphism $\pi_1(X, x_0) \to \pi_1(X, x_1)$, through $[\lambda] \mapsto [\bar{\mu} + \lambda + \mu]$.

There is an important link between the theory of the fundamental groupoid of a space and the theory of coverings spaces over that space. The first result to be proved is known as "uniqueness of path lifting".

Proposition (5.21). *Let $p\colon X \to B$ be a covering map. Let x be a point of X and let μ be a path in B which starts at px. Then there exists a unique path λ in X which starts at x and which satisfies the lifting condition $\mu = p\lambda$.*

For consider the pull-back μ^*X of X with respect to μ. This is a covering space of I and so trivial by (3.54). Choose a section s of μ^*X over I so that

$s(0) = (0, x)$. The composition of s with the structure map $\mu^*X \to X$ over μ is one lifting of μ satisfying the initial condition. Since I is connected the lifting is unique. By a similar argument, with $I \times I$ in place of I, we obtain

Proposition (5.22). *Let $p: X \to B$ be a covering map. Let x be a point of X and let μ, μ' be paths in B such that $\mu(0) = px = \mu'(0)$ and $\mu(1) = \mu'(1)$. Let λ, λ' be liftings of μ, μ' which start at x. If $\mu \sim \mu'$ then $\lambda(1) = \lambda'(1)$ and $\lambda \sim \lambda'$.*

An important special case is when the covering is of the form $p: G \to G/H$, where H is a closed discrete subgroup of the topological group G. Recall from Chapter 1 that in this situation a groupoid π can be constructed as follows, without using the topology. The objects of π are points of the factor space $B = G/H$. We may regard G/H (right cosets) as a right G-space through translations, in the usual way. Then if $b, b' \in B$ the morphisms of π from b to b' are the elements $g \in G$ such that $bg = b'$. Composition of morphisms is given by the group operation in G.

Let us now compare π with the fundamental groupoid $\pi_1(B)$ of the factor space. A functor $\partial: \pi_1(B) \to \pi$ may be defined as follows. On objects ∂ is the identity. As for morphisms, let $[\mu]$ be a path-class in B from b to b', say. Choose any element $x \in G$ such that $px = b$ and let x' be the end of a lift λ of μ starting at x. We define $\partial[\mu] = x^{-1}x'$. To check that this is well defined, observe that if $h \in H$ then $h\lambda$ is a lift of μ starting at hx and ending at hx', and then $(hx)^{-1}(hx') = x^{-1}x'$, as required. To check that ∂ transforms the path-class of the stationary loop at b into the neutral element e of G, observe that if $px = b$ then the stationary loop at x is a lift of the stationary loop at b. To check that ∂ satisfies the composition law for morphisms, let μ be a path in B from b to b' and let μ' be a path in B from b' to b''. Choose an element $x \in G$ such that $px = b$, lift μ to a path λ in G from x to x', say, and then lift μ' to a path λ' in G from x' to x'', say. Then $\lambda + \lambda'$ is a lift of $\mu + \mu'$ which starts at x and ends at x''. Write $x^{-1}x' = g$ and $(x')^{-1}x'' = g'$, so that $x^{-1}x'' = gg'$. Then

$$\partial([\mu] . [\mu']) = \partial[\mu + \mu'] = gg' = \partial[\mu] . \partial[\mu'],$$

as required. Thus $\partial: \pi_1(B) \to \pi$ is a functor, and we now prove

Proposition (5.23). *If G is path-connected then ∂ is surjective. If G is simply-connected then ∂ is injective.*

Thus if G is both path-connected and simply-connected (as will be the case if G is contractible) then $\pi_1(G/H) \approx \pi$, as a groupoid. To prove (5.23), first suppose that G is path-connected. Let $b, b' \in B$ and let $g \in G$ be an element such that $bg = b'$. Let v be a path in G from e to g. Then bv is a path in B from b to b', and if $x \in p^{-1}b$ then xv is a lift of bv which starts at x and ends at xg, so that $\partial[bv] = g$. Secondly suppose that G is simply-connected. Let μ, μ' be paths in B from b to b'. Let $x \in p^{-1}b$, $x' \in p^{-1}b'$ and let λ, λ' be lifts of μ, μ', respectively, to paths in G from x to x'. Then $\lambda \sim \lambda'$, since G is simply-connected, and so $\mu \sim \mu'$, by projection. Therefore ∂ is injective.

We at once deduce from (5.23) that if G is path-connected and simply-connected then the fundamental group of G/H is isomorphic to H. For example, taking $G = \mathbb{R}$ and $H = \mathbb{Z}$, the fundamental group of the circle $\mathbb{R}/\mathbb{Z}$ is cyclic infinite. If we only know that G is path-connected then (5.23) still shows that H is isomorphic to a factor group of the fundamental group; hence the fundamental group is non-abelian whenever H is non-abelian. For example, take $G = S^3$, represented as the group of quaternions of unit modulus (in fact S^3 is simply-connected, although we have not proved this). Take H to be the subgroup of order 8 consisting of the quaternions

$$\{\pm 1, \pm i, \pm j, \pm k\}.$$

We at once conclude that the factor space has non-abelian fundamental group.

These special cases we have been considering include a wide variety of examples where the fundamental groupoid can be calculated. Of course there are other, more general, methods of calculation which can be used but these are outside the scope of this book. One may also investigate, for spaces X and Y, the groupoid of which the objects are maps of X and Y and the morphisms are homotopies between such maps, classified relative to $\dot{I} \times X$.

Much of homotopy theory is concerned with the category $\mathscr{T}^*/\simeq$, and we discuss this next. As before, spaces under $*$ will be called pointed spaces or spaces with basepoints. Maps and homotopies under $*$ will be called pointed maps and pointed homotopies.

It follows at once from what we have already established about the relation between homotopy and quotients that if Φ is a continuous endofunctor of $\mathscr{T}$ then the corresponding reduced endofunctor Φ^* of $\mathscr{T}^*$ induces an endofunctor of $\mathscr{T}^*/\simeq$, and similarly for multiple functors. For example the reduced suspension Σ^* induces an endofunctor of $\mathscr{T}^*/\simeq$. More generally the smash product $\wedge$ induces a binary functor at the $\mathscr{T}^*/\simeq$ level.

Let us now consider binary systems consisting of a pointed space X and a pointed map

$$m: X \times X \to X,$$

called the multiplication. As we shall see, it is interesting to study systems which satisfy "up to homotopy" the axioms for a topological group. The following conditions turn out to be important.

Definition (5.24). The multiplication m is homotopy-commutative if $m \simeq^* mt$, where t is the switching equivalence as shown below.

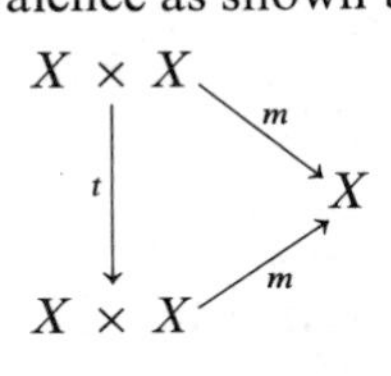

Definition (5.25). The multiplication m is homotopy-associative if

$$m(m \times \mathrm{id}) \simeq^* m(\mathrm{id} \times m),$$

as shown below

$$\begin{array}{ccc} X \times X \times X & \xrightarrow{\mathrm{id} \times m} & X \times X \\ {\scriptstyle m \times \mathrm{id}}\downarrow & & \downarrow{\scriptstyle m} \\ X \times X & \xrightarrow[m]{} & X \end{array}$$

Definition (5.26). The multiplication m satisfies the Hopf condition if

$$m(\mathrm{id} \times c)\Delta \simeq^* \mathrm{id} \simeq^* m(c \times \mathrm{id})\Delta,$$

where

$$X \xrightarrow{\Delta} X \times X \underset{c \times \mathrm{id}}{\xrightarrow{\mathrm{id} \times c}} X \times X \xrightarrow{m} X.$$

Here c, as usual, denotes the nul-map. A binary system where the multiplication satisfies the Hopf condition is called a Hopf space, or H-space.

Definition (5.27). The pointed map $v: X \to X$ is a homotopy inversion for the multiplication m if

$$m(\mathrm{id} \times v)\Delta \simeq^* c \simeq^* m(v \times \mathrm{id})\Delta,$$

where

$$X \xrightarrow{\Delta} X \times X \underset{v \times \mathrm{id}}{\xrightarrow{\mathrm{id} \times v}} X \times X \xrightarrow{m} X.$$

A homotopy-associative Hopf space for which the multiplication admits a homotopy inversion is called a group-like space; in that case the pointed homotopy class of the homotopy inverse is uniquely determined. Thus a group-like space is essentially a group object in the category $\mathscr{T}^*/\simeq$. Topological groups are group-like spaces, of course; further examples will be given in a moment.

Note that if X is a group-space, i.e. a space on which the structure of a topological group is defined, then various multiplications on X can be defined in terms of the group structure so as to satisfy the Hopf condition. For example one may define

$$m(x, y) = x^2yx^{-1}, x^3yx^{-2}, \ldots.$$

Such Hopf spaces are not, in general, group-like spaces.

Obviously a multiplication on the pointed space X determines a multiplication (binary operation) on the pointed set $\pi^*(A, X)$ for all spaces A. If the former is homotopy-commutative then the latter is commutative, and similarly with the other conditions. Thus $\pi^*(A, X)$ is a group if X is group-like.

Similar results hold for mapping spaces. Thus a multiplication on X determines a multiplication on map*(A, X), for all regular A. If the former is homotopy-commutative then so is the latter, and similarly with the other conditions. Thus map*(A, X) is group-like if X is group-like.

Proposition (5.28). *Let X be a group-like space. Then all the path-components of X have the same homotopy type.*

This is untrue for Hopf spaces in general, since $X' + *$ is a Hopf space for any space X' with multiplication given by the nul-map on $X' \times X'$.

To prove (5.28) first observe that any self-map of X determines, by restriction, maps of the path-components of X into the path-components of X, and similarly with homotopies. Let m be the multiplication on X and let v be a homotopy-inverse for m. Let X_0 be the path-component of the basepoint x_0 and X_1 the path-component of a given point x_1. Let $\phi: X \to X$ be given by $\phi(x) = m(x, x_1)$ and let $\psi: X \to X$ be given by $\psi(x) = m(x, v(x_1))$. Then ϕ maps X_0 into X_1 and ψ maps X_1 into X_0, since $m(x_1, v(x_1)) \in X_0$. Since ψ is a homotopy inverse of ϕ, by the group-like axioms, it follows at once that the map $X_1 \to X_0$ determined by ψ is a homotopy inverse of the map $X_0 \to X_1$ determined by ϕ. Therefore X_1 has the same homotopy type as X_0, which proves (5.28).

Next let us see what happens when we "dualize" the definitions (5.24)–(5.27). Consider "cobinary" systems consisting of a pointed space X and a pointed map

$$m: X \to X \vee X,$$

called the comultiplication.

Definition (5.29). The comultiplication m is homotopy-commutative if $m \simeq^* tm$, where t is the switching equivalence as shown below.

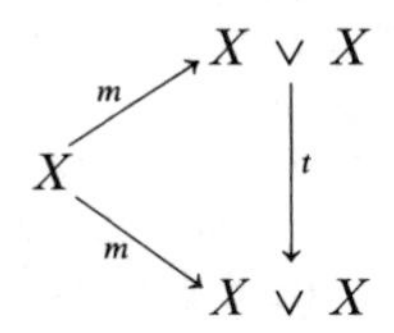

Definition (5.30). The comultiplication m is homotopy-associative if $(m \vee \text{id})m \simeq^* (\text{id} \vee m)m$, as shown below.

$$\begin{array}{ccc} X & \xrightarrow{m} & X \vee X \\ \downarrow{\scriptstyle m} & & \downarrow{\scriptstyle \text{id} \vee m} \\ X \vee X & \xrightarrow[m \vee \text{id}]{} & X \vee X \vee X \end{array}$$

Definition (5.31). The comultiplication m satisfies the coHopf condition if

$$\nabla(\mathrm{id} \vee c)m \simeq^* \mathrm{id} \simeq^* \nabla(c \vee \mathrm{id})m,$$

where

$$X \xrightarrow{m} X \vee X \underset{c \vee \mathrm{id}}{\xrightarrow{\mathrm{id} \vee c}} X \vee X \xrightarrow{\nabla} X.$$

A cobinary system where the comultiplication satisfies the coHopf condition is call a coHopf space.

Definition (5.32). The pointed map $v: X \to X$ is a homotopy-inversion for the comultiplication m if

$$\nabla(\mathrm{id} \vee v)m \simeq^* c \simeq^* \nabla(v \vee \mathrm{id})m,$$

where

$$X \xrightarrow{m} X \vee X \underset{v \vee \mathrm{id}}{\xrightarrow{\mathrm{id} \vee v}} X \vee X \xrightarrow{\nabla} X.$$

A homotopy-associative coHopf space for which the comultiplication admits a homotopy-inversion is called a cogroup-like space; in that case the pointed homotopy class of the homotopy inversion is uniquely determined. The term "cogroup-like" is not intended to suggest "like a cogroup", since cogroups are of no interest whatever, but rather to suggest a dual relationship to "group-like".

Obviously a comultiplication on X determines a multiplication on the pointed set $\pi^*(X, Y)$, for all pointed spaces Y. If the former is homotopy-commutative then the latter is commutative, and similarly with the other conditions. Thus $\pi^*(X, Y)$ becomes a group if X is cogroup-like. If X is a coHopf space and Y is a Hopf space then, for reasons explained in Chapter 1, the multiplication on $\pi^*(X, Y)$ determined by the comultiplication on X is the same as that determined by the multiplication on Y, and is both commutative and associative. The reader may wish to run through the proof of this in the present situation as an exercise.

Similar results hold for mapping spaces. Thus if X has closed basepoint a comultiplication on X determines a multiplication on $\mathrm{map}^*(X, Y)$, for all pointed spaces Y. If the former is homotopy-commutative then so is the latter, and similarly with the other conditions. Thus $\mathrm{map}^*(X, Y)$ is group-like if X is cogroup-like. If X is a coHopf space with closed basepoint and Y is a Hopf space the multiplication on $\mathrm{map}^*(X, Y)$ determined by the comultiplication on X is pointed homotopic to that determined by the multiplication on Y, and is homotopy-commutative.

The basic example of a cogroup-like space is the circle $S = I/\dot{I}$ (by a coincidence, S is also a group-space). The comultiplication is given by the map

$$I \to I_2 = [0, 1] \vee [1, 2],$$

defined by $t \mapsto 2t$, followed by the standard equivalence $I_2 \to I$. The three conditions which have to be satisfied follow at once from (5.18)–(5.20). For each pointed space Y the group-like space map*(S, Y) is called the space of loops on Y and denoted by ΩY. The group $\pi^*(S, Y)$ is just the fundamental group of Y. As we have seen in our discussion of the fundamental groupoid, fundamental groups are not, in general, commutative, and so the coHopf structure on S cannot be homotopy-commutative.

If X has closed basepoint then, by (3.2), a comultiplication on X determines a comultiplication on the smash product $X \wedge Y$ for all spaces Y with closed basepoint. If the former is homotopy-commutative then so is the latter, and similarly with the other conditions. Thus $X \wedge Y$ is cogroup-like if X is cogroup-like; in particular the suspension ΣY is cogroup-like.

Next let us extend the notion of homotopy to the category $\mathscr{T}_B$ of spaces over a given space B.

Definition (5.33). Let $\theta, \phi: X \to Y$ be maps over B. A homotopy over B of θ into ϕ is a homotopy in the ordinary sense which is a map over B at each stage of the deformation.

In fact the term fibre homotopy is customary rather than homotopy over B, as long as there can be no doubt as to what the base space is. If there exists a homotopy over B of θ into ϕ we say that θ is homotopic to ϕ over B and write $\theta \simeq_B \phi$. A fibre homotopy into a fibrewise nul-map is called a fibrewise nul-homotopy. The proof that homotopy over B is an equivalence relation between maps over B is the same as in the ordinary case. The set of equivalence classes is denoted by $\pi_B(X, Y)$. The operation of composition for homotopy classes over B is denoted by:

$$\pi_B(Y, Z) \times \pi_B(X,Y) \to \pi_B(X, Z).$$

Thus we may, when convenient, pass from the category $\mathscr{T}_B$ of spaces and maps over B to the category $\mathscr{T}_B/\simeq$ of spaces and homotopy classes over B. Maps over B which are equivalences in the latter category are called homotopy equivalences over B, or fibre homotopy equivalences, and spaces over B which are equivalent in this sense are said to be of the same fibre homotopy type. In particular spaces over B which are of the same fibre homotopy type as B itself are said to be fibrewise contractible. Clearly a space X over B is fibrewise contractible if and only if the identity id_X is fibrewise nul-homotopic. For example, if X is a space over B then the fibre-cone $\Gamma_B X$ is fibrewise contractible.

Definition (5.34). Let $\theta: X \to Y$ and $\phi: Y \to X$ be maps over B such that $\phi\theta \simeq_B \mathrm{id}_X$. Then ϕ is a left inverse of θ, up to homotopy over B, and θ is a right inverse of ϕ, up to homotopy over B.

Note that if θ admits a left inverse ϕ, up to homotopy over B, and a right inverse ϕ', up to homotopy over B, then $\phi \simeq_B \phi'$ and θ is a fibre homotopy equivalence. It is important to appreciate that a map over B may admit a homotopy inverse as an ordinary map but not as a map over B. However under certain conditions the existence of the former implies the existence of the latter, as we shall see in the next chapter.

From the formal point of view π_B constitutes a binary functor from $\mathscr{T}_B$ to $\mathscr{S}$, contravariant in the first entry and covariant in the second. There is a natural equivalence

$$\pi_B(X, Y_1 \times_B Y_2) \equiv \pi_B(X, Y_1) \times \pi_B(X, Y_2)$$

for all spaces X, Y_1, Y_2 over B, and a natural equivalence

$$\pi_B(X_1 + X_2, Y) \equiv \pi_B(X_1, Y) \times \pi_B(X_2, Y)$$

for all spaces X_1, X_2, Y over B. In fact the fibre product $\times_B$ and the sum $+$ induce binary functors at the $\mathscr{T}_B/\simeq$ level. Hence the fibre homotopy types of $X \times_B Y$ and $X + Y$ depend only on the fibre homotopy types of X and Y.

A homotopy $f_t\colon X \to Y$ over B determines a homotopy $\xi^*f_t\colon \xi^*X \to \xi^*Y$ over B' for each space B' and map $\xi\colon B' \to B$. Thus ξ induces a function

$$\xi^\#\colon \pi_B(X, Y) \to \pi_{B'}(\xi^*X, \xi^*Y)$$

and so determines a natural transformation of π_B into $\pi_{B'}$.

Definition (5.35). Let X be a space over B. If X has the same fibre homotopy type as $B \times T$ for some space T, then X is trivial, with fibre T, in the sense of the fibre homotopy type.

Local triviality, in the sense of fibre homotopy type, is defined similarly.

When the domain is the base space B itself, so that maps over B are sections, a different terminology is customary. A fibre homotopy $f_t\colon B \to X$, for any space X over B, is called a vertical homotopy or homotopy through sections.

Example (5.36). Let X and Y be spaces over B which admit sections. Then the fibre join $X *_B Y$ admits precisely one vertical homotopy class of section.

The proof is a fibrewise version of the proof of (5.1).

For another type of example consider a subspace Y of the space $B \times \mathbb{R}^n$ ($n \geq 0$) over B. Let us say that Y is fibrewise convex if for each point b of B the fibre Y_b is convex in $\{b\} \times \mathbb{R}^n$. If this condition is satisfied then for each space X over B there exists precisely one fibre homotopy class of maps of X into Y over B. Again, let us say that Y is fibrewise starlike from a given section $s\colon B \to Y$ if for each point b of B the fibre Y_b is starlike from $s(b)$ in $\{b\} \times \mathbb{R}^n$. If this condition is satisfied then for each space X over B each map of X into Y over B is nul-homotopic with nul-map given by s.

Let Ψ_B be a continuous endofunctor of $\mathscr{T}_B$. Then a fibre homotopy $f: I \times X \to Y$ determines a fibre homotopy $\hat{f}: I \times \Psi_B X \to \Psi_B Y$, according to the rule

$$\hat{f}_t = \Psi_B(f_t) \qquad (0 \leq t \leq 1).$$

Thus if $\theta, \phi: X \to Y$ are fibre homotopic then so are $\Psi_B \theta, \Psi_B \phi: \Psi_B X \to \Psi_B Y$. Consequently Ψ_B determines a function

$$\Psi_{B\#}: \pi_B(X, Y) \to \pi_B(\Psi_B X, \Psi_B Y),$$

and thereby induces an endofunctor of the quotient category $\mathscr{T}_B/\simeq$. Similarly in the case of multiple functors.

For example consider the fibre join $*_B$, with the fine topology. It follows from the above that the fibre homotopy type of $X *_B Y$ depends only on the fibre homotopy types of X and Y. Also the same argument as was used in the ordinary case shows that the fine fibre-join has the same fibre homotopy type as the coarse fibre-join, and hence that the fine fibre-join is associative up to fibre homotopy equivalence.

Next a result about the relation between fibre homotopy and fibrewise mapping spaces.

Proposition (5.37). *Let X, Y, Z be spaces over B. Then for any map $\theta: X \to Y$ over B the fibre homotopy class of the precomposition*

$$\theta^*: \operatorname{map}_B(Y, Z) \to \operatorname{map}_B(X, Z)$$

depends only on the fibre homotopy class of θ. Also for any map $\phi: Y \to Z$ over B the fibre homotopy class of the postcomposition

$$\phi_*: \operatorname{map}_B(X, Y) \to \operatorname{map}_B(X, Z)$$

depends only on the fibre homotopy class of ϕ.

The proof is a routine generalization of the corresponding result in the ordinary case and will therefore be left to serve as an exercise. We conclude that the fibre homotopy type of $\operatorname{map}_B(X, Y)$ depends only on the fibre homotopy types of X and Y. In particular $\operatorname{map}_B(X, Y)$ has the same fibre homotopy type as Y when X is fibrewise contractible, while $\operatorname{map}_B(X, Y)$ is fibrewise contractible when Y is fibrewise contractible.

Let us turn now to the category $\mathscr{T}_B^B$ of sectioned spaces over a given space B. Here too an appropriate notion of homotopy can be defined as follows.

Definition (5.38). Let $\theta, \phi: X \to Y$ be maps of sectioned spaces over B. A fibrewise pointed homotopy of θ into ϕ is a fibre homotopy in the ordinary sense which is a fibrewise pointed map at each stage of the deformation.

If there exists a fibrewise pointed homotopy of θ into ϕ we write $\theta \simeq_B^B \phi$. The proof that this is an equivalence relation between fibrewise pointed maps is the same as in the ordinary case. The pointed set of equivalence classes is denote by $\pi_B^B(X, Y)$. The operation of composition for maps of sectioned spaces induces an operation of composition for fibrewise pointed homotopy classes:

$$\pi_B^B(Y, Z) \times \pi_B^B(X, Y) \to \pi_B^B(X, Z).$$

Thus, we may, when convenient, pass from the category $\mathscr{T}_B^B$ of sectioned spaces and fibrewise pointed maps to the category $\mathscr{T}_B^B/\simeq$ of sectioned spaces and fibrewise pointed homotopy classes. Fibrewise pointed maps which are equivalences in the latter category are called fibrewise pointed homotopy equivalences, and sectioned spaces which are equivalent in this sense are said to have the same fibrewise pointed homotopy type.

From the formal point of view π_B^B constitutes a binary functor from $\mathscr{T}_B^B$ to $\mathscr{S}^*$, contravariant in the first entry and covariant in the second. There is a natural equivalence

$$\pi_B^B(X, Y_1 \times_B Y_2) \equiv \pi_B^B(X, Y_1) \times \pi_B^B(X, Y_2)$$

for all sectioned spaces X, Y_1, Y_2 over B, and a natural equivalence

$$\pi_B^B(X_1 \vee_B X_2, Y) \equiv \pi_B^B(X_1, Y) \times \pi_B^B(X_2, Y)$$

for all sectioned spaces X_1, X_2, Y over B.

A fibrewise pointed homotopy $f_t\colon X \to Y$ over B determines a fibrewise pointed homotopy $\xi^* f_t\colon \xi^* X \to \xi^* Y$ over B' for each space B' and map $\xi\colon B' \to B$. Thus ξ induces a function

$$\xi^{\#}: \pi_B^B(X, Y) \to \pi_{B'}^{B'}(\xi^* X, \xi^* Y)$$

and so determines a natural transformation of π_B^B into $\pi_{B'}^{B'}$.

Definition (5.39). Let X be a sectioned space over B. If there exists a pointed space T and a fibrewise pointed homotopy equivalence $X \to B \times T$ then X is trivial, with fibre T, in the sense of fibrewise pointed homotopy type.

Local triviality, in the same sense, is defined similarly.

It follows from what we have already established about the relation between homotopy and quotients that if Φ_B is a continuous endofunctor of $\mathscr{T}_B$ then the corresponding reduced endofunctor Φ_B^B of $\mathscr{T}_B^B$ induces an endofunctor of $\mathscr{T}_B^B/\simeq$, and similarly for multiple functors. For example the reduced fibre suspension Σ_B^B induces an endofunctor of $\mathscr{T}_B^B/\simeq$. More generally the fibre smash product $\wedge_B$ induces a binary functor at the $\mathscr{T}_B^B/\simeq$ level.

Our next result concerns the relation between fibrewise pointed homotopy and fibrewise mapping spaces.

Proposition (5.40). *Let X, Y, Z be sectioned spaces over B. Then for any map $\theta\colon X \to Y$ of sectioned spaces over B the fibrewise pointed homotopy class of*

$$\theta^*: \mathrm{map}_B^B(Y, Z) \to \mathrm{map}_B^B(X, Z)$$

depends only on the fibrewise pointed homotopy class of θ. *Also for any map* $\phi: Y \to Z$ *of sectioned spaces over* B *the fibrewise pointed homotopy class of*

$$\phi_*: \mathrm{map}_B^B(X, Y) \to \mathrm{map}_B^B(X, Z)$$

depends only on the homotopy class of ϕ.

The proof is a routine generalization of the corresponding result in the ordinary case and will therefore be left to serve as an exercise. We conclude that within the class of sectioned spaces over B the fibrewise pointed homotopy type of $\mathrm{map}_B^B(X, Y)$ depends only on the fibrewise pointed homotopy type of X and Y; in particular $\mathrm{map}_B^B(X, Y)$ has the same fibrewise pointed homotopy type as Y when X is fibrewise pointed contractible, while $\mathrm{map}_B^B(X, Y)$ is fibrewise pointed contractible when Y is fibrewise pointed contractible.

Let us now consider fibrewise binary systems consisting of a sectioned space X over B and a map

$$m: X \times_B X \to X$$

of sectioned spaces, called the fibrewise multiplication. Among the conditions which can be satisfied by such a system the following turn out to be important.

Definition (5.41). The fibrewise multiplication m is fibre homotopy-commutative if $m \simeq_B^B mt$, where t is the switching equivalence as shown below.

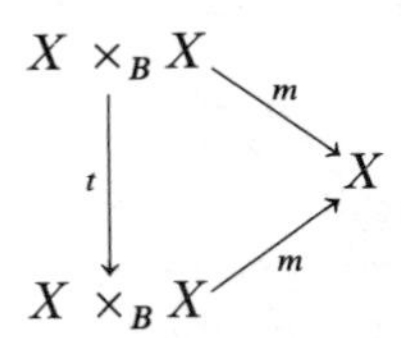

Definition (5.42). The fibrewise multiplication m is fibre homotopy-associative if

$$m(m \times \mathrm{id}) \simeq_B^B m(\mathrm{id} \times m),$$

as shown below.

$$\begin{array}{ccc} X \times_B X \times_B X & \xrightarrow{\mathrm{id} \times m} & X \times_B X \\ \downarrow{\scriptstyle m \times \mathrm{id}} & & \downarrow{\scriptstyle m} \\ X \times_B X & \xrightarrow[m]{} & X \end{array}$$

Definition (5.43). The fibrewise multiplication m satisfies the Hopf condition if

$$m(\mathrm{id} \times c)\Delta \simeq_B^B \mathrm{id} \simeq_B^B m(c \times \mathrm{id})\Delta,$$

where

$$X \xrightarrow{\Delta} X \times_B X \underset{c \times \mathrm{id}}{\xrightarrow{\mathrm{id} \times c}} X \times_B X \xrightarrow{m} X.$$

Here c, as usual, denotes the nul-map. A fibrewise binary system where the fibrewise multiplication satisfies the Hopf condition is called a fibrewise Hopf space.

Definition (5.44). The map $v: X \to X$ of sectioned spaces is a fibre homotopy-inversion for the fibrewise multiplication m if

$$m(\mathrm{id} \times v)\Delta \simeq_B^B c \simeq_B^B m(v \times \mathrm{id})\Delta$$

where

$$X \xrightarrow{\Delta} X \times_B X \underset{v \times \mathrm{id}}{\xrightarrow{\mathrm{id} \times v}} X \times_B X \xrightarrow{m} X.$$

A fibre homotopy-associative fibrewise Hopf space for which the fibrewise multiplication admits a fibre homotopy-inversion is called a fibrewise group-like space; in that case the fibrewise pointed homotopy class of the fibre homotopy-inversion is uniquely determined. Thus a fibrewise group-like space is essentially a group object in the category $\mathscr{T}_B^B/\simeq$.

Note that if T is a group-like space then $B \times T$ (with the obvious structure) is a fibrewise group-like space over B, for any B. More generally if X is a fibrewise group-like space over B then $\xi^* X$ is a fibrewise group-like space space over B' for each space B' and map $\xi: B' \to B$.

Obviously a fibrewise multiplication on the sectioned space X over B determines a multiplication on the pointed set $\pi_B(A, X)$, for all spaces A over B. If the former is fibre homotopy-commutative then the latter is commutative, and similarly with the other conditions. Thus $\pi_B(A, X)$ is a group if X is fibrewise group-like.

Similar results hold for fibrewise mapping spaces. Thus a fibrewise multiplication on X determines a fibrewise multiplication on $\mathrm{map}_B(A, X)$ for all regular A over B. If the former is fibre homotopy-commutative then so is the latter, and similarly with the other conditions. Thus $\mathrm{map}_B(A, X)$ is fibrewise group-like if X is fibrewise group-like.

In particular if the sectioned space X over B is fibrewise group-like then the space $\mathrm{sec}_B X$ of sections of X is group-like and the set $\pi^* \mathrm{sec}_B X$ of vertical homotopy classes of sections is a group.

Dually let us also consider the fibrewise cobinary systems consisting of a sectioned space X over B and a map

$$m: X \to X \vee_B X$$

of sectioned spaces, called the fibrewise comultiplication.

Definition (5.45). The fibrewise comultiplication m is fibre homotopy-commutative if $m \simeq_B^B tm$, where t is the switching equivalence as shown below.

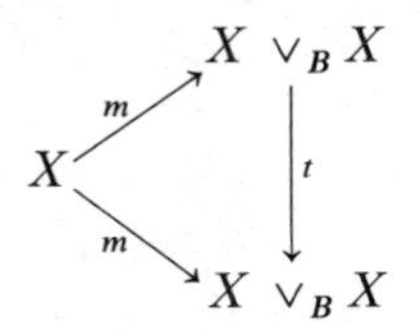

Definition (5.46). The fibrewise comultiplication m is fibre homotopy-associative if

$$(m \vee \mathrm{id})m \simeq_B^B (\mathrm{id} \vee m)m,$$

as shown below.

$$\begin{array}{ccc} X & \xrightarrow{m} & X \vee_B X \\ {\scriptstyle m}\downarrow & & \downarrow{\scriptstyle \mathrm{id} \vee m} \\ X \vee_B X & \xrightarrow[m \vee \mathrm{id}]{} & X \vee_B X \vee_B X \end{array}$$

Definition (5.47). The fibrewise comultiplication m satisfies the coHopf condition if

$$\nabla(\mathrm{id} \vee c)m \simeq_B^B \mathrm{id} \simeq_B^B \nabla(c \vee \mathrm{id})m,$$

where

$$X \xrightarrow{m} X \vee_B X \xrightarrow[c \vee \mathrm{id}]{\mathrm{id} \vee c} X \vee_B X \xrightarrow{\nabla} X.$$

A fibrewise cobinary system where the fibrewise comultiplication satisfies the coHopf condition is called a fibrewise coHopf space.

Definition (5.48). The map $v: X \to X$ of sectioned spaces over B is a fibre homotopy-inversion for the fibrewise comultiplication m if

$$\nabla(\mathrm{id} \vee v)m \simeq_B^B c \simeq_B^B \nabla(v \vee \mathrm{id})m,$$

where

$$X \xrightarrow{m} X \vee_B X \xrightarrow[v \vee \mathrm{id}]{\mathrm{id} \vee v} X \vee_B X \xrightarrow{\nabla} X.$$

A fibre homotopy-associative fibrewise coHopf space which admits a fibre homotopy-inversion is called a fibrewise cogroup-like space; in that case the fibrewise pointed homotopy class of the fibre homotopy-inversion is uniquely determined.

Note that if T is a cogroup-like space then $B \vee T$ (with the obvious structure) is a fibrewise cogroup-like space over B, for any B. More generally if X is a fibrewise cogroup-like space over B then ξ^*X is a fibrewise cogroup-like space over B' for each space B' and map $\xi: B' \to B$.

Obviously a fibrewise comultiplication on the sectioned space X over B determines a multiplication on the pointed set $\pi_B^B(X, Y)$, for all sectioned spaces Y over B. If the former is fibre homotopy-commutative then the latter is commutative, and similarly with the other conditions. Thus $\pi_B^B(X, Y)$ is a group when X is fibrewise cogroup-like. If X is a fibrewise coHopf space and Y is a fibrewise Hopf space then, for reasons explained in Chapter 1, the multiplication on $\pi_B^B(X, Y)$ determined by the fibrewise comultiplication on X is the same as that determined by the fibrewise multiplication on Y, and is both commutative and associative.

Similar results hold for fibrewise mapping spaces. Thus if X has closed section a fibrewise comultiplication on X determines a fibrewise multiplication on $\mathrm{map}_B^B(X, Y)$ for all sectioned spaces Y. If the former is fibre homotopy-commutative then so is the latter, and similarly with the other conditions. Thus $\mathrm{map}_B^B(X, Y)$ is fibrewise group-like if X is fibrewise cogroup-like. If X is a fibrewise coHopf space with closed section and Y is a fibrewise Hopf space the fibrewise multiplication on $\mathrm{map}_B^B(X, Y)$ determined by the fibrewise comultiplication on X is fibrewise pointed homotopic to that determined by the fibrewise multiplication on Y, and is both fibre homotopy-commutative and fibre homotopy-associative.

If X has closed section then, by (3.80), a fibrewise comultiplication on X determines a fibrewise comultiplication on $X \wedge_B Y$ for all spaces Y over B with closed section. If the former is fibre homotopy-commutative then so is the latter, and similarly with the other conditions. Thus $X \wedge_B Y$ is fibrewise cogroup-like if X is fibrewise cogrouplike; in particular the fibre-suspension $\Sigma_B Y$ is fibrewise cogroup-like, for all spaces Y over B with closed section.

We turn next to the category G–$\mathscr{T}$ of G-spaces, where G is a topological group. Here too an appropriate notion of homotopy can be defined as follows.

Definition (5.49). Let $\theta, \phi: X \to Y$ be G-maps. A G-homotopy of θ into ϕ is a homotopy in the ordinary sense which is a G-map at each stage in the deformation.

The term equivariant homotopy is also used when there can be no doubt as to the group concerned. A G-homotopy into a constant G-map is called a G-nul-homotopy. The proof that the existence of a G-homotopy is an equivalence relation between G-maps is the same as in the ordinary case. The set of equivalence classes is denoted by $\pi_G(X, Y)$. The operation of composition for G-maps induces an operation of composition for G-homotopy classes:

$$\pi_G(Y, Z) \times \pi_G(X, Y) \to \pi_G(X, Z).$$

Thus we may, when convenient, pass from the category G–$\mathscr{T}$ of G-spaces and G-maps to the category G–$\mathscr{T}/\simeq$ of G-spaces and G-homotopy classes. The G-maps which are equivalences in the latter category are called G-homotopy equivalences, and G-spaces which are equivalent in this sense are said to have the same G-homotopy type. In particular G-spaces which are of the same G-homotopy type as the point-space are said to be G-contractible. Clearly a G-space X is G-contractible if and only if the identity id_X is G-nul-homotopic. For example, if X is a G-space then the cone ΓX is G-contractible.

Definition (5.50). Let $\theta: X \to Y$ and $\phi: Y \to X$ be G-maps such that $\phi\theta$ is G-homotopic to id_X. Then ϕ is a G-homotopy left inverse of θ, and θ is a G-homotopy right inverse of ϕ.

Note that if θ admits a G-homotopy left inverse ϕ and a G-homotopy right inverse ϕ' then ϕ and ϕ' are G-homotopic and θ is a G-homotopy equivalence. It is important to appreciate that a G-map may admit a homotopy inverse as an ordinary map but not as a G-map. However, under certain conditions the existence of the one implies the existence of the other, as we shall see in Chapter 8.

From the formal point of view π_G constitutes a binary functor from G–$\mathscr{T}$ to $\mathscr{S}$, contravariant in the first entry and covariant in the second. There is a natural equivalence

$$\pi_G(X, Y_1 \times Y_2) \equiv \pi_G(X, Y_1) \times \pi_G(X, Y_2)$$

for all G-spaces X, Y_1, Y_2 and a natural equivalence

$$\pi_G(X_1 + X_2, Y) \equiv \pi_G(X_1, Y) \times \pi_G(X_2, Y)$$

for all G-spaces X_1, X_2, Y. In fact the product $\times$ and the sum $+$ induce binary functors at the G–$\mathscr{T}/\simeq$ level. Hence the G-homotopy types of $X \times Y$ and $X + Y$ depend only on the G-homotopy types of X and Y.

Recall that a homomorphism $\alpha: G' \to G$ of topological groups determines a functor

$$\alpha^{\#}: G\text{–}\mathscr{T} \to G'\text{–}\mathscr{T}.$$

Evidently a G-homotopy $f: I \times X \to Y$ determines a G'-homotopy $\alpha^{\#}f: I \times \alpha^{\#}X \to \alpha^{\#}Y$, where $(\alpha^{\#}f)_t = \alpha^{\#}(f_t)$ $(0 \leq t \leq 1)$. Thus α determines a natural transformation of π_G into $\pi_{G'}$.

It is important to note that both the fixed-point space functor and the orbit space functor respect the relation of homotopy. Specifically a G-homotopy $f: I \times X \to Y$ determines a homotopy $f^G: I \times X^G \to Y^G$ and a homotopy $f/G: I \times X/G \to Y/G$, where

$$(f^G)_t = (f_t)^G, \qquad (f/G)_t = f_t/G \qquad (0 \leq t \leq 1)$$

Thus the fixed-point space functor and the orbit space functor induce functors G–$\mathscr{T}/\simeq \to \mathscr{T}/\simeq$.

Homotopy-theoretic versions of the results of Chapter 4 can be obtained wherever appropriate without difficulty. For example, take (4.40), where X and Y are G-spaces with G compact and X Hausdorff. According to that result, applied to $I \times X$ instead of X, the G-maps $I \times X \to Y$ correspond precisely to the sections of $(I \times X : Y)$ over $(I \times X)/G$, i.e. to the sections of $I \times (X : Y)$ over $I \times X/G$, in the way described. Hence it follows at once that the G-homotopy classes of G-maps $X \to Y$ correspond precisely to the vertical homotopy classes of sections of $(X : Y)$ over X/G.

I hope that the reader will accept that by proceeding in this way it is perfectly straightforward to produce equivariant versions of everything we have done earlier in this chapter. Specifically one can introduce a notion of G-homotopy under a G-space A, in particular a notion of pointed G-homotopy, and a notion of G-homotopy over a G-space B, also a notion of fibrewise pointed G-homotopy over a G-space B. It seems quite unnecessary to spell all this out. Instead, whenever I need to refer to a result of this nature I shall refer to "the equivariant version" of whichever is the appropriate result in the ordinary theory.

Exercise (5.51). Let H be a normal subgroup of the path-connected group G. Show that the inner automorphisms of G/H are pointed G-homotopic to the identity.

Looking ahead to the next chapter I would like to return to a couple of topics considered earlier in the present chapter but from a somewhat different point of view. Let us work primarily in the basic category $\mathscr{T}$.

By the mapping cylinder $M = M(\phi)$ of a map $\phi: X \to Y$ I mean the push-out of the cotriad

$$I \times X \overset{\sigma_0}{\leftarrow} X \overset{\phi}{\rightarrow} Y.$$

Thus M comes equipped with maps

$$I \times X \to M \leftarrow Y.$$

Moreover a map $\rho: M \to Y$ is defined by $\phi\pi_2$ on $I \times X$ and by the identity on Y. We can interpret M, with the above section, as the fibre cone on X, as a space over Y. Thus M, as a sectioned space over Y, is fibrewise contractible.

Proposition (5.52). *The map $\phi: X \to Y$ admits a homotopy inverse on the left if and only if the mapping cylinder $M(\phi)$ admits a cosection under X.*

For let $\theta: Y \to X$ be a map and let $h: I \times X \to X$ be a homotopy of $\theta\phi$ into id_X. Then a cosection $M(\phi) \to X$ is given by the push-out of θ and h. Conversely a cosection determines a homotopy inverse on the left by restriction to Y.

In case A is a subspace of the space X one must be careful not to confuse the mapping cylinder $M(u)$ of the inclusion $u: A \to X$ with the subspace $I \times A \cup \{0\} \times X$ of the cylinder $I \times X$. The continuous bijection

$$M(u) \to I \times A \cup \{0\} \times X,$$

defined in the obvious way, is not in general a closed map. If, however, A is closed in X then $(I \times A) \cap (\{0\} \times X) = \{0\} \times A$ is closed in $I \times X$ and the bijection is a homeomorphism.

There is another construction, known as the double mapping cylinder, which is also useful. Here we start with a cotriad

$$X_0 \overset{\phi_0}{\leftarrow} X \overset{\phi_1}{\rightarrow} X_1.$$

The double mapping cylinder $M = M(\phi_0, \phi_1)$ is defined to be the quotient space of $X_0 + X_1 + (I \times X)$ with respect to the relations

$$(0, x) \sim \phi_0 x, \qquad (1, x) \sim \phi_1 x \qquad (x \in X).$$

Thus M comes equipped with maps

$$X_0 \rightarrow M \leftarrow X_1.$$

Note that $M(\phi_0, \phi_1)$ is equivalent to $M(\phi_1, \phi_0)$. We have already met an example of this construction in the join $X_0 * X_1$ (fine topology), which is just the double mapping cylinder of the cotriad

$$X_0 \leftarrow X_0 \times X_1 \rightarrow X_1.$$

The reason for the term "double mapping cylinder" is that $M(\phi_0, \phi_1)$ can be regarded as the push-out of the cotriad

$$M(\phi_0, \mathrm{id}_X) \equiv M(\phi_0) \leftarrow X \rightarrow M(\phi_1) \equiv M(\mathrm{id}_X, \phi_1).$$

The double mapping cylinder is also known as the homotopy push-out, for the following reason. Suppose that we have a homotopy-commutative diagram as shown below.

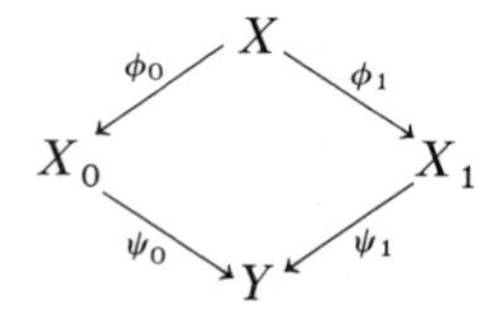

To each homotopy $f: I \times X \rightarrow Y$ of $\psi_0 \phi_0$ into $\psi_1 \phi_1$ there is associated a map of $M(\phi_0, \phi_1)$ into Y, given by ψ_0 on X_0, by ψ_1 on X_1 and by f on $I \times X$.

Now suppose that one of ϕ_0, ϕ_1 is injective, so that the (topological) push-out of the cotriad

$$X_0 \overset{\phi_0}{\leftarrow} X \overset{\phi_1}{\rightarrow} X_1$$

is defined. By taking the push-out to be Y, and the associated maps to be ψ_0, ψ_1, we obtain a diagram as above which is commutative. Then by taking f to be the stationary homotopy we obtain a map of the homotopy push-out into the topological push-out. Conditions under which this map is a homotopy equivalence will be obtained in the next chapter.

Exercise (5.53). Show that for all cotriads

$$X_0 \overset{\phi_0}{\leftarrow} X \overset{\phi_1}{\rightarrow} X_1$$

the homotopy type of the homotopy push-out $M(\phi_0, \phi_1)$ depends only on the homotopy classes of ϕ_0 and ϕ_1.

All this can be dualized as follows. By the mapping path-space $W = W(\phi)$ of a map $\phi: X \to Y$ I mean the pull-back of the triad

$$PY \overset{\rho_0}{\rightarrow} Y \overset{\phi}{\leftarrow} X.$$

Thus W comes equipped with maps

$$PY \leftarrow W \rightarrow X.$$

Moreover a map $\sigma: X \to W$ is defined, where the first component of σ sends $x \in X$ into the stationary path at ϕx, while the second component is id_X. Regarding W as a space under X, with this insertion, we see that W can be interpreted as the space $P^X Y$ of paths in Y based in X. Thus W, as a sectioned space over X, is fibrewise contractible.

Proposition (5.54). *The map $\phi: X \to Y$ admits a homotopy inverse on the right if and only if the mapping path-space $W(\phi)$ admits a section over Y.*

For let $\theta: Y \to X$ be a map and let $h: I \times Y \to Y$ be a homotopy of $\phi\theta$ into id_Y. Then a section $Y \to W(\phi)$ is given by the pull-back of θ and h. Conversely a section determines a homotopy inverse on the right by projection to X.

There is another construction, known as the double mapping path-space, which is also useful. Here we start with a triad

$$Y_0 \overset{\psi_0}{\rightarrow} Y \overset{\psi_1}{\leftarrow} Y_1.$$

The double mapping path-space $W = W(\psi_0, \psi_1)$ is defined to be the subspace of $Y_0 \times Y_1 \times PY$ consisting of triples (y_0, y_1, λ) such that $\rho_0\lambda = y_0$ and $\rho_1\lambda = y_1$. Thus W comes equipped with maps

$$Y_0 \leftarrow W \rightarrow Y_1.$$

Note that $W(\psi_0, \psi_1)$ is equivalent to $W(\psi_1, \psi_0)$. The reason for the term "double mapping path-space" is that $W(\psi_0, \psi_1)$ may be regarded as the pull-back of the triad

$$W(\psi_0, \mathrm{id}_Y) \equiv W(\psi_0) \rightarrow Y \leftarrow W(\psi_1) \equiv W(\mathrm{id}_Y, \psi_1).$$

The double mapping path-space is also known as the homotopy pull-back, for the following reason. Suppose that we have a homotopy-commutative diagram as shown below.

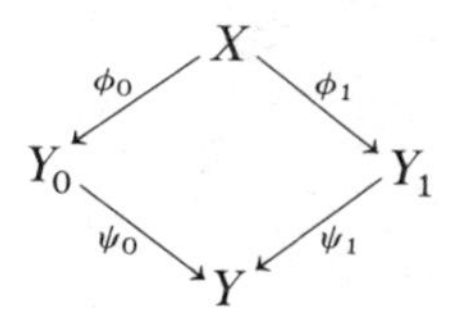

To each homotopy $f: X \to PY$ of $\psi_0 \phi_0$ into $\psi_1 \phi_1$ there is associated a map of X into $W(\psi_0, \psi_1)$, with first component $\psi_0 \phi_0$, second component $\psi_1 \phi_1$, and third component f.

In particular take the (topological) pull-back of the triad

$$Y_0 \overset{\psi_0}{\to} Y \overset{\psi_1}{\leftarrow} Y_1$$

to be X, and the associated maps to be ϕ_0, ϕ_1. We obtain a diagram as above which is commutative. Then by taking f to be the stationary homotopy we obtain a map of the topological pull-back into the homotopy pull-back. Conditions under which this map is a homotopy equivalence will be given in the next chapter.

Exercise (5.55). Show that for all triads

$$Y_0 \overset{\psi_0}{\to} Y \overset{\psi_1}{\leftarrow} Y_1$$

the homotopy type of the homotopy pull-back $W(\psi_0, \psi_1)$ depends only on the homotopy classes of ψ_0 and ψ_1.

References

R. Brown. *Elements of Modern Topology*, McGraw-Hill, London, 1968.

R. Brown and P. R. Heath. Coglueing homotopy equivalences. *Math. Zeit.* **113** (1970), 313–325.

P. J. Hilton. *Homotopy Theory and Duality*. Gordon and Breach, New York, 1965.

I. M. James. Ex-homotopy theory. *Ill. J. Math.* **15** (1971), 324–327.

M. Mather. Pull-backs in homotopy theory. *Can. J. Math.* **28** (1976), 225–263.

E. H. Spanier. *Algebraic Topology*. McGraw-Hill, New York, 1966.

R. M. Vogt. Convenient categories of topological spaces for homotopy theory. *Arch. der Math.* **22** (1971), 545–555.

R. M. Vogt. A note on homotopy equivalences. *Proc. Amer. Math. Soc.* **32** (1972), 627–629.

R. M. Vogt. Homotopy limits and colimits. *Math. Zeit.* **134** (1973), 11–52.

R. M. Vogt. Commuting homotopy limits and colimits. *Math. Zeit.* **175** (1980), 77–80.

M. Walker. Homotopy pull-backs and applications to duality. *Can. J. Math.* **29** (1977), 45–46.

G. W. Whitehead. *Elements of Homotopy Theory*. Springer-Verlag, New York, 1978.

CHAPTER 6

Cofibrations and Fibrations

Problems concerning the extension of continuous functions are central to topology. One is given a space X and a subspace A of X. One is also given a space E and a map $f: A \to E$. The question is: does there exist an extension of f over X, i.e. a map $g: X \to E$ such that $g|A = f$?

For example, take $E = A$ and f the identity: does there exist a retraction $X \to A$? For another example, take $X = I \times Y$ and $A = \dot{I} \times Y$: given maps $\theta, \phi: Y \to E$ does there exists a homotopy of θ into ϕ?

Consideration of these problems is greatly facilitated if the inclusion $A \to X$ satisfies a condition called the cofibration condition, to be defined in a moment. Discussion of cofibrations occupies the first part of this chapter. In the second part we discuss the dual notion of fibrations; covering projections are examples of fibrations.

Definition (6.1). The map $u: A \to X$ is a cofibration if u has the following property for all spaces E. Let $f: X \to E$ be a map and let $g: A \to PE$ be a homotopy such that $\rho_0 g = fu$. Then there exists a homotopy $h: X \to PE$ such that $\rho_0 h = f$ and $hu = g$.

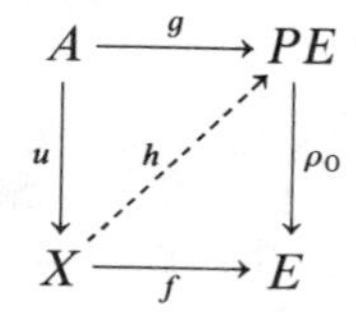

The property involved here is called the homotopy extension property (HEP); the homotopy of fu is "extended" to a homotopy h of f itself (the terminology is derived from the case when u is injective which is, as we shall see, typical). Instead of saying that u is a cofibration we may occasionally say that X is a cofibre space under A: for example the sum of cofibre spaces under A is a cofibre space under A. An important special case is when A is a subspace of X. In that case we describe (X, A) as a cofibred pair when the inclusion $A \to X$ is a cofibration.

Homeomorphisms are cofibrations. Maps with empty domain are cofibrations. The composition of cofibrations is a cofibration.

Proposition (6.2). *Let $u: A \to X$ and $u': A \to X'$ be spaces under A and let $\phi: X \to X'$ be a map under A. Suppose that ϕ admits a left inverse. Then u is a cofibration if u' is a cofibration.*

For let $\psi: X' \to X$ be a left inverse of ϕ. For any space E let $f: X \to E$ and $g: A \to PE$ be such that $\rho_0 g = fu$. Then $\rho_0 g = f\psi u'$, since $\psi u' = u$. Since u' is a cofibration there exists a map $k: X' \to PE$ such that $ku' = g$ and such that $\rho_0 k = f\psi$. Write $k\phi = h$. Then $hu = k\phi u = ku' = g$, while $\rho_0 h = \rho_0 k\phi = f\psi\phi = f$. Thus u is a cofibration as asserted.

In cofibration theory an important part is played by the mapping cylinder. Given a map $u: A \to X$ the mapping cylinder $M = M(u)$ is defined, we recall, as the push-out of the cotriad

$$I \times A \xleftarrow{\sigma_0} A \xrightarrow{u} X$$

Thus M comes equipped with canonical maps

$$I \times A \xrightarrow{H} M \xleftarrow{j} X,$$

so that the adjoint $\hat{H}$ of H makes the diagram shown below commutative.

$$\begin{array}{ccc} A & \xrightarrow{\hat{H}} & PM \\ {\scriptstyle u}\downarrow & & \downarrow{\scriptstyle \rho_1} \\ X & \xrightarrow[j]{} & M \end{array}$$

If u is a cofibration there exists a map $h: X \to PM$ such that $hu = \hat{H}$. Now $\rho_1 hu = \rho_1 \hat{H}$ is an embedding, hence u is an embedding. Thus we have

Proposition (6.3). *If the map $u: A \to X$ is a cofibration then u is an embedding.*

This result clarifies the situation greatly. It means that, without real loss of generality, we can restrict attention to the case when A is a subspace of X and u is the inclusion. In that case there is a continuous bijection

$$M \to (\{0\} \times X) \cup (I \times A)$$

determined by the triad

$$I \times A \to (\{0\} \times X) \cup (I \times A) \leftarrow \{0\} \times X \equiv X.$$

If, moreover, A is closed in X, the continuous bijection is a homeomorphism and we may identify M with $(\{0\} \times X) \cup (I \times A)$, as we have seen in the previous chapter.

The proof of our next result is purely formal and will be left to serve as an exercise.

Proposition (6.4). *Let $u: A \to X$ be a cofibration. Then the push-out $u': A' \to \xi_* X$ is a cofibration for each space A' and map $\xi: A \to A'$.*

For each map $u: A \to X$ the mapping cylinder $M = M(u)$ may be interpreted as the fibre-cone of A, regarded as a space over X. From the present point of view the fibrewise contractibility of the fibre-cone is equivalent to the statement that $j: X \to M$ is a homotopy equivalence over X. Consider now the map $k: M \to I \times X$ which is derived from the diagram

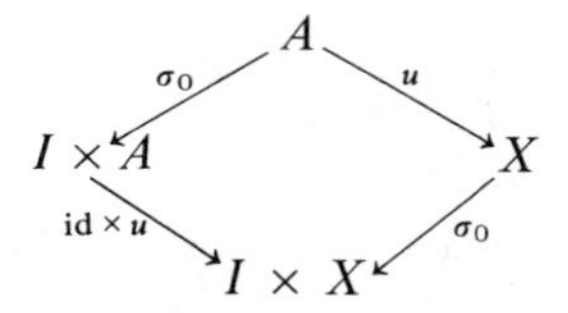

Proposition (6.5). *The map u is a cofibration if and only if the map k admits a left inverse $l: I \times X \to M$.*

For suppose that l exists. Given maps $f: X \to E$ and $g: A \to PE$, related as before, we consider the triad

$$I \times A \xrightarrow{\hat{g}} E \xleftarrow{f} X,$$

where $\hat{g}$ is the adjoint of g. Precomposition of l with the push-out $M \to E$ of the triad yields a map $I \times X \to E$, of which the adjoint $X \to PE$ provides a filler as required.

Conversely, suppose that u is a cofibration. Take $E = M = M(u)$, in the definition, and consider the commutative diagram:

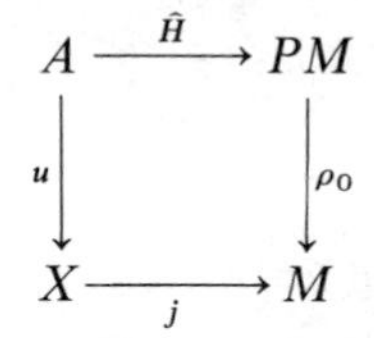

The adjoint of the filler $X \to PM$ is the required left inverse of k.

In view of our previous remarks we can at once deduce

Proposition (6.6). *Let X be a space and let A be a subspace of X. If the pair (X, A) is cofibred then $\{0\} \times X \cup I \times A$ is a retract of $I \times X$. Conversely if $\{0\} \times X \cup I \times A$ is a retract of $I \times X$, and if A is closed in X, then the pair (X, A) is cofibred.*

Strøm has shown that, rather surprisingly, the converse is true even if A is not closed in X, but we shall not be making any use of this fact.

An an application of (6.6), take $(X, A) = (I, \{0\})$. We have already seen, in Chapter 2 above, that $\{0\} \times I \cup I \times \{0\}$ is a retract of $I \times I$. Therefore $(I, \{0\})$ is a cofibred pair. More generally $(I, \{t\})$ is a cofibred pair for each point $t \in I$. Similarly $(I^n, \{t\})$ is a cofibred pair for each point $t \in I^n$, where $n = 1, 2, \ldots$.

For another application, recall from Chapter 2 that $(\{0\} \times I^n) \cup (I \times \dot{I}^n)$ is a retract of $I \times I^n$, for $n = 1, 2, \ldots$. Hence the pair $(I^n, \dot{I}^n)$ is cofibred, and hence the homeomorphic pair (B^n, S^{n-1}) is also cofibred. More generally, for any space A and map $\phi\colon S^{n-1} \to A$, the inclusion $A \to X$ is a cofibration, where X is the push-out of the cotriad

$$B^n \overset{\supseteq}{\leftarrow} S^{n-1} \overset{\phi}{\rightarrow} A.$$

This is the source of many important examples.

Although a cofibration $u\colon A \to X$ is always an embedding, as we have seen, the embedding is not necessarily closed. For example, take X to be the pair $\{a, b\}$, where $\{a\} = A$ is open and $\{b\}$ is not. But using (6.5) one can easily show that the inclusion is a cofibration in this case. It is true, however, that a cofibration $u\colon A \to X$ is a closed embedding whenever X is Hausdorff. For then $I \times X$ is Hausdorff and so the graph Γ of the self-map kl of $I \times X$ is closed. However uA coincides with $\sigma_1^{-1}\Delta^{-1}\Gamma$, where

$$X \overset{\sigma_1}{\rightarrow} I \times X \overset{\Delta}{\rightarrow} I \times X \times I \times X.$$

Hence u is a closed embedding, as asserted.

Recall that for any cotriad

$$X \overset{u}{\leftarrow} A \overset{v}{\rightarrow} Y$$

the homotopy push-out $M(u, v)$ is defined. Moreover when u is injective there is a canonical map $\lambda\colon M(u, v) \to X +^A Y$, the topological push-out. As another application of (6.5) we have

Proposition (6.7). *If u is a cofibration (and hence injective) the canonical map*

$$\lambda\colon M(u, v) \to X +^A Y$$

is a homotopy equivalence.

For let $l\colon I \times X \to M(u)$ be a left inverse of the natural map k, as in (6.5). Define

$$\mu\colon X + Y \to M(u, v)$$

to be l_0 on X and to be id_Y on Y. Then $\lambda\mu = \mathrm{id}$, while $\mu\lambda \simeq \mathrm{id}$, using the homotopy defined by l.

Corollary (6.8). *Let (X, A) be a cofibred pair, where A is contractible. Then the natural projection $X \to X/A$ is a homotopy equivalence.*

For the inclusion $X \to X \cup \Gamma A$ is a homotopy equivalence, as we have seen in the previous chapter. Moreover $X \cup \Gamma A$ is the homotopy push-out of the cotriad

$$X \leftarrow A \to *,$$

and so the natural projection

$$X \cup \Gamma A \to \frac{X \cup \Gamma A}{\Gamma A} \equiv \frac{X}{A}$$

is a homotopy equivalence. By composing these two homotopy equivalences we obtain (6.8). Another straightforward consequence of (6.7) is

Corollary (6.9). *Suppose that $u \simeq u'$ and $v \simeq v'$, where*

$$X \xleftarrow{u} A \xrightarrow{v} Y, \qquad X \xleftarrow{u'} A \xrightarrow{v'} Y.$$

If u and u' are cofibrations then the homotopy type under A of $X +^A Y$ is the same for (u, v) as it is for (u', v').

In particular we have

Corollary (6.10). *Let $u: A \to X$ be a cofibration. Let A' be a space and let $\xi_0, \xi_1: A \to A'$ be maps. If ξ_0 and ξ_1 are homotopic then $\xi_{0*}X$ and $\xi_{1*}X$ have the same homotopy type under A'.*

We already know that any map $u: A \to X$ can be factored into the composition of a closed embedding and a homotopy equivalence. In fact, writing $M = M(u)$, we have that $u = \rho\sigma_0$, where

$$A \xrightarrow{\sigma_0} M \xrightarrow{\rho} X.$$

We are now going to show that σ_0 here is a cofibration, so that u can be factored into the composition of a closed cofibration and a homotopy equivalence. This is a most useful result.

Since $\{0\} \times A = \sigma_0 A$ is a closed subset of M we may apply (6.6) (without needing Strøm's refinement) and prove our assertion by constructing a retraction of $I \times M$ onto the subspace $(\{0\} \times M) \cup (I \times \{0\} \times A)$. First consider the retraction

$$r: I \times I \to (\{0\} \times I) \cup (I \times \{0\})$$

given by projection from the point $(2, 1)$.

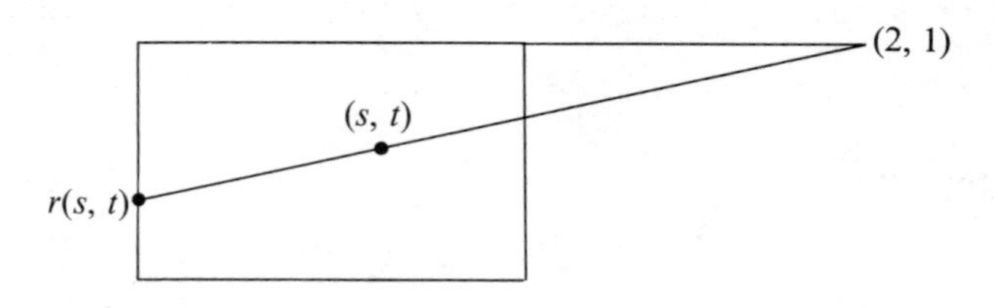

The retraction of $I \times M$ on $(\{0\} \times M) \cup (I \times \{0\} \times A)$ is induced by the map

$$R: I \times (X + I \times A) \to (\{0\} \times M) \cup (I \times \{0\} \times A)$$

which is given by

$$(t, x) \mapsto (0, x) \qquad (x \in X),$$

$$(s, t, a) \mapsto (r(s, t), a) \qquad (s, t \in I, a \in A).$$

Since $r(s, 0) = (0, 0)$ for all s we have

$$R(s, 0, a) = (r(s, 0), a) = (0, 0, a) = (0, ua) = R(0, ua).$$

Thus R induces the required retraction.

Consider a pair (X, A). By a Strøm structure on (X, A) I mean a pair (α, h) consisting of a map $\alpha: X \to I$ such that $\alpha = 0$ throughout A and a homotopy $h: I \times X \to X$ rel A of id_X such that $h(t, x) \in A$ whenever $t > \alpha(x)$. In case A is closed in X these conditions imply that $A = \alpha^{-1}(0)$. For suppose that $\alpha(x) = 0$, where $x \in X$. Then $x = h(0, x) \in A$, since $h(1/n, x) \in A$ for all n.

For pairs (X, A) which admit a Strøm structure, the existence of a Strøm structure (α, h) such that $\alpha < 1$ throughout X is equivalent to the existence of a deformation retraction of X onto A. For if such a Strøm structure (α, h) exists then $h_1 X \subset A$ and so h_t is a deformation retraction. Conversely if A is a deformation retract of X, with deformation retraction $H: I \times X \to X$, then a Strøm structure (α, h) on (X, A) can be replaced by a Strøm structure (α', h'), where

$$\alpha'(x) = \min(\alpha(x), \tfrac{1}{2}),$$

$$h'(t, x) = H(\min(2t, 1), h(t, x)),$$

and then $\alpha' < 1$ throughout X.

The relation between the cofibration condition and the existence of a Strøm structure is shown in

Proposition (6.11). *Suppose that the pair (X, A) is cofibred. Then there exists a Strøm structure on (X, A). Conversely suppose that there exists a Strøm structure on (X, A). Then the pair $(X, \mathcal{C}\ell\, A)$ is cofibred; in particular the pair (X, A) is cofibred when A is closed in X.*

For suppose that (X, A) is cofibred. By (6.6) there exists a retraction

$$r: I \times X \to (\{0\} \times X) \cup (I \times A).$$

Consider the projections

$$I \xleftarrow{r_1} I \times X \xrightarrow{r_2} X.$$

Since I is compact a map $\alpha: X \to I$ is given by

$$\alpha(x) = \sup_{t \in I} |r_1(t, x) - t|.$$

Then (α, r_2) constitute a Strøm structure on (X, A).

Conversely suppose that (α, h) is a Strøm structure on (X, A). Then a retraction

$$r: I \times X \to (\{0\} \times X) \cup (I \times A)$$

is defined by

$$r(t, x) = \begin{cases} (0, h(t, x)) & (t \leq \alpha(x)) \\ (t - \alpha(x), h(t, x)) & (t \geq \alpha(x)). \end{cases}$$

So if A is closed in X then, by (6.6), the pair (X, A) is cofibred. In any case the pair $(X, \mathscr{C}\ell\ A)$ is cofibred, by the same argument but with h replaced by h', where

$$h'(t, x) = h(\min(t, \alpha(x)), x).$$

It is a formality to show that if (X, A) is a cofibred pair then so is $(X \times Y, A \times Y)$ for all spaces Y. However the next result is not just a formality.

Proposition (6.12). *Let (X, A) be a cofibred pair and let (Y, B) be a closed cofibred pair. Then*

$$(X, A) \times (Y, B) = (X \times Y, A \times Y \cup X \times B)$$

is a cofibred pair. Moreover if B is a deformation retract of Y then

$$A \times Y \cup X \times B$$

is a deformation retract of $X \times Y$.

For let (α, h) and (β, k) be Strøm structures for the pairs (X, A) and (Y, B). Define $\gamma: X \times Y \to I$ by

$$\gamma(x, y) = \min(\alpha(x), \beta(y))$$

and define $l: I \times X \times Y \to X \times Y$ by

$$l(t, x, y) = (h(\min(t, \beta(y)), x), k(\min(t, \alpha(x)), y).$$

Then (γ, l) constitutes a Strøm structure for the pair $(X, A) \times (Y, B)$. Moreover if $\beta < 1$ throughout Y then $\gamma < 1$ throughout $X \times Y$, from which (6.12) follows at once.

Thus if (X, A) is a cofibred pair then $\{0\} \times X \cup I \times A$ is not just a retract of $I \times X$ but actually a deformation retract.

For another example, let T_0 be a closed subspace of the space T and let Φ be the endofunctor of $\mathscr{T}$ which assigns to each space X the push-out of the cotriad

$$T \times X \leftarrow T_0 \times X \rightarrow T_0,$$

and similarly with maps.

Corollary (6.13). *Suppose that the pair (T, T_0) is cofibred. Then the inclusion $\Phi A \rightarrow \Phi X$ is a cofibration whenever the inclusion $A \rightarrow X$ is a cofibration. In particular the inclusions $\Gamma A \rightarrow \Gamma X$ and $\Sigma A \rightarrow \Sigma X$ are cofibrations.*

For since $(I, \{0\})$ and (T, T_0) are cofibred pairs so is

$$(I, \{0\}) \times (T, T_0) = (I \times T, \{0\} \times T \cup I \times T_0).$$

Moreover $\{0\} \times T \cup I \times T_0$ is a deformation retract of $I \times T$, since $\{0\}$ is a deformation retract of I. Since (X, A) is also a cofibred pair so is

$$\begin{aligned}(I \times T, \{0\} \times T \cup I \times T_0) \times (X, A)\\ = (I \times T \times X, \{0\} \times T \times X \cup I \times T_0 \times X \cup I \times T \times A).\end{aligned}$$

Moreover $\{0\} \times T \times X \cup I \times T_0 \times X \cup I \times T \times A$ is a deformation retract of $I \times T \times X$ since $\{0\} \times T \cup I \times T_0$ is a deformation retract of $I \times T$. Hence and from (2.101), (6.13) follows at once. A similar argument shows that if (X, A) and (Y, B) are closed cofibred pairs then so is

$$(X * Y, A * Y \cup X * B),$$

in the fine topology.

Perhaps the most useful result of this type, due to Lillig, is the following:

Proposition (6.14). *Let X be a space and let A, B be closed subspaces of X. Write $A \cap B = C$. If (X, A), (X, B) and (X, C) are cofibred then so is*

$$(X, A \cup B).$$

Before proving this we need a lemma.

Lemma (6.15). *Let (X, A) be a closed cofibred pair. For any space E let $K, L \colon I \times X \rightarrow E$ be maps which agree on $\{0\} \times X \cup I \times A$. Then*

$$K \simeq L \operatorname{rel}(\{0\} \times X \cup I \times A).$$

For since (X, A) is cofibred so is

$$\begin{aligned}(I, \dot{I}) \times (I, \{0\}) \times (X, A)\\ = (I \times I \times X, \dot{I} \times I \times X \cup I \times \{0\} \times X \cup I \times I \times A).\end{aligned}$$

Now K, L determine a map $\dot{I} \times I \times X \rightarrow E$, while the stationary homotopy of $K|(\{0\} \times X \cup I \times A)$ into $L|(\{0\} \times X \cup I \times A)$ determines a map

$I \times (\{0\} \times X \cup I \times A) \to E$. These maps coincide on the intersection of their domains and so determine a map

$$\dot{I} \times I \times X \cup I \times \{0\} \times X \cup I \times I \times A \to E.$$

Since $\{0\}$ is a deformation retract of I it follows by (6.12) that

$$\dot{I} \times I \times X \cup I \times \{0\} \times X \cup I \times I \times A$$

is a deformation retract of $I \times I \times X$. Hence we obtain (6.15).

We are now ready for (6.14) itself. The proof involves the push-out $\tilde{X}$, say, of the cotriad

$$I \times X \leftarrow I \times C \to C.$$

Since (X, A) and (Y, B) are closed cofibred pairs there exist maps $\alpha, \beta: X \to I$ such that $A = \alpha^{-1}(0)$ and $B = \beta^{-1}(0)$. Then a right inverse $l: X \to \tilde{X}$ of the natural projection $\tilde{X} \to X$ is given by

$$l(x) = \begin{cases} \left[\dfrac{\alpha(x)}{\alpha(x) + \beta(x)}, x\right] & (x \notin C) \\ [0, x] = [t, x] & (x \in C). \end{cases}$$

The continuity argument is straightforward. Note that $l(x) = [0, x]$ on A, while $l(x) = [1, x]$ on B.

Let $f: X \to E$ be a map and let $g: I \times (A \cup B) \to E$ be a homotopy of $f|(A \cup B)$. Let $h_A, h_B: I \times X \to E$ be homotopies of f which extend $g|I \times A$, $g|I \times B$ respectively. Applying (6.15) to the pair (X, C) we conclude that there exists a homotopy H, rel$(\{0\} \times X \cup I \times C)$, of h_A into h_B. So now let h be a map which makes the following diagram commutative

$$\begin{array}{ccc} I \times I \times X & \xrightarrow{T \times \mathrm{id}} & I \times I \times X \\ \Big\downarrow{\scriptstyle \mathrm{id} \times p} & & \Big\downarrow{\scriptstyle H} \\ I \times \tilde{X} & \xrightarrow[h]{} & E \end{array}$$

Here p is the quotient map and T switches factors. Then the composition

$$I \times X \xrightarrow{\mathrm{id} \times l} I \times \tilde{X} \xrightarrow{h} E$$

is a homotopy of f which extends g, as required.

Proposition (6.16). *Let $u: A \to X$ and $v: A \to Y$ be spaces under the given space A. Let $\phi: X \to Y$ be a map such that $\phi u \simeq v$. If u is a cofibration then $\phi \simeq \psi$ for some map $\psi: X \to Y$ under A.*

For let $g: I \times A \to Y$ be a homotopy of ϕu into v. Since $g_0 = \phi u$ and since u is a cofibration there exists an extension $f: I \times X \to Y$ of g such that $f_0 \simeq \phi$ under A. Take ψ to be f_1; then $\psi u = v$.

Proposition (6.17). *Let $u: A \to X$ be a cofibration. Let $\theta: X \to X$ be a map under A, and suppose that $\theta \simeq \mathrm{id}_X$. Then there exists a map $\theta': X \to X$ under A such that $\theta'\theta \simeq \mathrm{id}_X$ under A.*

For let f_t be a homotopy of θ into id_X. Then $f_t u$ is a homotopy of u into itself. Since u is a cofibration there exists a map $\phi: X \to X$ under A such that $\phi \simeq \mathrm{id}_X$ under A, also a homotopy $g_t: X \to X$ of ϕ under $f_t u$. Take $\theta' = g_1$; we shall prove that $\theta'\theta \simeq \mathrm{id}_X$, under A.

The juxtaposition k_s of $g_{1-s}\theta$ and f_s is a homotopy of $\theta'\theta$ into id_X. Now $f_t u = g_t u$, and hence $H_{(s,0)} = k_s u$, where $H: I \times I \times A \to X$ is given by

$$H(s, t, a) = \begin{cases} \theta'(1 - 2s(1 - t), ua) & (s \le \frac{1}{2}) \\ \theta'(1 - 2(1 - s)(1 - t), ua) & (s \ge \frac{1}{2}). \end{cases}$$

Since u is a cofibration we can extend H to a map $K: I \times I \times X \to X$ such that $K_{(s,0)} \simeq k_s$ under A. Then

$$\theta'\theta = k_0 \simeq K_{(0,0)} \simeq K_{(0,1)} \simeq K_{(1,1)} \simeq K_{(1,0)} \simeq k_1 = \phi,$$

all homotopies being under A.

The main use of (6.17) is to prove

Proposition (6.18). *Let $u: A \to X$ and $v: A \to Y$ be cofibrations. Let $\phi: X \to Y$ be a map under A. Suppose that ϕ, as an ordinary map, is a homotopy equivalence. Then ϕ is a homotopy equivalence under A.*

For let $\psi: Y \to X$ be a homotopy inverse of ϕ, as an ordinary map. Then $\psi v = \psi\phi u \simeq u$. Hence $\psi \simeq \psi'$ for some map ψ' under A. Since $\psi'\phi \simeq \mathrm{id}_X$ and since $\psi'\phi$ is under A, there exists, by (6.17) a map $\psi'': X \to Y$ under A such that $\psi''\psi'\phi \simeq \mathrm{id}_X$ under A. Thus ϕ admits a homotopy left inverse $\phi' = \psi''\psi'$ under A.

Now ϕ' is a homotopy equivalence, since ϕ is a homotopy equivalence, and so the same argument, applied to ϕ' instead of ϕ, shows that ϕ' admits a homotopy left inverse ϕ'' under A. Thus ϕ' admits both a homotopy right inverse ϕ under A and a homotopy left inverse ϕ'' under A. Hence ϕ' is a homotopy equivalence under A, and so ϕ itself is a homotopy equivalence under A, as asserted.

By way of application let us prove a result originally due to Fox.

Proposition (6.19). *Let $u: A \to X$ be a homotopy equivalence. Then A is a deformation retract of the mapping cylinder $M = M(u)$.*

Here we regard A as a space under A with insertion id_A and M as a space under A with insertion σ_0, so that σ_0 is a map under A. Now $u = \rho\sigma_0$, where $\rho: M \to X$ is the usual retraction. Since u and ρ are homotopy equivalences, so is σ_0. Also id_A and σ_0 are cofibrations, and so σ_0 is a homotopy equivalence under A, by (6.18). This proves (6.19).

For another application, suppose that $u: A \to X$ is a cofibration, and hence that the canonical map $k: M \to I \times X$ admits a left inverse $l: I \times X \to M$, as in (6.5). Then l is a cofibration, by (6.2). Also k is obviously a homotopy equivalence and so l is a homotopy equivalence, as an ordinary map. Hence l is a homotopy equivalence under $I \times X$, by (6.18), where M is regarded as a space under $I \times X$ using l and $I \times X$ as a space under $I \times X$ using the identity.

Proposition (6.20). *Let $u: A \to X$ be a cofibration.*
(i) *If u admits a left inverse up to homotopy then u admits a left inverse.*
(ii) *If u admits a left inverse $r: X \to A$ and is a homotopy equivalence then u is a homotopy equivalence under A.*

The first assertion is just a special case of (6.16) with $Y = A$. For the second we apply (6.18) to obtain that r is a homotopy equivalence under A. For formal reasons the homotopy inverse, under A, of r can only be u, and the result follows.

The property of being a cofibration is not an invariant of homotopy type, under the domain. Specifically if X and Y are spaces of the same homotopy type under A, and if the insertion $u: A \to X$ is a cofibration, then the insertion $v: A \to Y$ may fail to be a cofibration. This motivates

Definition (6.21). The map $u: A \to X$ is a weak cofibration if there exists a cofibration $v: A \to Y$ such that X and Y have the same homotopy type, as spaces under A.

Weak cofibrations satisfy a weak form of the homotopy extension property (WHEP). This property can be formulated in two alternative ways which are easily seen to be equivalent. Either way we are given a space E, a map $f: X \to E$ and a map $g: A \to PE$ such that $\rho_0 g = fu$. In the first formulation the property is that there exists a map $h: X \to PE$ such that $\rho_0 h \simeq f$ under A and such that $hu = g$. In the second formulation we further require that g is semi-stationary, in the sense that $g(x)$ is stationary on $[0, \frac{1}{2}]$ for each point $x \in A$, and then the property is that there exists a map $h: X \to PE$ such that $\rho_0 h = f$ and such that $hu = g$. The proof that weak cofibrations have the WHEP is not difficult; the converse is a strenuous exercise. The reader may wish to establish this fact and to investigate which properties of cofibrations are also properties of weak cofibrations.

The category $\mathscr{T}(2)$ of pairs of spaces has been discussed in Chapter 1. The relation of homotopy between the maps of the category is defined in the obvious way, and hence the notion of homotopy equivalence. Specifically, let (θ, ϕ) be a map of the pair, as indicated in the diagram on the left

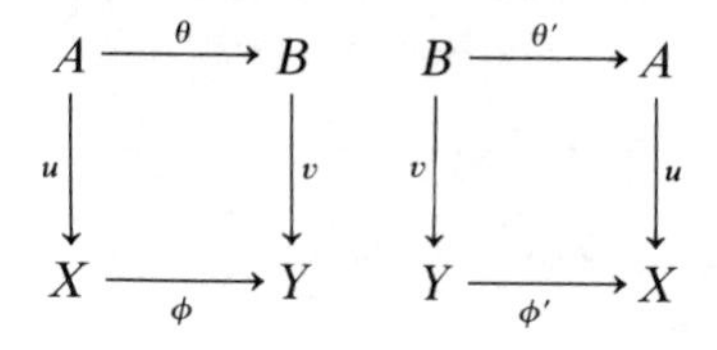

Then (θ, ϕ) is a homotopy equivalence of the pair if there exists a map (θ', ϕ') of the pair, as indicated in the diagram on the right, such that $(\theta'\theta, \phi'\phi) \simeq (\mathrm{id}_A, \mathrm{id}_X)$, $(\theta\theta', \phi\phi') \simeq (\mathrm{id}_B, \mathrm{id}_Y)$, through homotopies of maps of the pair. Clearly if (θ, ϕ) is a homotopy equivalence of the pair then θ and ϕ are both homotopy equivalences. In general the converse is false, but we prove

Proposition (6.22). *Suppose that u and v are cofibrations. If θ and ϕ are homotopy equivalences, then (θ, ϕ) is a homotopy equivalence of the pair.*

We shall show that (θ, ϕ) admits a left inverse, up to homotopy of the pair. Then the same argument, applied to the left inverse, will show that the left inverse admits a left inverse, in the same sense, and so is a homotopy equivalence of the pair. The result will then follow at once.

So let $\theta' : B \to A$ be a homotopy inverse of θ and let $\phi' : Y \to X$ be a homotopy inverse of ϕ. Then

$$\phi' v \simeq \phi' v \theta\theta' \simeq \phi'\phi u \theta' \simeq u\theta'.$$

Since v is a cofibration we can deform ϕ' into a map ϕ'' such that $\phi'' v = u\theta'$. So without real loss of generality we may suppose that ϕ' is such that (θ', ϕ') is a map of the pair.

Now choose a homotopy $g: I \times A \to A$ of $\theta'\theta$ into id_A. Since $ug_0 = u\theta'\theta = \phi' v\theta = \phi'\phi u$, and since u is a cofibration, there exists a homotopy $h: I \times X \to X$ extending ug such that $h_0 \simeq \phi'\phi$ under A. (Here ug_0 is the insertion of X). Write $h_1 = k$. Then $ku = h_1 u = ug_1 = u$, so that k is a map under A. Also k is a homotopy equivalence, since $k = h_1 \simeq h_0 \simeq \phi'\phi \simeq \mathrm{id}_X$. So by (6.17) there exists a map $k' : X \to X$ under A such that $k'k \simeq \mathrm{id}_X$ under A.

To complete the proof we show that $(\theta', k'\phi')$ is a left inverse of (θ, ϕ), up to homotopy of the pair, as follows. Choose a homotopy $H: I \times X \to X$ under A of $k'k$ into id_X and then define $K: I \times X \to X$ by

$$K(t, x) = \begin{cases} k'h(2t, x) & (0 \le t \le \frac{1}{2}) \\ H(2t - 1, x) & (\frac{1}{2} \le t \le 1). \end{cases}$$

Then K is a homotopy of $k'h_0$ into id_X. Since H is a homotopy under A and since $ug = k'h(\mathrm{id} \times u)$ we have that

$$K(t, ua) = u(\min(2t, 1), a).$$

So define $g' : I \times A \to A$ by

$$g'(t, a) = g(\min(2t, 1), a).$$

Then g' is a homotopy of $\theta'\theta$ into id_A such that $K(\mathrm{id} \times u) = ug'$. This proves that $(\theta'\theta, k'h_0) \simeq (\mathrm{id}_A, \mathrm{id}_X)$, by a homotopy of the pair, and since

$$(\theta'\theta, k'\phi'\phi) \simeq (\theta'\theta, k'h_0),$$

also by a homotopy of the pair, the result follows.

We turn next to the category of pointed spaces. One might expect that it would be necessary to introduce a notion of pointed cofibration and go through everything again. However, a glance at the diagram at the beginning of this chapter suggests why this can be avoided. For if $u: A \to X$ is a map of pointed spaces, and if E is a pointed space, then an extension to X of a pointed map $A \to PE$ is automatically a pointed map. Of course the homotopy extension property in the category of pointed spaces is, a priori, a weaker property than the homotopy extension property in the category of spaces, but this seems to be of little practical importance.

Definition (6.23). The pointed space X is well-pointed if the map $* \to X$ to the basepoint is a cofibration.

For example if $u: A \to X$ is a cofibration then the space X/A obtained by collapsing A is well-pointed. In particular the sphere $S^n = B^n/S^{n-1}$ well-pointed.

Products of well-pointed spaces are well-pointed. It follows, for example, that a well-pointed space X admits a Hopf structure if and only if the co-diagonal $X \vee X \to X$ can be extended to a map $X \times X \to X$.

We have already discussed, in Chapter 3, the general procedure for "reducing" an endofunctor of $\mathscr{T}$. Let T_0 be a closed subspace of the space T and let Φ be the endofunctor which assigns to each space X the push-out of the cotriad

$$T \times X \leftarrow T_0 \times X \to T_0,$$

and similarly with maps. If X is well-pointed then the inclusion $\Phi * \to \Phi X$ is a cofibration, by (6.13). Hence if $T = \Phi *$ is contractible the natural projection

$$\Phi X \to \Phi^* X = \Phi X / \Phi *$$

is a homotopy equivalence. In particular the natural projections $\Gamma X \to \Gamma^* X$ and $\Sigma X \to \Sigma^* X$ are homotopy equivalences.

Similarly if X and Y are well-pointed spaces then the natural projection

$$X * Y \to \Sigma^*(X \wedge Y),$$

given by $(t, x, y) \mapsto (t, (x, y))$, is a homotopy equivalence; here Σ^* denotes the reduced suspension. More generally consider the cotriad

$$X \overset{u}{\leftarrow} A \overset{v}{\rightarrow} Y$$

where A, X and Y are pointed spaces and where u, v are pointed maps. The homotopy push-out $M(u, v)$ of the above cotriad contains, in a natural way, the homotopy push-out of the cotriad

$$* \leftarrow * \rightarrow *$$

which we may identify with $I \times *$. If A is well-pointed the inclusion of $I \times *$ in $M(u, v)$ is a cofibration and so, since $I \times *$ is contractible, we have a homotopy equivalence

$$M(u, v) \to M(u, v)/M(*, *).$$

The codomain here is called the reduced homotopy push-out; we shall be making use of it later.

More generally, consider the category $\mathscr{T}_B^B$ of sectioned spaces over a given space B.

Definition (6.24). The sectioned space X over B is well-sectioned if the section $B \to X$ is a cofibration.

For example, if T is well-pointed then $B \times T$ is well-sectioned as a sectioned space over B. Moreover if X is well-sectioned over B then $\xi^* X$ is well-sectioned over B' for each space B' and map $\xi\colon B' \to B$.

Let B be a space and let X, Y be sectioned spaces over B. Given a fibrewise pointed map $\phi\colon X \to Y$ over B we denote by $\Gamma_B(\phi)$ the reduced homotopy push-out of the cotriad

$$Y \xleftarrow{\phi} X \xrightarrow{p} B$$

We refer to $\Gamma_B(\phi)$ as the homotopy cofibre, particularly when B is a point. Evidently the homotopy type of $\Gamma_B(\phi)$, as a sectioned space over B, depends only on the fibrewise pointed homotopy class of ϕ. In particular if ϕ is nulhomotopic, in this sense, then $\Gamma_B(\phi)$ is homotopy equivalent to $\Gamma_B(c)$ as a sectioned space, and $\Gamma_B(c) = Y \vee_B \Sigma_B^B X$.

Moreover if ϕ is a cofibration then, by the fibrewise pointed version of (6.9), the homotopy push-out of the cotriad

$$Y \xleftarrow{\phi} X \xrightarrow{p} B$$

is homotopy equivalent to the topological push-out, $p_* Y$, as a sectioned space over B, indeed as a space under X and over B.

Returning to the general case, consider the homotopy push-out $\Gamma_B(\phi)$ of the given cotriad, where $\phi\colon X \to Y$ is a fibrewise pointed map over B. We can identify $\Gamma_B(\phi)$ with the push-out $\phi_* \Gamma_B(X)$, where $\Gamma_B(X)$ is the homotopy push-out of the cotriad

$$X \xleftarrow{\text{id}} X \xrightarrow{p} B.$$

Since $\Gamma_B(X)$ cofibres over X so $\Gamma_B(\phi) = \phi_* \Gamma_B(X)$ cofibres over Y, using the first insertion $\phi_1\colon Y \to \Gamma_B(\phi)$. So it follows from the remark in the previous paragraph, that $\Gamma_B(\phi_1)$ is homotopy equivalent to the topological push-out $q_* \Gamma_B(\phi)$ of the cotriad

$$\Gamma_B(\phi) \underset{\phi_1}{\leftarrow} Y \underset{q}{\rightarrow} B.$$

Now $q_* \Gamma_B(\phi)$ is equivalent to the reduced fibre suspension $\Sigma_B^B(X)$, as a sectioned space over B, and so we conclude that $\Gamma_B(\phi_1)$ is homotopy equivalent to $\Sigma_B^B(X)$, in the same sense. Consequently if ϕ_1 is nulhomotopic over B then $\Sigma_B^B X \simeq_B \Gamma_B(\phi) \vee_B \Sigma_B^B Y$; in particular $\Gamma_B(\phi)$ is a coHopf space, over B.

Proposition (6.25). *Let $\phi: X \to Y$ be a fibrewise pointed map over B. Then the sequence*

$$\pi_B^B(\Gamma_B(\phi), E) \overset{\phi_1^*}{\to} \pi_B^B(Y, E) \overset{\phi^*}{\to} \pi_B^B(X, E)$$

of pointed sets is exact for all sectioned spaces E over B.

Since $\phi_1 \circ \phi \simeq c$ we have $\phi^* \circ \phi_1^* = 0$ immediately. To complete the proof, let $f: Y \to E$ be a fibrewise pointed map such that $\phi \circ f \simeq c$. Let $h: I \times X \to E$ be a fibrewise pointed homotopy such that $h_0 = \phi \circ f$ and $h_1 = c$. Then h and ϕ together form a fibrewise pointed map $g: \Gamma_B(\phi) \to E$ such that $g \circ \phi_1 = f$.

To conclude the first part of this chapter we discuss, for the pointed theory, an obvious generalization of the notion of retract which plays a role in certain results.

Definition (6.26). The closed subspace A of the pointed space X is retractile if the suspension $\Sigma^* A$ is a retract of $\Sigma^* X$.

Proposition (6.27). *Let (X, A) be a closed cofibred pair. If A is retractile in X then $\Sigma^* X$ has the same fibre homotopy type as $\Sigma^* A \vee \Sigma^*(X/A)$.*

For consider the sequence

$$\pi^*(X/A, \Sigma^* A) \leftarrow \pi^*(\Sigma^* A, \Sigma^* A) \leftarrow \pi^*(\Sigma^* X, \Sigma^* A)$$

of pointed sets. Since $\Sigma^* A$ is a retract of $\Sigma^* X$ the function on the right is surjective and so the function on the left is trivial. Applying our previous analysis to the projection $X \to X/A$ the conclusion follows at once.

Proposition (6.28) (G. W. Whitehead). *Let $A_1, \ldots, A_n$ be retracts of the pointed space X. Suppose that the pair (X, A) is cofibred, where $A = A_1 \cup \cdots \cup A_n$. Also suppose that the pair (A_J, A_I) is cofibred, for each subset $J \subset \{1, \ldots, n\}$ and subset $I \subset J$. Finally suppose that there exist retractions $r_i: X \to A_i$ $(i = 1, \ldots, n)$ such that $r_i A_j \subset A_j$ $(j = 1, \ldots, n)$. Then A is retractile in X.*

Here A_J denotes the intersection of the A_j for all $j \in J$. To prove (6.28), first observe that for each subset $J \subset \{1, \ldots, n\}$ a retraction $r_J: X \to A_J$ is defined by composing the retractions r_j, in some order, for all $j \in J$. We regard the r_J as maps into A, by composition with the inclusion, so that

$$\Sigma^* r_J: \Sigma^* X \to \Sigma^* A.$$

We use these retractions $\Sigma^* r_J$ to define a map $R\colon \Sigma^* X \to \Sigma^* A$, where

$$R = \Sigma(-1)^{|J|}\Sigma^* r_J,$$

the summation being taken over all subsets J of $\{1, \ldots, n\}$, in any order. It is not hard to show that R is a retraction up to homotopy, and so is homotopic to a retraction. Thus A is retractile in X, as asserted.

For an example where the conditions of (6.28) are fulfilled, take X to be the product

$$X_1 \times \cdots \times X_t$$

of well-pointed spaces $X_1, \ldots, X_t$. For each subset $K \subset \{1, \ldots, t\}$ define $X_K \subset X$ to be the subspace consisting of points $(x_1, \ldots, x_k, \ldots, x_t)$ $(x_k \in X_k)$ such that $x_k = *$ whenever $k \notin K$. The X_K satisfy the conditions of (6.28), with $n = 2^t$; the retraction $X \to X_K$ is defined by dropping the components x_k for which $k \notin K$.

We now turn from cofibrations to fibrations. As the terminology suggests, the latter theory is to some extent dual to the former. However the duality is somewhat formal in character and, although often suggestive, does not extend to the deeper properties.

Definition (6.29). The map $p\colon E \to B$ is a fibration if p has the following property for all spaces X. Let $f\colon X \to E$ be a map and let $g\colon I \times X \to B$ be a homotopy such that $g\sigma_0 = pf$. Then there exists a homotopy $h\colon I \times X \to E$ such that $h\sigma_0 = f$ and $ph = g$.

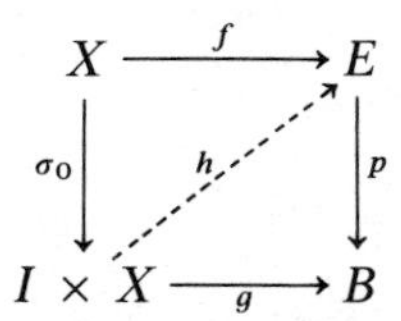

The property involved here is called the homotopy lifting property (HLP): the homotopy g of pf is lifted to a homotopy h of f itself. The term fibre mapping used to be employed instead of fibration but this usage seems to be dying out, possibly because it is liable to be confused with the term fibre-preserving map. Another term in common use is that of fibre space; to say that E is a fibre space over B is synonymous with saying that the projection $p\colon E \to B$ is a fibration.

Clearly the composition of fibrations is a fibration. Also the fibre product of fibre spaces over B is a fibre space over B. The map $T \to *$ is a fibration for each space T. More generally the map $B \times T \to B$ is a fibration for each space T. Also we have

Proposition (6.30). *Let $p\colon E \to B$ and $p'\colon E' \to B$ be spaces over B, and let $\phi\colon E \to E'$ be a map over B. Suppose that ϕ admits a left inverse. Then p is a fibration if p' is a fibration.*

For let ψ be a left inverse of ϕ. For any space X let $f: X \to E$ and $g: I \times X \to B$ be maps such that $g\sigma_0 = pf$. Then $g\sigma_0 = p'\phi f$, since $p'\phi = p$. If p' is a fibration there exists a map $k: I \times X \to E'$ such that $p'k = g$ and such that $k\sigma_0 = \phi f$. Write $\psi k = h: I \times X \to E$. Then $ph = p\psi k = p'k = g$, while

$$h\sigma_0 = \psi k\sigma_0 = \psi\phi f = f.$$

Thus p is a fibration, as asserted.

The proofs of our next two results are purely formal in character and will be left to serve as exercises

Proposition (6.31). *Let $p: E \to B$ be a fibration. Then the pull-back*

$$p': \xi^*E \to B'$$

is a fibration for each space B' and map $\xi: B' \to B$.

Proposition (6.32). *Let $p: E \to B$ be a fibration. Then*

$$p^*: \mathrm{map}(Y, E) \to \mathrm{map}(Y, B)$$

is a fibration for all locally compact regular Y.

Proposition (6.33). *Let $p: E \to B$ be a covering map. Then p is a fibration.*

For let X be a space, let $f: X \to E$ be a map, and let $g: I \times X \to B$ be a homotopy of pf. For each point $x \in X$, the path $g_x: I \to B$ from $pf(x)$ can be lifted to a unique path $h_x: I \to E$ from $f(x)$. The maps h_x, for $x \in X$, determine a function $h: I \times X \to E$. We shall prove that h is continuous and so constitutes a homotopy of f, as required.

Suppose, to obtain a contradiction, that h fails to be continuous on $I \times \{x_0\}$, for some point $x_0 \in X$. Let t_0 be the greatest lower bound of the points $t \in I$ such that h is not continuous at (t, x_0). Let U be a neighbourhood of the point $g(t_0, x_0) = b_0$ such that E_U is trivial. Since g is continuous there exists a convex neighbourhood V of t_0 in I and a neighbourhood W of x_0 in X such that $V \times W \subset g^{-1}U$. We distinguish two cases.

First suppose that $t_0 = 0$. Let N be a neighbourhood of $f(x_0)$ in E which projects topologically onto U. Since f is continuous there exists a neighbourhood of x_0 contained in $f^{-1}(U)$. Clearly we can choose W in the above so that this further condition is satisfied. Now sg and h coincide on $V \times W$, where s is given by $(p|N)^{-1}$. Since $sg|(V \times W)$ is continuous, so therefore is $h|(V \times W)$. Since $V \times W$ is open this contradicts the definition of t_0.

On the other hand suppose that $t_0 > 0$. Let $t' \in [0, t_0)$, with $t' \in V$. Let N be a neighbourhood of $h(t', x_0)$ which projects topologically onto U, and let $s: U \to N$ be the inverse of $p|N$. Now h is continuous at (t', x_0), by definition of t_0. Hence there exists a neighbourhood V' of t' and a neighbourhood W' of x_0 such that $V' \times W' \subset h^{-1}N$. Clearly we can choose these neighbourhoods so that $V' \subset V$ and $W' \subset W$. Now for each point $x \in W'$ the path in B

given by $g(t, x)$, for $t \geq t'$ and $t \in V$, can be lifted to the path $h(t, x)$ and to the path $sg(t, x)$ in E. Since $h(t', x) = sg(t', x)$ it follows that $h(t, x) = sg(t, x)$ for $t \geq t'$ and $t \in V$, by uniqueness of pathlifting. However, sg is continuous on a neighbourhood of (t_0, x_0) and so h is continuous on that neighbourhood also, contrary to the definition of t_0. This completes the proof.

As we shall see in the next chapter, the same conclusion holds for fibre bundles of which the fibres are not necessarily discrete, subject to a numerability restriction which can be avoided when, as here, the fibres are discrete.

The following result, which is known as Borsuk's fibre theorem, assures us of a plentiful supply of fibre spaces.

Proposition (6.34). *Let $u: A \to X$ be a closed cofibration, where X is locally compact regular. Then*

$$u^*: \operatorname{map}(X, E) \to \operatorname{map}(A, E)$$

is a fibration for all spaces E.

For let Y be a space, let $f: Y \to \operatorname{map}(X, E)$ be a map, and let $g: I \times Y \to \operatorname{map}(A, E)$ be a homotopy such that $g\sigma_0 = u^*f$. Since u is a cofibration so is

$$u \times \mathrm{id}: A \times Y \to X \times Y.$$

Hence there exists a filler $\hat{h}$ of the diagram shown below, where $\hat{f}$, $\hat{g}$ are the adjoints of f, g respectively.

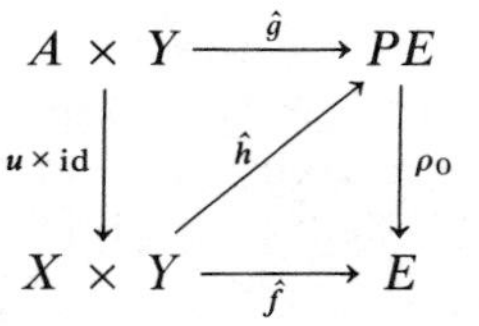

$$\begin{array}{ccc} A \times Y & \xrightarrow{\hat{g}} & PE \\ {\scriptstyle u \times \mathrm{id}} \downarrow & \overset{\hat{h}}{\nearrow} & \downarrow {\scriptstyle \rho_0} \\ X \times Y & \xrightarrow[\hat{f}]{} & E \end{array}$$

Then the adjoint $h: I \times Y \to \operatorname{map}(X, E)$ of $\hat{h}$ constitutes a homotopy of f such that $u^*h = g$.

It follows from (3.61) that when the projection $X \to X/A$ is compact the fibre of the above fibration, over the constant map of A to a basepoint $*$ of E, can be identified with the space $\operatorname{map}^*(X/A, E)$ of pointed maps.

An important special case of (6.34) occurs when $X = I$ and $A = *$, with u mapping $*$ to $0 \in I$. The result shows that $\rho_0: PE \to E$ is a fibration for all spaces E. More generally, let E' be a space and let $\xi: E' \to E$ be a map. Recall that the mapping path-space of ξ is defined as the pull-back of the triad

$$PE \xrightarrow{\rho_0} E \xleftarrow{\xi} E'$$

Since PE is a fibre space over E the mapping path-space $\xi^*PE = W(\xi)$ is a fibre space over E'.

Changing the notation slightly, consider the mapping path-space $W = W(p)$ of a map $p: E \to B$. We have seen in the previous chapter that W is contractible as a space over E. Since W is the pull-back of the triad

$$PB \xrightarrow{\rho_0} B \xleftarrow{p} E$$

there is a canonical map $k\colon PE \to W$ determined by the cotriad

$$PB \underset{Pp}{\longleftarrow} PE \underset{\rho_0}{\longrightarrow} E.$$

Proposition (6.35). *The map $p\colon E \to B$ is a fibration if and only if the map $k\colon PE \to W$ admits a right inverse.*

For suppose that p is a fibration. Take W as the domain in the definition (6.29), take f to be k and take g to be given by

$$I \times W \xrightarrow{\mathrm{id} \times \pi_1} I \times PB \xrightarrow{\pi_1} PB \overset{\rho_0}{\to} B.$$

We obtain a homotopy $h\colon I \times W \to E$ of which the adjoint $\hat{h}\colon W \to PE$ is a right inverse of k.

Conversely suppose that $l\colon W \to PE$ is a right inverse of k. Let X be a space, let $f\colon X \to E$ be a map, and let $g_t\colon X \to B$ be a homotopy such that $g_0 = pf$. A map $\xi\colon X \to W$ is determined by the cotriad

$$PB \overset{\hat{g}}{\leftarrow} X \overset{f}{\to} E,$$

and the composition $l\xi\colon X \to PE$ is a lifting of $\hat{g}$ to a homotopy of f, as required.

We already know that any map $p\colon E \to B$ can be factored into the composition

$$E \overset{\sigma}{\to} W \xrightarrow{\rho_1} B,$$

where $W = W(p)$. We are now going to show that ρ_1 here is a fibration, so that p can be factored into the composition of a homotopy equivalence σ and a fibration ρ_1.

In the proof we express maps into W in the usual way by giving their components in E and PB; the latter we at once convert into maps of the cylinder on the domain into B. So let X be a space, let $f\colon X \to W$ be a map, and let $g_t\colon X \to B$ be a homotopy such that $g_0 = \rho_1 f = f''\sigma_1$, where $f'\colon X \to E$ and $f''\colon I \times X \to B$ are the components of f. Then a map

$$h''\colon I \times I \times X \to B$$

is defined by

$$h''(s, t, x) = \begin{cases} f''(2s(2-t)^{-1}, x) & (s \leq 1 - t/2) \\ g(2s - 2 + t, x) & (s \geq 1 - t/2). \end{cases}$$

Take h'' as the second component of a deformation h of f in which the first component h' remains stationary at f'. Then $h_0 = f$ and $\rho_1 h_t = g_t$, as required.

All cofibrations are embeddings, as we have seen. Not all fibrations are quotient maps however: some condition is necessary, as in

Proposition (6.36) (Strøm). *Let E be a fibre space over the locally path-connected space B. Then pE is a quotient space of E, where $p\colon E \to B$ is the projection.*

We may suppose, without real loss of generality, that p is surjective. We have that $p\rho_1 = \rho_1 k$, as shown below.

$$\begin{array}{ccc} PE & \xrightarrow{k} & W \\ {\scriptstyle \rho_1}\downarrow & & \downarrow{\scriptstyle \rho_1} \\ E & \xrightarrow[p]{} & B \end{array}$$

Now k admits a right inverse, by (6.35), and so is a quotient map. We shall show that $\rho_1: W \to B$ is a quotient map, hence $\rho_1 k$ is a quotient map, from which it follows at once that p is a quotient map.

Suppose then that $\rho_1^{-1}V$ is open in W, where V is a subset of B. For each point b of V we have that $(e, c_b) \in W_V$, where $e \in E_b$ and c_b is the stationary path at b. Hence there exists a neighbourhood U of c_b in PB such that $(e \times U) \cap W \subset W_V$. Without real loss of generality we may assume that $U = PN$, where N is a neighbourhood of b in B. Since B is locally path-connected the path component N_0 of N containing b is open. For any point $b' \in N_0$ there exists a path λ in N from b to b'. Then $(e, \lambda) \in W_V$, since $\lambda \in U$, and so $b' = \rho_1(e, \lambda) \in V$. Thus $N_0 \subset V$ and so $b \in \mathscr{Int}\ V$, since N_0 is open. Therefore V is open and so $\rho_1: W \to B$ is a quotient map, as asserted. Now (6.36) follows at once.

It can be shown that the homotopy push-out of a cotriad of fibrations is again a fibration. Unfortunately the proof of this striking result is both lengthy and complicated. There is, however, a simple proof in the following special case.

Proposition (6.37) (Hall). *Let $p_i: E_i \to B$ $(i = 1, 2, \ldots)$ be a fibration. Then*

$$p: E_1 *_B E_2 \to B$$

is a fibration, where the fibre-join is taken with the coarse topology.

Of course it follows as a special case that the fibre-cone $\Gamma_B E$ of a fibre space E over B is again a fibre space over B, with the coarse topology, and similarly with the fibre-suspension $\Sigma_B E$.

To prove (6.37) consider the mapping path-spaces $W_i = W(p_i)$ $(i = 1, 2)$ and $W = W(p)$ with their canonical maps

$$k_i: PE_i \to W_i\ (i = 1, 2), \qquad k: P(E_1 *_B E_2) \to W.$$

Since p_i is a fibration there exists, by (6.35), a right inverse $l_i: W_i \to PE_i$ of k_i. So we can construct a right inverse $l: W \to P(E_1 *_B E_2)$ of k by writing

$$l(s)(te_1, (1 - t)e_2, \lambda) = (tl_1(e_1, \lambda)(s), (1 - t)l_2(e_2, \lambda)(s)),$$

where $e_1 \in E_1$, $e_2 \in E_2$, $\lambda \in PB$ and $s, t \in I$.

In the previous chapter we discussed the homotopy pull-back $W(p, q)$ of a triad

$$X \xrightarrow{p} B \xleftarrow{q} Y$$

and pointed out the existence of a natural map $\xi: X \times_B Y \to W(p, q)$ of the topological pull-back. In general ξ is not a homotopy equivalence but now we prove

Proposition (6.38). *If one of the maps p, q (say p) is a fibration then the natural map*

$$\xi: X \times_B Y \to W(p, q)$$

is a homotopy equivalence.

For let $l: W = W(p) \to PX$ be a right inverse of the canonical map k, as in (6.35), and let

$$\eta: W(p, q) \to X \times_B Y$$

be given by (π_2, π_3). Then $\eta\xi = \mathrm{id}$ and a homotopy $h: I \times W \to W$ of $\xi\eta$ into the identity is given by

$$h(x, y, \lambda)(t) = (l(x, \lambda)(t), y, \lambda).$$

Corollary (6.39). *Suppose that $p \simeq p'$ and $q \simeq q'$, where*

$$X \xrightarrow{p} B \xleftarrow{q} Y, \qquad X \xrightarrow{p'} B \xleftarrow{q'} Y.$$

If p and p' are fibrations then $X \times_B Y$ has the same homotopy type over B in both cases.

In particular we have

Corollary (6.40). *Let $p: X \to B$ be a fibration. Let B' be a space and let ξ_0, $\xi_1: B' \to B$ be maps. If ξ_0 and ξ_1 are homotopic then $\xi_0^* X$ and $\xi_1^* X$ have the same homotopy type over B.*

There are some important results due to Strøm which combine features of fibration and cofibration theory. For example

Proposition (6.41). *Let X be a space and let $\alpha: X \to I$ be a map such that $A = \alpha^{-1}(0)$ is a deformation retract of X. Let $p: E \to B$ be a fibration. Then for each pair of maps $f: A \to E$ and $g: X \to B$ such that $pf = g|A$ there exists an extension $h: X \to E$ of f such that $ph = g$.*

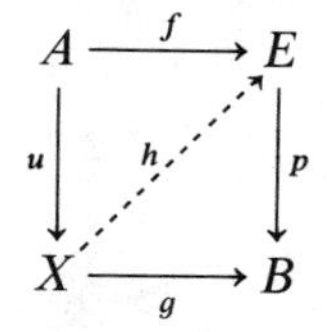

For let $r: X \to A$ be a retraction and let H be a homotopy rel A of ur into id_X. Define $D: I \times X \to X$ by

$$D(t, x) = H\left(\min\left(1, \frac{t}{\alpha(x)}\right), x\right).$$

Since p is a fibration there exists a homotopy $K: I \times X \to E$ such that $pK = gD$ and such that $K(0, x) = fr(x)$. Then

$$h(x) = K(\alpha(x), x) \qquad (x \in X)$$

defines a filler as required. Notice, incidentally, that h is unique up to homotopy rel A.

The main use of this result is to prove

Proposition (6.42). *Let $p: E \to X$ be a fibration and let A be a closed subspace of X. If the pair (X, A) is cofibred then so is the pair (E, E_A).*

For let $\alpha: X \to I$ and $h: I \times X \to X$ be a Strøm structure for the pair (X, A). Since $E \equiv \{0\} \times E$ is a deformation retract of $I \times E$, and since $E = \pi_2^{-1}0$, we can lift the deformation $h_t p: E \to X$ of p to a deformation $k_t: E \to E$ of id_E. Then $(\alpha p, l)$ is a Strøm structure for the pair (E, E_A), where

$$l(t, e) = k(\min(t, \alpha p(e)), e).$$

For example, let X be a pointed space. Let ΛX be the space of based paths in X and let ΩX be the space of based loops. For each point (t, λ) of the cone $\Gamma \Omega X$, a path μ in X is given by $\mu(s) = \lambda(st)$. In this way a map $\Gamma \Omega X \to \Lambda X$ is defined, and is the identity on ΩX. Using (6.42) we obtain

Corollary (6.43). *Let X be a well-pointed space. Then the above map*

$$(\Gamma \Omega X, \Omega X) \to (\Lambda X, \Omega X)$$

is a homotopy equivalence of the pair.

For since X is well-pointed the inclusion $\Omega X \to \Lambda X$ is a cofibration, by (6.42). Also $(\Gamma Y, Y)$ is a cofibred pair for any space Y, in particular for $Y = \Omega X$. Since the map in (6.43) is the identity on ΩX, and since both $\Gamma \Omega X$ and ΛX are contractible, the result follows from (6.22) above.

One often needs a result which combines both homotopy extension and homotopy lifting, such as

Proposition (6.44). *Let $p: E \to B$ be a fibration and let (X, A) be a closed cofibred pair. Let $f: X \to E$ be a map and let $g_t: X \to B$, $l_t: A \to E$ be homotopies such that $pl_t = g_t | A$ and such that $g_0 = pf$, $l_0 = f | A$. Then there exists a homotopy $l_t: X \to E$ of f extending l_t and lifting g_t.*

For $\{0\} \times X \cup I \times A$ is a deformation retract of $I \times X$, by (6.11). Moreover there exists a map $\beta: X \to I$ such that $A = \beta^{-1}\{0\}$. Hence $\alpha^{-1}(0) = \{0\} \times X \cup I \times A$, where $\alpha: I \times X \to I$ is given by

$$\alpha(t, x) = t . \beta(x).$$

This proves (6.44), using (6.41), and we deduce

Corollary (6.45). *Let $p: E \to B$ be a fibration. Let A be a closed subset of B such that the pair (B, A) is cofibred. If s, s' are sections of E over B which coincide on A, and if s, s' are homotopic as maps then s, s' are vertically homotopic* rel A.

An important special case is when A is empty. Then (6.45) shows that the classification of sections by ordinary homotopy is the same as the classification of sections by vertical homotopy, in the case of a fibration.

To prove (6.45) let $h_t: B \to E$ be a homotopy of s into s'. Let $k_t: B \to E$ be given by

$$k_t(b) = \begin{cases} h_{2t}(b) & (0 \le t \le \frac{1}{2}) \\ h_1 p h_{2-2t}(b) & (\frac{1}{2} \le t \le 1). \end{cases}$$

Since $pk_t = pk_{1-t}$ the map $pk: I \times B \to B$ is homotopic, rel $\dot{I} \times B$, to the projection π_1. Now

$$(I, \dot{I}) \times (B, A) = (I \times B, \dot{I} \times B \cup I \times A)$$

is a closed cofibred pair, by (6.12). Applying (6.44), therefore, it follows that k is homotopic rel$(\dot{I} \times B \cup I \times A)$ to a map l which constitutes a vertical homotopy of s into s'.

Proposition (6.46). *Let $p: E \to B$ be a fibration, where B is locally compact regular. Let A be a closed subset of B such that the pair (B, A) is cofibred. Then the map*

$$\sec_B(E) \to \sec_A(E_A),$$

given by restriction to the subset, is a fibration.

This also follows from (6.44) together with the observation that maps $Y \to \sec_B(E)$, for any space Y, correspond precisely to sections of $Y \times E$ as a space over $Y \times B$, and similarly for maps $Y \to \sec_A(E_A)$.

Proposition (6.47). *Let $p: E \to B$ be a fibration. Let $u: A \to X$ be a closed cofibration , where X is locally compact regular. Then the map*

$$(u^*, p_*): \mathrm{map}(X, E) \to \mathrm{map}(A, E) \times_{\mathrm{map}(A, B)} \mathrm{map}(X, B)$$

is a fibration.

Here map(A, E) and map(X, B) are regarded as spaces over map(A, B) using p_* and u^* respectively in order to form the fibre product. To prove the result we have to find a filler as in the diagram shown below.

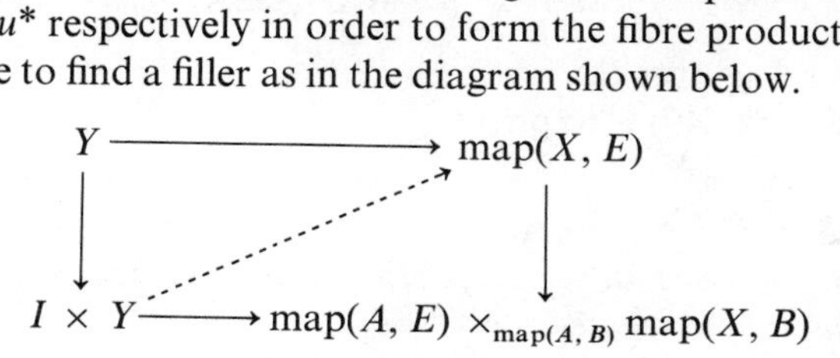

By taking adjoints and patching we convert this into a diagram as follows.

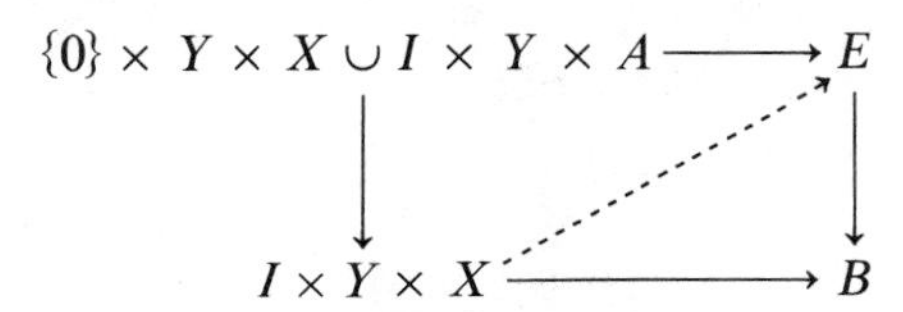

The existence of the filler shown here is established by (6.44) and its adjoint provides a filler of the original diagram.

We now turn to a series of results which the reader will at once recognize as duals of the corresponding results for cofibrations given in the first part of this chapter.

Proposition (6.48). *Let $p: E \to B$ and $q: F \to B$ be spaces over the given space B. Let $\phi: E \to F$ be a map such that $q\phi \simeq p$. If q is a fibration then $\phi \simeq \psi$ for some map $\psi: E \to F$ over B.*

For let $g: I \times E \to B$ be a homotopy of $q\phi$ into p. Since $g_0 = q\phi$ and since q is a fibration there exists a lifting $h: I \times E \to F$ of g such that $h_0 = \phi$ over B. Take ψ to be h_1; then $q\psi = p$, as asserted.

Proposition (6.49). *Let $p: E \to B$ be a fibration. Let $\theta: E \to E$ be a map over B, and suppose that $\theta \simeq \mathrm{id}_E$. Then there exists a map $\theta': E \to E$ over B such that $\theta\theta' \simeq \mathrm{id}_E$ over B.*

For let f_t be a homotopy of θ into id_E. Then pf_t is a homotopy of p into itself. Since p is a fibration there exists a map $\phi: E \to E$ over B such that $\phi \simeq \mathrm{id}_E$ over B, also a homotopy $g_t: E \to E$ of ϕ over pf_t. Take $\theta' = g_1$; we shall prove that $\theta\theta' \simeq \mathrm{id}_E$ over B.

The juxtaposition k_s of θg_{1-s} and f_s is a homotopy of $\theta\theta'$ into id_E. Now $pf_t = pg_t$ and hence $H_{(s,0)} = pk_s$, where $H: I \times I \times E \to B$ is given by

$$H(s, t, e) = \begin{cases} p\theta'(1 - 2s(1 - t), e) & (0 \le s \le \frac{1}{2}) \\ p\theta'(1 - 2(1 - s)(1 - t), e) & (\frac{1}{2} \le s \le 1). \end{cases}$$

Since p is a fibration we can lift H to a map $K: I \times I \times E \to E$ such that $K_{(s,0)} \simeq k_s$ over B. Then

$$\theta\theta' = k_0 \simeq K_{(0,0)} \simeq K_{(0,1)} \simeq K_{(1,1)} \simeq K_{(1,0)} \simeq k_1 = \phi,$$

all homotopies being over B.

The main use of (6.49) is to prove

Proposition (6.50). *Let $p: E \to B$ and $q: F \to B$ be fibrations. Let $\phi: E \to F$ be a map over B. Suppose that ϕ, as an ordinary map, is a homotopy equivalence. Then ϕ is a homotopy equivalence over B.*

For let $\psi: F \to E$ be a homotopy inverse of ϕ, as an ordinary map. Then $p\psi = q\phi\psi \simeq q$. Hence $\psi \simeq \psi'$ for some map ψ' over B. Since $\phi\psi' \simeq \mathrm{id}_E$ and since $\phi\psi'$ is over B there exists, by (6.49), a map $\psi'': E \to F$ over B such that $\phi\psi'\psi'' \simeq \mathrm{id}_E$ over B. Thus ϕ admits a homotopy right inverse $\phi' = \psi'\psi''$ over B.

Now ϕ' is a homotopy equivalence, since ϕ is a homotopy equivalence, and so the same argument, applied to ϕ' instead of ϕ, shows that ϕ' admits a homotopy right inverse ϕ'' over B. Thus ϕ' admits both a homotopy left inverse ϕ over B and a homotopy right inverse ϕ'' over B. Hence ϕ' is a homotopy equivalence over B and so ϕ itself is a homotopy equivalence over B, as asserted.

Proposition (6.51). *Let $p: E \to B$ be a homotopy equivalence. Then the mapping path-space $W = W(p)$ is contractible over B.*

Here we regard B as a space over B with projection id_B, and W as a space over B with projection ρ_0, so that ρ_0 is a map over B. Now $p = \rho_0 \sigma$, where $\sigma: E \to W$ is the standard embedding. Since σ and p are homotopy equivalences, so is ρ_0. Also id_B and ρ_0 are fibrations, and so ρ_0 is a homotopy equivalence over B, by (6.50). This proves (6.51).

Proposition (6.52). *If $p: E \to B$ is a fibration then PE is contractible over $W(p)$.*

Of course the converse is obvious from (6.35). First observe that the projection $k: PE \to W$ is a fibration. This follows at once from the definition (6.29) and the observation that $(I \times I, I \times \{0\} \cup \{0\} \times I)$ is homeomorphic to $(I \times I, I \times \{0\})$. Then observe that k is a homotopy equivalence and hence, by (6.50), a homotopy equivalence over W. This completes the proof.

Proposition (6.53). *Let $p: E \to B$ be a fibration.*

(i) *If p admits a right inverse up to homotopy then p admits a right inverse.*
(ii) *If p admits a right inverse $s: B \to E$ and is a homotopy equivalence then p is a homotopy equivalence over B.*

The first assertion is just a special case of (6.48) with $F = B$. For the second we apply (6.50) to obtain that s is a homotopy equivalence over B. For formal reasons the homotopy inverse, over B, of s can only be p and the result follows.

The property of being a fibration is not an invariant of homotopy type, over the domain. For example, take $B = I$, $E = I \times I$ and

$$E' = (\{0\} \times I) \cup (I \times \{0\}),$$

as shown below.

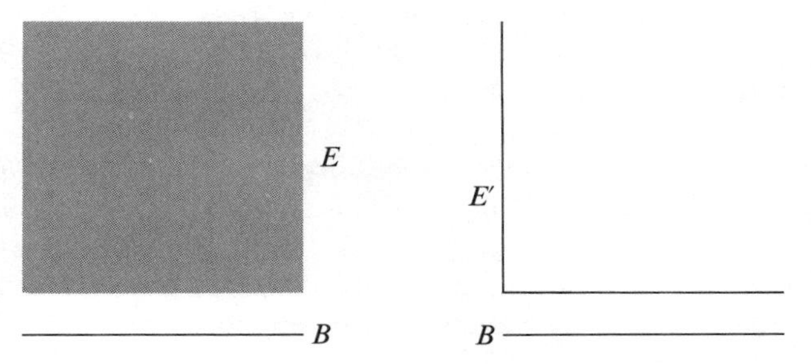

Here the projection is the standard one, given by $(\alpha, \beta) \mapsto \beta$, i.e. vertical projection in terms of the diagram. Taking $X = *$ one sees that although E is a fibre space over B, E' is not. However, both E and E' are contractible over B and so have the same homotopy type over B. This example helps to motivate

Definition (6.54). The map $p: E \to B$ is a weak fibration if there exists a fibration $q: F \to B$ such that E and F have the same homotopy type, as spaces over B.

Weak fibrations satisfy a weak form of the homotopy lifting property (WHLP). This property can be formulated in either of two ways which are easily seen to be equivalent. Either way we are given a space X, a map $f: X \to E$ and a map $g: I \times X \to B$ such that $g\sigma_0 = pf$. In the first formulation the property is that there exists a map $h: I \times X \to E$ such that $h\sigma_0 \simeq f$ over B and such that $ph = g$. In the second formulation we require that g is semistationary and then the property is that there exists a map $h: I \times X \to E$ such that $h\sigma_0 = f$ and such that $ph = g$. The proof that weak fibrations have the WHLP is not difficult; the converse is a strenuous exercise. The reader may wish to establish this fact and to investigate which properties of fibrations are also properties of weak fibrations.

We now come to a set of results which have no obvious parallel in cofibration theory. Let A be a space and let D be a space over the cylinder $I \times A$. We regard D as a space over A by composing with the projection $I \times A \to A$. We also regard $D_t = D|(\{t\} \times A)$ $(0 \leq t \leq 1)$ as a space over A in the obvious way so that the inclusion $u_t: D_t \to D$ is a map over A. We prove

Proposition (6.55). *If the projection $p: D \to I \times A$ is a fibration then D_0 is a deformation retract of D over A.*

For consider the map $\xi: I \times I \times A \to I \times A$ which is given by $\xi(s, t, x) = ((1 - s)t, x)$. We can lift $\xi(\mathrm{id}_I \times p): I \times D \to I \times A$ to a homotopy $h: I \times D \to D$, rel D_0, such that $h_0 = \mathrm{id}_D$. Now $h_1 D \subset D_0$ and so h_1 determines a retraction $r_0: D \to D_0$ over A. We have $r_0 u_0 = \mathrm{id}_{D_0}$ and $u_0 r_0 = h_1 \simeq h_0 = \mathrm{id}_D$ over A, as required.

Similarly, D_1 is a deformation retract of D over A, under the same hypothesis, and so D_0 has the same homotopy type as D_1, over A. More specifically, there exists a homotopy equivalence $r_1 u_0 : D_0 \to D_1$ over A such that

$$u_1(r_1 u_0) = (u_1 r_1) u_0 \simeq u_0,$$

over A. We use this to obtain

Proposition (6.56). *Let $p: E \to B$ be a fibration. Let $\phi_0, \phi_1 : A \to B$ be homotopic maps. Then $\phi_0^* E$ and $\phi_1^* E$ have the same homotopy type over A. More specifically there exists a homotopy equivalence ψ over A such that $\tilde{\phi}_1 \psi \simeq \tilde{\phi}_0$ over ϕ_0, as shown below:*

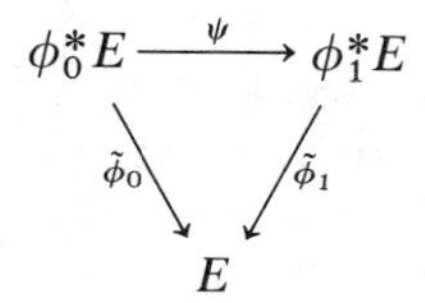

Here $\tilde{\phi}_t$ $(t = 0, 1)$ is the canonical map over ϕ_t. In fact (6.56) follows at once from (6.55), and the remark we have just made, on taking $D = \phi^* E$ where $\phi: I \times A \to B$ is a homotopy of ϕ_0 into ϕ_1. An alternative proof is given in (6.40).

Corollary (6.57). *Let $p: E \to B$ be a fibration, where B is contractible. Then E is trivial in the sense of fibre homotopy type over B.*

To deduce the corollary take $A = B$, take ϕ_0 to be constant and take $\phi_1 = \mathrm{id}_B$.

This has the following consequence. Suppose that $\theta, \theta' : E \to F$ are maps over B and that

$$\phi_1^* \theta \simeq \phi_1^* \theta' : \phi_1^* E \to \phi_1^* F$$

over A. Then

$$\phi_0^* \theta \simeq \phi_0^* \theta' : \phi_0^* E \to \phi_0^* F$$

over A. For example we have

Corollary (6.58). *Let E, F be fibre spaces over B. Let A be a subspace of B which is contractible in B. Let $\theta, \phi : E \to F$ be maps over B which are homotopic over some point of A. Then $\theta_A, \phi_A : E_A \to F_A$ are homotopic over A.*

Finally we prove

Proposition (6.59). *Let E be a fibre space over B, and let $\phi: A \to B$ be a homotopy equivalence. Then the standard map $\tilde{\phi}: \phi^* E \to E$ is a homotopy equivalence over ϕ.*

For let $\psi: B \to A$ be a homotopy inverse of ϕ. We have the commutative diagram:

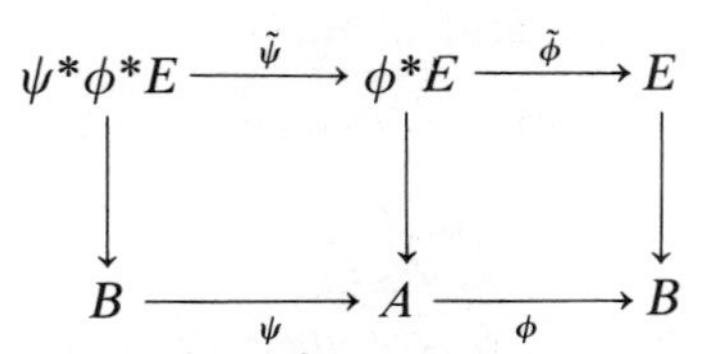

Now $\psi^*\phi^*E$ is equivalent to $(\phi\psi)^*E$, as a space over B, and $\phi\psi \simeq \mathrm{id}_B$. By (6.56), therefore, there exists a homotopy equivalence

$$\eta: \psi^*\phi^*E \to E$$

over B such that $\eta \simeq \widetilde{\phi\psi} = \tilde{\phi}\tilde{\psi}$ over $\phi\psi$. Thus $\tilde{\phi}\tilde{\psi}$ is a homotopy equivalence over $\phi\psi$; in particular $\tilde{\phi}$ has a right homotopy inverse $\tilde{\psi}$ over ψ. Now apply the same argument with $\phi^*E \to A$ in place of $E \to B$ and with ψ in place of ϕ; we conclude that $\tilde{\psi}$ has a left homotopy inverse over ϕ. Therefore $\tilde{\psi}$ has both a left and right homotopy inverse over ϕ and so is a homotopy equivalence over ϕ. Consequently $\tilde{\phi}$ is a homotopy equivalence over ϕ, as asserted.

Proposition (6.60). *Let E, F be fibre spaces over B. If E is locally compact regular over B then* $\mathrm{map}_B(E, F)$ *is a fibre space over B.*

We work directly from the definition (6.29) of a fibration. Let X be any space and let $\phi: I \times X \to B$ be a homotopy. Let

$$\hat{f}: X \to \mathrm{map}_B(E, F)$$

be a map over ϕ_0. Since E is locally compact regular over B the adjoint

$$f: X \times_B E \to F$$

of $\hat{f}$ is defined where X is regarded as a space over B through ϕ_0. Now consider the fibre product

$$(I \times X) \times_B E,$$

where $I \times X$ is regarded as a space over B through ϕ. Recall that $X \times_B E$ is a deformation retract of $(I \times X) \times_B E$ over X, where X is embedded in $I \times X$ through u_0. If $r_0: (I \times X) \times_B E \to X \times_B E$ is a retraction then

$$fr_0: (I \times X) \times_B E \to F$$

agrees with f on $X \times_B E$. Now let $h_t: I \times X \to I \times X$ be defined by $h_t(s, x) = (st, x)$. Then

$$(I \times X) \times_B E \xrightarrow{\pi_1} I \times X \xrightarrow{h_t} I \times X \xrightarrow{\phi} B$$

is a homotopy of qfr_0 into $\phi\pi_1$, $\mathrm{rel}(X \times_B E)$. Since $(I \times X, X)$ is always a cofibration we can lift this to a homotopy, $\mathrm{rel}(X \times_B E)$, of fr_0 into a map

$$g: (I \times X) \times_B E \to F$$

over $\phi\pi_1$. Now the adjoint

$$\hat{g}: I \times X \to \text{map}_B(E, F)$$

is over ϕ and satisfies $\hat{g}u_0 = \hat{f}$. Therefore $\text{map}_B(E, F)$ is a fibre space over B, as asserted.

Let B be a space and let X, Y be sectioned spaces over B. Given a map $\phi: X \to Y$ of sectioned spaces over B we denote by $\Lambda_B(\phi)$ the homotopy pull-back of the triad

$$X \xrightarrow{\phi} Y \xleftarrow{t} B,$$

where t is the section of Y. When $B = *$ it is common to refer to $\Lambda_*(\phi)$ as the homotopy fibre of ϕ. Evidently the homotopy type of $\Lambda_B(\phi)$, as a sectioned space over B, depends only on the homotopy class of ϕ, as a map of sectioned spaces over B. In particular if ϕ is nul-homotopic, as a map of sectioned spaces, then $\Lambda_B(\phi)$ is homotopy equivalent to $\Lambda_B(c)$, as a sectioned space, and

$$\Lambda_B(c) = X \times_B \Omega_B Y.$$

Moreover, if ϕ is a fibration then, by the section-preserving version of (6.38), the homotopy pull-back of the triad

$$X \xrightarrow{\phi} Y \xleftarrow{t} B$$

is homotopy-equivalent to the topological pull-back t^*X, as a sectioned space over B, indeed as a space over X with section over B.

Returning to the general case, consider the homotopy pull-back $\Lambda_B(\phi)$ of the given triad, where $\phi: X \to Y$ is a map of sectioned spaces over B. We can identify $\Lambda_B(\phi)$ with the pull-back $\phi^*\Lambda_B(Y)$, where $\Lambda_B(Y)$ is the homotopy pull-back of the triad

$$Y \xrightarrow{\text{id}} Y \xleftarrow{t} B.$$

Since $\Lambda_B(Y)$ fibres over Y so $\Lambda_B(\phi) = \phi^*\Lambda_B(Y)$ fibres over X, using the first projection $\phi^1: \Lambda_B(\phi) \to X$. So it follows, from the remark in the previous paragraph, that $\Lambda_B(\phi^1)$ is homotopy equivalent to the topological pull-back $s^*\Lambda_B(\phi)$ of the triad

$$\Lambda_B(\phi) \xrightarrow{\phi^1} X \xleftarrow{s} B.$$

Now $s^*\Lambda_B(\phi)$ is equivalent to $\Omega_B(Y)$, as a sectioned space over B, and so we conclude that $\Lambda_B(\phi^1)$ is homotopy equivalent to $\Omega_B(Y)$, as a sectioned space over B.

We illustrate these remarks, with $B = *$, by proving

Proposition (6.61). *Let $\phi: X \to Y$ be a fibration, where X and Y are pointed spaces. Suppose that the fibre Z over the basepoint is contractible in X. Then ΩY has the same pointed homotopy type as $Z \times \Omega X$; consequently Z is a Hopf space.*

To prove this result, which is a version of a theorem of Spanier and Whitehead, we apply the argument of the previous paragraph to the inclusion $u: Z \to X$. We see that $\Lambda(u)$ has the pointed homotopy type of ΩY, since ϕ is a fibration, and has the pointed homotopy type of $\Lambda(c) = Z \times \Omega X$, since Z is contractible in X. Thus $Z \times \Omega X$ is a Hopf space, since ΩY is a Hopf space, and so Z is a Hopf space, since $Z \times *$ is a retract of $Z \times \Omega X$.

Proposition (6.62). *For each well-pointed space X the homotopy fibre of the diagonal $\Delta: X \to X \times X$ has the same pointed homotopy type as ΩX.*

For the homotopy fibre $\Lambda(\Delta)$ can be identified with the fibre product $\Lambda(X) \times_X \Lambda(X)$, where $\Lambda(X)$ is regarded as a space over X in the usual way. Also a pointed map $\xi: \Omega(X) \to \Lambda(X) \times_X \Lambda(X)$ is given by $\xi(\lambda) = (\lambda', \lambda'')$, where

$$\lambda'(t) = \lambda(t/2), \qquad \lambda''(t) = \lambda(1 - t/2) \qquad (0 \leq t \leq 1),$$

and a pointed homotopy inverse η of ξ is given by $\eta(\mu, v) = \mu - v$, the track difference.

Proposition (6.63). *For well-pointed spaces X, Y the homotopy fibre of the inclusion*

$$X \vee Y \to X \times Y$$

has the same pointed homotopy type as the join $\Omega X * \Omega Y$ with fine topology.

For the homotopy fibre can be identified with the push-out of the cotriad

$$\Lambda(X) \times \Omega(Y) \leftarrow \Omega(X) \times \Omega(Y) \rightarrow \Omega(X) \times \Lambda(Y).$$

As in (6.43) we can replace Λ by $\Gamma\Omega$, without affecting the pointed homotopy type, so that the homotopy fibre is equivalent to the push-out of the cotriad

$$\Gamma\Omega(X) \times \Omega(Y) \leftarrow \Omega(X) \times \Omega(Y) \rightarrow \Omega(X) \times \Gamma\Omega(Y),$$

which is one of the representations of the join.

Proposition (6.64). *For each well-pointed space X the homotopy fibre of the codiagonal*

$$\nabla: X \vee X \to X$$

has the same pointed homotopy type as $\Sigma\Omega(X)$.

Here the homotopy fibre can be identified with the push-out of the cotriad

$$\Lambda(X) \leftarrow \Omega(X) \rightarrow \Lambda(X),$$

which has the same pointed homotopy type as the push-out of the cotriad

$$\Gamma\Omega(X) \leftarrow \Omega(X) \rightarrow \Gamma\Omega(X).$$

This is one of the representations of $\Sigma\Omega(X)$.

Proposition (6.65). *For each well-pointed space X the homotopy fibre of the adjunction $\Sigma\Omega(X) \to X$ has the same pointed homotopy type as the join $\Omega(X) * \Omega(X)$ with fine topology.*

I leave this last illustration as an exercise for the reader.

REFERENCES

P. I. Booth. The exponential law of maps, I. *Proc. London Math. Soc.* (3) **20** (1970), 179–192.

R. Brown. *Elements of Modern Topology*. McGraw-Hill, London, 1968.

T. tom Dieck, K. H. Kamps and D. Puppe. *Homotopietheorie*. Springer Lecture Notes, Vol. 157, 1970.

B. Eckmann and P. J. Hilton. Operators and cooperators in homotopy theory. *Math. Zeit.* **141** (1960), 1–21.

B. Eckmann and P. J. Hilton. A natural transformation in homotopy theory and a theorem of G. W. Whitehead. *Math. Zeit.* **82** (1963), 115–124.

I. M. Hall. The generalized Whitney sum. *Quart. J. Math. Oxford* **16** (1965), 360–384.

H. M. Hastings. Fibrations of compactly-generated spaces. *Michigan Math. J.* **21** (1974), 243–251.

W. Hurewicz. On the concept of fibre spaces. *Proc. Nat. Acad. Sci.* **41** (1955), 956–961.

J. Lillig. A union theorem for cofibrations. *Arch. Math.* **24** (1973), 410–415.

D. Puppe. Homotopiemengen und ihre induzierten Abbildungen, I. *Math. Zeit.* **69** (1958), 299–344.

D. Puppe. Bemerkungen über die Erweiterung von Homotopien. *Arch. Math.* **18** (1967), 81–88.

D. G. Quillen. *Homotopical algebra*. Springer Lecture Notes, Vol. 43, 1967.

J. W. Rutter. A coclassifying map for the inclusion of the wedge in the product. *Math. Zeit.* **129** (1972), 173–183.

J. W. Rutter. Fibred joins of fibrations and maps I. *Bull. London Math. Soc.* **4** (1972), 187–190.

J. W. Rutter. Fibred joins of fibrations and maps, II. *J. London Math. Soc.* **8** (1974), 453–459.

E. H. Spanier. *Algebraic Topology*. McGraw-Hill, New York, 1966.

A. Strøm. Note on cofibrations. *Math. Scand.* **19** (1966), 11–14.

A. Strøm. Note on cofibrations, II. *Math. Scand.* **22** (1969), 130–142.

A. Strøm. The homotopy category is a homotopy category. *Arch. Math.* **23** (1972), 435–441.

P. Tulley McAuley. A note on paired fibrations. *Proc. Amer. Math. Soc.* **34** (1972), 534–540.

G. W. Whitehead. On mappings into group-like spaces. *Comment. Math. Helv.* **28** (1954), 320–328.

G. W. Whitehead. *Elements of Homotopy Theory*. Springer-Verlag, New York, 1978.

CHAPTER 7

Numerable Coverings

In what we have done so far the Hausdorff and regularity axioms have played an important part. We now need to introduce two more separation axioms and to discuss various concepts which are associated with them.

Definition (7.1). The space X is normal if for each disjoint pair H, K of closed sets of X there exists a disjoint pair U, V of open sets of X such that $H \subset U$ and $K \subset V$.

In fact the conclusion can be strengthened. For if U is as in (7.1) then $\mathscr{C}\ell\, U \subset X - V \subset X - K$; thus $\mathscr{C}\ell\, U$, K are a disjoint pair of closed sets and so there exists a neighbourhood W of K such that $\mathscr{C}\ell\, W$ does not meet $\mathscr{C}\ell\, U$.

Obviously discrete spaces and indiscrete spaces are normal. The point-pair with the indiscrete topology is an example of a normal space which is not Hausdorff. The point-pair with the topology in which only one point is closed is an example of a normal space which is not regular.

Metric spaces are normal. For suppose that X is a metric space with metric d. For any $x \in X$ define

$$d(x, H) = \inf\{d(x, h) \mid h \in H\}$$

and define $d(x, K)$ similarly. Then the open sets

$$U = \{x \in X \mid d(x, H) < d(x, K)\}, \qquad V = \{x \in X \mid d(x, K) < d(x, H)\}$$

satisfy the requirements of (7.1).

Proposition (7.2). *If X is compact regular (and hence if X is compact Hausdorff) then X is normal.*

For by regularity there exists, for each point $x \in H$, a disjoint pair U_x, V_x of open sets of X such that $x \in U_x$ and $K \subset V_x$. The family U_x $(x \in H)$ forms an open covering of the compact H. Extract a finite subcovering indexed by $x_1, \ldots, x_n$, say. Then the union

$$U = U_{x_1} \cup \cdots \cup U_{x_n}$$

is a neighbourhood of H which does not meet the neighbourhood

$$V = V_{x_1} \cap \cdots \cap V_{x_n}$$

of K.

Proposition (7.3). *Let $\phi: X \to Y$ be a closed surjection. If X is normal then so is Y.*

For let H, K be a disjoint pair of closed subsets of Y. Since X is normal there exist disjoint open sets U, V of X such that $\phi^{-1}H \subset U$ and $\phi^{-1}K \subset V$. Since ϕ is closed there exist, by (2.27), open sets U', V' of Y such that $H \subset U'$, $K \subset V'$ and such that $\phi^{-1}U' \subset U$, $\phi^{-1}V' \subset V$. Then U', V' are disjoint, since ϕ is surjective, and the conclusion follows.

Proposition (7.4). *Let A be a closed subspace of the normal space X. Then A is normal.*

For let H, K be a disjoint pair of closed sets of A. Then H, K are also closed in X, since A is closed in X, and so there exists a disjoint pair U, V of open sets of X such that $H \subset U$, $K \subset V$. Then $U \cap A$, $V \cap A$ is a disjoint pair of open sets of A such that $H \subset U \cap A$, $K \subset V \cap A$.

Non-closed subspaces of a normal space are not, in general, normal. This gives point to

Definition (7.5). The space X is completely normal if each subspace of X is normal.

For example metric spaces are completely normal since each subspace of a metric space is again a metric space.

Definition (7.6). Subsets H, K of the space X are mutually separated if $\mathscr{C}\ell\, H$ does not meet K and $\mathscr{C}\ell\, K$ does not meet H.

For example disjoint open subsets are always mutually separated, and disjoint closed subsets are also mutually separated.

Proposition (7.7). *The space X is completely normal if and only if for each mutually separated pair H, K of subsets of X there exists a disjoint pair U, V of open sets of X such that $H \subset U$, $K \subset V$.*

To prove sufficiency, let A be a subspace of X and let H, K be a disjoint pair of closed subsets of A. Then $K = L \cap A$, where L is closed in X, and so

$$H \cap \mathscr{Cl}\, K \subset H \cap L = H \cap K = \varnothing.$$

Thus H does not meet $\mathscr{Cl}\, K$ and similarly K does not meet $\mathscr{Cl}\, H$. Since H, K are mutually separated there exists, by hypothesis, a disjoint pair U, V of open sets of X such that $H \subset U$, $K \subset V$. Then $U \cap A$, $V \cap A$ are open in A and $H \subset U \cap A$, $K \subset V \cap A$.

To prove necessity, let H, K be mutually separated subsets of X. Consider the open set $Y = X - \mathscr{Cl}\, H \cap \mathscr{Cl}\, K$. We have

$$H \subset Y \cap \mathscr{Cl}\, H, \qquad K \subset Y \cap \mathscr{Cl}\, K,$$

since Y contains both H and K. Now $Y \cap \mathscr{Cl}\, H$ does not meet $Y \cap \mathscr{Cl}\, K$, since

$$(Y \cap \mathscr{Cl}\, H) \cap (Y \cap \mathscr{Cl}\, K) = Y \cap (\mathscr{Cl}\, H \cap \mathscr{Cl}\, K) = \varnothing.$$

Also Y is normal, since X is completely normal and so there exists a disjoint pair U, V of open sets of Y such that $Y \cap \mathscr{Cl}\, H \subset U$, $Y \cap \mathscr{Cl}\, K \subset V$. But U, V are open in X, since Y is open in X, and so the proof is complete.

In previous chapters we have discussed Lusternik–Schnirelmann category and its generalizations. For each of these numerical invariants one is given a collection Γ of open sets, covering the given space X, and then seeks the least number of members of Γ required to cover X. Let us say that such a collection is admissible if

(i) any open subset of a member of Γ is also a member of Γ, and
(ii) the disjoint union of members of Γ is also a member of Γ.

Examples where these two conditions are satisfied are given in Chapter 3 above. We now prove

Proposition (7.8). *Let Γ, Γ', Γ'' be admissible open coverings of the completely normal space X. Suppose that if $U \in \Gamma'$ and $V \in \Gamma''$ then $U \cap V \in \Gamma$. Then the covering number of X with respect to Γ is less than the sum of the covering numbers of X with respect to Γ' and with respect to Γ''.*

We may suppose that the covering numbers with respect to Γ' and Γ'' are finite, say m and n respectively, since otherwise there is nothing to be proved. So let $\{U_1, \ldots, U_m\}$ be a covering by members of Γ' and let $\{V_1, .\ , V_n\}$ be a covering by members of Γ''. Consider the subsets

$$U_i' = U_i \backslash (U_1 \cup \cdots \cup U_{i-1}) \qquad (i = 1, \ldots, m)$$

$$V_j' = V_j \backslash (V_1 \cup \cdots \cup V_{j-1}) \qquad (j = 1, \ldots, n).$$

Clearly $\mathscr{Cl}\, U'_i \subset X - (U_1 \cup \cdots \cup U_{i-1})$ and so does not meet any of the sets $U'_1, \ldots, U'_{i-1}$. Similarly $\mathscr{Cl}\, V'_j$ does not meet any of the sets $V'_1, \ldots, V_{j-1}$. For $k = 2, \ldots, m + n$ write

$$W_k = \bigcup_{i+j=k} (U'_i \cap V'_j).$$

Then the closure of one component $U'_i \cap V'_j$ of W_k cannot meet any other component. In other words the components of W_k are mutually separated. Since X is completely normal there exist neighbourhoods N_{ij} of each $U'_i \cap V'_j$ which are mutually disjoint. Now $U_i \cap V_j$ is a member of Γ, hence $N_{ij} \cap U_i \cap V_j$ is a member of Γ. Hence the union

$$W'_k = \bigcup_{i+j=k} (N_{ij} \cap U_i \cap V_j)$$

is a member of Γ. However $\{U'_1, \ldots, U'_m\}$ and $\{V'_1, \ldots, V'_n\}$ both cover X, hence the family

$$\{U'_i \cap V'_j\} \qquad (i = 1, \ldots, m; j = 1, \ldots, n)$$

also covers X. Both N_{ij} and $U_i \cap V_j$ contain $U'_i \cap V'_j$, hence the family

$$\{N_{ij} \cap U_i \cap V_j\} \qquad (i = 1, \ldots, m; j = 1, \ldots, n)$$

also covers X. Therefore $\{W'_2, \ldots, W'_m\}$ covers X and (7.8) is established.

Recall that the fibre product of locally sectionable spaces over B is also locally sectionable over B.

Corollary (7.9). *Let B be a space such that the product $B \times B$ is completely normal. Let E, F be locally sectionable spaces over B. Then the fibre product $E \times_B F$ is also locally sectionable and*

$$\operatorname{secat}_B(E \times_B F) \leq \operatorname{secat}_B E + \operatorname{secat}_B F - 1.$$

Also recall that the product of locally categorical spaces is locally categorical.

Corollary (7.10). *Let X, Y be locally categorical path-connected spaces. Suppose that $X \times Y$ is completely normal. Then*

$$\operatorname{cat}(X \times Y) \leq \operatorname{cat} X + \operatorname{cat} Y - 1.$$

Proposition (7.11). *Let X, Y be normal spaces, let A be closed in X and let $\phi: A \to Y$ be continuous. Then the push-out of the cotriad*

$$X \overset{\supseteq}{\leftarrow} A \overset{\phi}{\rightarrow} Y$$

is a normal space.

For let $H_i\,(i = 1, 2)$ be a disjoint pair of closed sets of the push-out $X +^A Y = Z$. Since Y is normal there exists a pair $V_i\,(i = 1, 2)$ of neighbourhoods of $Y \cap \pi^{-1}(H_i)$ in Y such that $\mathscr{Cl}\, V_1$ does not meet $\mathscr{Cl}\, V_2$. Since X is normal, and since $\pi(\mathscr{C}\ell\ V_i)$ is closed, there exists a disjoint pair $U_i\ (i = 1, 2)$ of neighbourhoods of $X \cap \pi^{-1}(H_i \cap \pi(\mathscr{C}\ell\ V_i))$ in X. Now H_i is contained in $\pi((U_i - A) \cup V_i)$, moreover the pair $\pi((U_i - A) \cup V_i)$ are disjoint. We complete the proof by showing that these sets are open.

For this purpose we can drop the suffix i. To show that $\pi((U - A) \cup V)$ is open in Z we need to show that $\pi^{-1}\pi((U - A) \cup V)$ meets X and Y in open sets. Now

$$Y \cap \pi^{-1}\pi((U - A) \cup V) = V,$$

which is open in Y. Also

$$X \cap \pi^{-1}\pi((U - A) \cup V) = (U - A) \cup \phi^{-1}V.$$

Moreover $\phi^{-1}V = W \cap A$, where W is open in X, and $\phi^{-1}V \subset U$ so that

$$\begin{aligned}(U - A) \cup \phi^{-1}V &= (U - A) \cup (U \cap W \cap A)\\ &= U \cap (W \cup (X - A)),\end{aligned}$$

which is open in X. This completes the proof.

Definition (7.12). Let $\{X_j\}$ $(j \in J)$ be an open covering of the set X. A refinement of $\{X_j\}$ is an open covering $\{X'_k\}$ $(k \in K\}$ of X such that each member X'_k of the new covering is contained in some member X_j of the old covering.

Definition (7.13). The space X is paracompact if X is regular and each open covering of X has a locally finite refinement.

For example, compact regular spaces are paracompact since each covering admits a finite subcovering, which is a special case of a refinement.

Proposition (7.14). Let A be a closed subspace of the paracompact space X. Then A is also paracompact.

For let $\{U_j\}$ $(j \in J)$ be an open covering of A. Then $U_j = V_j \cap A$, for each index j, where V_j is open in X. The family $\{V_j\}$ $(j \in J)$ together with the open set $X - A$ forms an open covering of X. Since X is paracompact there exists a locally finite refinement $\{W_k\}$ $(k \in K)$. Consider the open covering $\{W_k \cap A\}$ of A. Each point $x \in A$ admits a neighbourhood N in X which meets W_k for at most a finite number of indices k. Then $N \cap A$ is a neighbourhood of x in A which meets $W_k \cap A$ for at most a finite number of indices k. Therefore $\{W_k \cap A\}$ is locally finite and (7.14) is proved.

Non-closed subspaces of a paracompact space are not, in general, paracompact. This gives point to

Definition (7.15). The space X is completely paracompact if each subspace of X is paracompact.

We have already seen that metric spaces are normal. In fact they are also paracompact, and hence completely paracompact. This important result is due to Stone, but unfortunately the proof is somewhat too technical to be given here.

Proposition (7.16). *Let X be paracompact. Then X is normal.*

For let H, K be a disjoint pair of closed subsets of X. By regularity there exists for each point $x \in H$ a neighbourhood U_x of which the closure does not meet K. Together with $X - K$ these neighbourhoods U_x $(x \in H)$ form an open covering of X. Since X is paracompact there exists a locally finite refinement. The union V of the members of the refinement which meet H is a neighbourhood of H. Moreover $\mathscr{Cl}\, V$ is the union of the closures of these members, since the refinement is locally finite, and so does not meet K. Thus $X - \mathscr{Cl}\, V$ is a neighbourhood of K which does not meet the neighbourhood V of H, and so (7.16) is proved.

Proposition (7.17). *Let $\phi: X \to Y$ be a compact surjection. If Y is paracompact then so is X.*

For let Γ be an open covering of X. Since ϕ is a compact surjection each point $y \in Y$ admits a neighbourhood $N(y)$ such that $\phi^{-1}N(y)$ can be covered by a finite subfamily Γ_y of Γ. The neighbourhoods $N(y)$, for $y \in Y$, form an open covering of Y. Since Y is paracompact there exists a locally finite refinement $\{V(y)\}$ $(y \in Y)$. Now Γ_y covers $\phi^{-1}N(y)$ and so covers $\phi^{-1}V(y)$. Consider the intersection Γ'_y of Γ_y and $\phi^{-1}V(y)$. The union Γ' of the Γ'_y for all $y \in Y$ constitutes a refinement of Γ. Moreover Γ' is locally finite since for each point $x \in X$ there exists a neighbourhood N of $\phi(x)$ which meets at most finitely many of the sets $V(y)$, and then the neighbourhood $\phi^{-1}N$ of x meets at most finitely many of the members of Γ'.

Exercise (7.18). Let X, Y be paracompact spaces, let A be closed in X and let $\phi: A \to Y$ be a closed map. Then the push-out of the cotriad

$$X \xleftarrow{\supseteq} A \xrightarrow{\phi} Y$$

is also paracompact.

After these preliminaries I turn now to the main subject of this chapter, i.e. numerable coverings. First a few remarks about partitions of unity.

For any space X and map $\pi: X \to \mathbb{R}$ the complement of $\pi^{-1}(0)$ is called the cozero set of π. Similarly if $\{\pi_j\}$ $(j \in J)$ is a family of maps $\pi_j: X \to \mathbb{R}$ then the family $\{V_j\}$, where $V_j = X - \pi_j^{-1}(0)$, is called the family of cozero sets of $\{\pi_j\}$. We describe the family $\{\pi_j\}$ as point-finite (resp. locally-finite) if the family $\{V_j\}$ of cozero sets is point-finite (resp. locally-finite). Thus $\{\pi_j\}$ is point-finite if for each point $x \in X$ we have $\pi_j(x) = 0$ for all but a finite number of indices j, while $\{\pi_j\}$ is locally finite if for each point $x \in X$ there exists a neighbourhood U such that $\pi_j = 0$ on U for all but a finite number of indices j.

In what follows the main interest is in non-negative maps $\pi: X \to \mathbb{R}$, i.e. maps such that $\pi(x) \geq 0$ for all $x \in X$. Any map $\pi: X \to \mathbb{R}$ determines a non-negative map $\pi^+: X \to \mathbb{R}$, where

$$\pi^+(x) = \max(0, \pi(x)).$$

Definition (7.19). A partition of unity on the space X is a point-finite family $\{\pi_j\}$ ($j \in J$ of non-negative maps $\pi_j: X \to \mathbb{R}$ such that $\Sigma\pi_j = 1$ throughout X.

For example, consider the join, with coarse topology, of a collection of spaces $\{X_j\}$ ($j \in J$). The family of coordinate functions $\{t_j\}$ ($j \in J$) constitutes a partition of unity. In general the family is not locally finite.

Note that if $\{\pi_j\}$ ($j \in J$) is a point-finite family of non-negative maps $\pi_j: X \to \mathbb{R}$ such that $\Sigma\pi_j$ is positive throughout X then $\{\pi'_j\}$ is a partition of unity, with the same cozero sets, where

$$\pi'_j = \pi_j/\Sigma\pi_i \qquad (j \in J).$$

We refer to $\{\pi'_j\}$ as the partition of unity obtained from $\{\pi_j\}$ by normalization.

Note also that if $\{\pi_j\}$ is a partition of unity of the space X then π_j may be regarded as a map of X into I, for each index j, and the cozero set of π_j is just $\pi_j^{-1}(0, 1]$.

Proposition (7.20). *Let $\{\pi_j\}$ ($j \in J$) be a partition of unity on the space X. Let $\varepsilon > 0$. Then each point $x \in X$ has a neighbourhood U such that $\pi_j < \varepsilon$ throughout U for all but a finite number of indices j.*

For given x, there exists a finite subset K of J such that $\Sigma_{k \in K}\pi_k(x) > 1 - \varepsilon$. Take U to be the set of points $y \in X$ such that $\Sigma_{k \in K}\pi_k(y) > 1 - \varepsilon$. If $\pi_j(y) \geq \varepsilon$, for some point $y \in U$, then $j \in K$ since $\Sigma\pi_i = 1$. This proves (7.20) and we deduce

Corollary (7.21). *Let $\{\pi_j\}$ ($j \in J$) be a partition of unity on the space X. Then the non-negative function $\mu: X \to \mathbb{R}$, given by*

$$\mu(x) = \max_{j \in J} \{\pi_j(x)\},$$

is continuous.

For by (7.20) $\mu(x)$ agrees locally with the maximum of finitely many of the π_j and is therefore continuous.

Definition (7.22). An envelope of unity on the space X is a point-finite family $\{\mu_j\}$ ($j \in J$) of non-negative maps $\mu_j: X \to \mathbb{R}$ such that $\max\{\mu_j\} = 1$ throughout X.

Note that if $\{\pi_j\}$ is a partition of unity then $\{\mu_j\}$ is an envelope of unity, where

$$\mu_j(x) = \frac{\pi_j(x)}{\mu(x)}.$$

Proposition (7.23). *Let* $\{\pi_j\}$ $(j \in J)$ *be a partition of unity on the space* X. *Then there exists a locally finite partition of unity* $\{\rho_j\}$ $(j \in J)$ *such that*

$$\rho_j^{-1}(0, 1] \subset \pi_j^{-1}(0, 1]$$

for all indices j.

For consider the family $\{\sigma_j\}$ of non-negative maps $\sigma_j \colon X \to \mathbb{R}$ where

$$\sigma_j(x) = \max{}^+(2\pi_j(x) - \mu(x)).$$

Here $\mu(x)$ means the same as in (7.21). Given a point $x \in X$ write $\varepsilon = \frac{1}{4}\mu(x)$. By (7.20) and (7.21) there exists a neighbourhood U of x and a finite subset $K \subset J$ such that $\mu > 2\varepsilon$ throughout U and such that $\pi_j < \varepsilon$ throughout U whenever $j \in J - K$. Hence $\sigma_j = 0$ throughout U whenever $j \in J - K$ and so the family $\{\sigma_j\}$ $(j \in J)$ is locally finite. However there exists an index $k \in K$ such that $\mu(x) = \pi_k(x)$ and then $\sigma_k(x) = \pi_k(x) > 0$. Thus $\Sigma\sigma_j$ is positive throughout X and a partition of unity $\{\rho_j\}$ is defined by normalizing the family $\{\sigma_j\}$. Moreover $\{\rho_j\}$ is locally finite, since $\{\sigma_j\}$ is locally finite, and

$$\rho_j^{-1}(0, 1] = \sigma_j^{-1}(0, 1] \subset \pi_j^{-1}(0, 1]$$

for each index j. This proves (7.23).

Our next result is reminiscent of the shrinking theorems of Chapter 2. Following E. Michael we prove

Proposition (7.24). *Let* $\{\pi_j\}$ $(j \in J)$ *be a locally finite partition of unity on the space* X. *Then there exists a partition of unity* $\{\rho_j\}$ $(j \in J)$ *such that*

$$\mathscr{Cl}\, \rho_j^{-1}(0, 1] \subset \pi_j^{-1}(0, 1]$$

for each index j.

For consider the family $\{\sigma_j\}$ $(j \in J)$ of non-negative functions $\sigma_j \colon X \to \mathbb{R}$ given by

$$\sigma_j = (\pi_j - \tfrac{1}{2}\mu_j)^+;$$

where μ_j means the same as before. Given $x \in X$ there exists an index j such that $\pi_j(x) = \mu(x)$ and hence $\sigma_j(x) > 0$. Therefore $\Sigma\sigma_i$ is positive throughout X and we define the family $\{\rho_j\}$ by normalizing the $\{\sigma_j\}$. Since $\{\pi_j\}$ is locally finite it follows that each σ_j and hence each ρ_j is continuous. Finally

$$\begin{aligned}\mathscr{Cl}\, \rho_j^{-1}(0, 1] &= \mathscr{Cl}\, \sigma_j^{-1}(0, 1] \\ &\subset \{x \in X \mid \pi_j(x) \geq \tfrac{1}{2}\mu_j(x)\} \\ &\subset \pi_j^{-1}(0, 1].\end{aligned}$$

This completes the proof of (7.24).

Definition (7.25). Let $\{X_j\}$ $(j \in J)$ be a covering of the space X. A numeration of $\{X_j\}$ is a locally finite partition of unity $\{\pi_j\}$ such that $\pi_j^{-1}(0, 1] \subset X_j$ for each index j. If there exists a numeration the covering is said to be numerable. A covering which is the family of cozero set of a locally finite partition of unity is said to be numerically defined.

In view of (7.24) we can strengthen the condition in (7.25) to $\mathscr{C}\ell\ \pi_j^{-1}(0, 1] \subset X_j$; if this further condition is satisfied we describe $\{\pi_j\}$ as subordinated to the covering. The sets $\mathscr{C}\ell\ \pi_j^{-1}(0, 1]$ are known as the supports of the π_j.

Example (7.26). Let X be a metric space. Then each point-finite open covering of X is numerable.

For let $\{V_j\}$ $(j \in J)$ be such a covering. Define $\pi_j \colon X \to \mathbb{R}$ by $\pi_j(x) = d(x, X - V_j)$, where d is the metric. Then $\Sigma \pi_j$ is defined and is positive throughout X. Thus a numeration of the given covering is obtained by normalization.

Proposition (7.27) (Milnor). *Let $\{V_j\}$ $(j \in J)$ be a numerable open covering of the space X. Then there exists a countable numerable open covering $\{W_n\}$ $(n = 1, 2, \ldots)$ of X such that each W_n is the disjoint union of open sets, each of which is contained in some member of the original covering $\{V_j\}$.*

For let $\{\pi_j\}$ $(j \in J)$ be a numeration of the given covering. For each point $x \in X$ let $K(x)$ denote the finite set of indices $j \in J$ for which $\pi_j(x) > 0$. For each finite subset $K \subset J$ let $W(K)$ denote the open set of X consisting of the points x such that:

$$\pi_i(x) < \pi_j(x) \quad \text{whenever} \quad i \in J - K \text{ and } j \in K.$$

If K and K' are two distinct subsets of J with the same finite number of elements then there exists an index i which is contained in K' but not in K, and an index j which is contained in K but not in K'; therefore $W(K)$ and $W(K')$ are disjoint. Define W_n $(n = 1, 2, \ldots)$ to be the union of all the sets $W(K(x))$ such that $K(x)$ has n elements. Each of the sets $W(K(x))$ is open and is contained in U_j whenever $j \in K(x)$.

To see that $\{W_n\}$ is numerable consider for each finite subset $K \subset J$ the non-negative map $\sigma_K \colon X \to \mathbb{R}$ given by

$$\sigma_K(x) = \min_{i \in K,\, j \in J - K}{}^{+} \ (\pi_i(x) - \pi_j(x)).$$

Then $W(K)$ is the cozero set of σ_K, and hence W_n is the cozero set of ρ_n, where

$$\rho_n(x) = \sum_{|K| = n} \sigma_K(x).$$

Also $\{\rho_n\}$ $(n = 1, 2, \ldots)$ is locally finite since for each point $x \in X$ there exists a neighbourhood V such that $\pi_i(x) > 0$ only for indices i in a finite subset $J' \subset J$, and hence $\rho_n(x) > 0$ only for indices $n \leq |J'|$. Finally we normalize in the usual way, and obtain a numeration of $\{W_n\}$ as required.

The last result in this series is somewhat harder to prove; however it will play an important part in what we are going to do later.

Proposition (7.28). *Let X be a space. Let $\{U_j\}$ $(j \in J)$ be a numerable open covering of the cylinder $I \times X$. Then there exists a numerable open covering $\{V_k\}$ $(k \in K)$ of X, and a family $\{\varepsilon_k\}$ $(k \in K)$ of positive real numbers, such that for all α, β with $|\alpha - \beta| < \varepsilon_k$ there exists an index j for which $[\alpha, \beta] \times V_k \subset U_j$.*

Let $\{\pi_j\}$ $(j \in J)$ be a numeration of the given covering of $I \times X$. For each r-tuple $k = (j_1, \ldots, j_r) \in J^r$ a map $\rho_k \colon X \to I$ is given by

$$\rho_k(x) = \min_{i=1,\ldots,r} \{\pi_{j_i}(t, x) | i - 1 \leq (r + 1)t \leq i + 1\}.$$

Define $K = \bigcup_{r=1}^{\infty} J^r$. I assert that $\{V_k\}$ $(k \in K)$ is a numerable open covering of X, where V_k is the cozero set of ρ_k.

The first step is to show that each point $x \in X$ is contained in V_k for some $k \in K$. To see this, observe that each point $(t, x) \in I \times X$ has a neighbourhood contained in some member U_j, say, of the given covering, and hence has a neighbourhood of the form $U(t, x) \times V(t, x)$ with this property. Moreover the neighbourhood can also be chosen, at the same time, so as to meet only a finite number of members of the covering. By compactness there exist real numbers $t_1, \ldots, t_n$ such that the open sets

$$U(t_1, x), \ldots, U(t_n, x)$$

cover the segment $I \times \{x\}$. Choose $r \geq 1$ so that $2/(r + 1)$ is a Lebesgue number of the covering. Consider the intersection

$$V = V(t_1, x) \cap \cdots \cap V(t_n, x).$$

Each of the sets

$$\left[\frac{i-1}{r+1}, \frac{i+1}{r+1}\right] \times V \qquad (i = 1, \ldots, r)$$

is contained in the set U_{j_i} for some $j_i \in J$. Hence x is contained in V_k for $k = (j_1, \ldots, j_r)$.

It remains to be shown that the covering $\{V_k\}$ $(k \in K)$ is locally finite. To see this let V be the neighbourhood of the given point $x \in X$ constructed in the previous paragraph. Then $I \times V$ meets U_j for at most a finite number of indices $j \in J$. But the segment $I \times \{x\}$ meets U_{j_i} for $i = 1, \ldots, r$, and so the family $\{V_k\}$ $(k \in K_r)$ is locally finite, where

$$K_r = J \cup J^2 \cup \cdots \cup J^r.$$

Now consider the family $\{\sigma_r\}$ $(r = 1, 2, \ldots)$ of maps $\sigma_r \colon X \to I$, where $\sigma_1 = 0$ and where

$$\sigma_r(x) = \sum_{k \in K_{r-1}} \rho_k(x) \qquad (r > 1).$$

For $k \in K_r - K_{r-1}$ we write $\tau_k = (\rho_k - r\sigma_k)^+$. Choose $k' = \{j_1, \ldots, j_r\} \in K$ with the least integer r such that $x \in V_{k'}$. Then $\sigma_r(x) = 0$ and so $\tau_{k'}(x) = \rho_{k'}(x)$. I now assert that the cozero sets of the family $\{\tau_k\}$ $(k \in K)$ cover X.

To see this, choose $m > r$ so that $\rho_{k'}(x) > 1/m$. Then $\sigma_m(x) > 1/m$ and so $m\sigma_m(y) > 1$ for all y in some neighbourhood N of x. However $\tau_k = 0$ on N for all $k = (j_1, \ldots, j_s)$ such that $s \geq m$. Therefore $\{\tau_k\}$ $(k \in K)$ is locally finite and so, after normalization, is a numeration of $\{V_k\}$. Finally we take $\varepsilon_k = 1/2r$, where $k = (j_1, \ldots, j_r)$ and the proof of (7.28) is complete.

The next thing we need to do is to characterize normal spaces and paracompact spaces in terms of numerable coverings, beginning with

Proposition (7.29). *The space X is normal if and only if each finite open covering of X is numerable.*

The "if" direction is obvious; in the other direction it is the result known as Urysohn's lemma, to be found in all the standard texts. In fact we can strengthen the conclusion as follows:

Proposition (7.30). *Let $\{U_j\}$ $(j \in J)$ be a locally finite open covering of the normal space X. Then $\{U_j\}$ is numerable.*

The first step is to shrink the given covering to a covering V_j $(j \in J)$ of X such that $\mathscr{C}\ell\ V_j \subset U_j$. By (7.29) there exists, for each index j, a map $\sigma_j \colon X \to I$ such that $\sigma_j = 0$ on $X - U_j$ and $\sigma_j = 1$ on $\mathscr{C}\ell\ V_j$. Since $\{U_j\}$ is locally finite so is $\{\sigma_j\}$. Since $\{V_j\}$ covers X and since $\sigma_j = 1$ on V_j we have $\Sigma_{j \in J}\, \sigma_j > 0$. Hence a numeration of the given covering is obtained by normalization.

Proposition (7.31). *Let $\{V_j\}$ be an open covering of the paracompact space X. Then $\{V_j\}$ is numerable.*

For by definition $\{V_j\}$ has a locally finite refinement $\{U_j\}$, say. Since X is normal, by (7.16), the refinement is numerable and so the original covering is numerable.

Earlier in this chapter we discussed Lusternik–Schnirelmann category and its generalizations. For each of these numerical invariants one is given a collection Γ of open sets, covering the given space X, and then seeks the least number of members of Γ required to cover X. If we require the latter covering to be numerable then the number, which may be greater than the previous one, is called the numerable category. In case X is paracompact the numbers are equal, of course. In any case the numerable category is countable, by (7.27), since each open subset of a member of Γ is a member of Γ and since the union of a disjoint family of members of Γ is a member of Γ.

Proposition (7.32). *Let Γ be a numerable open covering of the space X. Suppose that*

(i) *each open subset of a member of Γ is a member of Γ,*
(ii) *the union of each disjoint family of members of Γ is a member of Γ, and*
(iii) *the union of each finite family of members of Γ is a member of Γ.*

Then X is a member of Γ.

The first two "admissibility" conditions are satisfied by all the examples of category we have previously considered. However the third condition is not, in general satisfied; we refer to it as the "domino" condition.

By (7.27) and the first two conditions we may assume, without real loss of generality, that Γ is a countable numerable open covering $\{W_n\}$ ($n = 1, 2, \ldots$). By the domino condition the finite union $V_n = W_1 \cup \cdots \cup W_n$ is a member of Γ. Let $\{\pi_n\}$ ($n = 1, 2, \ldots$) be a numeration of $\{W_n\}$ and write $U_n = (\pi_1 + \cdots + \pi_n)^{-1}(1/n, 1]$. Then $\mathscr{C}\ell\, U_{n-1} \subset U_n$, for $n \geq 2$, and the family $\{U_n\}$ ($n = 1, 2, \ldots$) covers X. Now consider the family $N_1, N_2, \ldots$ of open sets of X, where $N_n = U_n$ for $n = 1, 2$, and where $N_n = U_n - \mathscr{C}\ell\, U_{n-2}$ for $n = 3, 4, \ldots$. Each N_n is an open set of V_n and so is a member of Γ. Also $N_1, N_3, \ldots$ are disjoint and $N_2, N_4, \ldots$ are disjoint. Therefore

$$X_1 = \bigcup_i N_{2i-1}, \qquad X_2 = \bigcup_i N_{2i} \qquad (i = 1, 2, \ldots)$$

are both members of Γ and so X is a member of Γ, as asserted.

Proposition (7.33). *Let E be a space over X. Suppose that E is numerably sectionable. Then the numerable sectional category* $\operatorname{secat}_X E$ *does not exceed n if the n-fold fibre join $E^{(n)}$ with coarse topology admits a section over X.*

In the proof we represent points of the fibre join as in Chapter 5. Suppose that $E^{(n)}$ admits a section s, say. Let $U_i \subset X$ ($i = 1, \ldots, n$) be the set of points x such that $t_i s(x) > 0$. A section $s_i: U_i \to E$ is then given by $s_i(x) = x_i s(x)$, where $x_i: t_i^{-1}(0, 1] \to E$ is the ith coordinate function.

Conversely let $\{U_i\}$ ($i = 1, \ldots, n$) be an open covering of X with numeration $\{t_i\}$, and let $\{s_i\}$ be a family of sections $s_i : U_i \to E$. Then a section $s: X \to E^{(n)}$ is given at $x \in X$ by

$$s(x) = \langle t_1(x)\, s_1(x), \ldots, t_n(x)\, s_n(x)\rangle.$$

This completes the proof.

Our next set of results will demonstrate the utility of numerations in homotopy theory. In fact we shall only be considering numerations with two functions $\{\pi, \rho\}$; since $\pi + \rho = 1$ the formulae can be expressed entirely in terms of ρ which is then termed the Urysohn function. To extend the results to finite numerations is just a matter of induction but to extend them to infinite (but still locally finite) numerations requires techniques which we have not developed.

In what follows it is often convenient to assume that a given homotopy h_t is bordered, in the sense that h_t is independent of t for $0 \leq t \leq \frac{1}{4}$ and for $\frac{3}{4} \leq t \leq 1$. Of course any homotopy can be converted into a bordered homotopy through reparametrization.

Proposition (7.34). *Let X and Y be spaces. Let $\{X_0, X_1\}$ be a numerable covering of X and let $\{Y_0, Y_1\}$ be a family of subsets of Y. Let $\phi_i: X_i \to Y_i$ $(i = 0, 1)$ be maps. Suppose that there exists a homotopy*

$$h_t: X_0 \cap X_1 \to Y_0 \cap Y_1$$

of the map determined by ϕ_0 into the map determined by ϕ_1. Then there exists a map $\phi: X \to Y$ such that $\phi X_i \subset Y_i$ $(i = 0, 1)$ and homotopies

$$f_t: X_0 \to Y_0, \qquad g_t = X_1 \to Y_1$$

of ϕ_0 into the map determined by ϕ and of ϕ_1 into the map determined by ϕ, respectively.

Let $\{\pi, \rho\}$ be a numeration of $\{X_0, X_1\}$, so that $\pi = 0$ away from X_0 and $\rho = 0$ away from X_1. We assume h_t to be bordered and then define ϕ by

$$\phi(x) = \begin{cases} h_{\rho(x)}(x) & (x \in X_0 \cap X_1) \\ \phi_0(x) & (0 \leq \rho(x) < \frac{1}{4}) \\ \phi_1(x) & (\frac{3}{4} < \rho(x) \leq 1). \end{cases}$$

Also we define f_t by

$$f_t(x) = \begin{cases} h_{t.\rho(x)}(x) & (x \in X_0 \cap X_1) \\ \phi_0(x) & (0 \leq \rho(x) < \frac{1}{4}) \end{cases}$$

and define g_t similarly, using π instead of ρ. Since $\rho^{-1}(0, 1) \subset X_0 \cap X_1$, and since X is covered by

$$\{\rho^{-1}(0, 1), \rho^{-1}[0, \tfrac{1}{4}), \rho^{-1}(\tfrac{3}{4}, 1]\}$$

the continuity of ϕ follows at once, and the continuity of f_t, g_t follows similarly.

Proposition (7.35). *Let B be a space and let X, Y be spaces over B. Let $\{X_0, X_1\}$ be a numerable covering of X and let $\{Y_0, Y_1\}$ be a family of subsets of Y. Let $\phi_i: X_i \to Y_i$ $(i = 0, 1)$ be fibre-preserving maps. Suppose that there exists a fibre-preserving homotopy*

$$h_t: X_0 \cap X_1 \to Y_0 \cap Y_1$$

of the map determined by ϕ_0 into the map determined by ϕ_1. Then there exists a fibre-preserving map $\phi: X \to Y$ such that $\phi X_i \subset Y_i$ $(i = 0, 1)$. Moreover *there exist fibre-preserving homotopies*

$$X_0 \to Y_0, \qquad X_1 \to Y_1$$

of ϕ_0 *into the map determined by* ϕ, *and of* ϕ_1 *into the map determined by* ϕ, *respectively.*

This is just a fibrewise version of (7.34) and to prove it we use precisely the same formulae as we used in the case of (7.34). As an application we prove

Proposition (7.36). *Let* B *be a space and let* X *be a space over* B *with projection* p. *Let* $\{X_0, X_1\}$ *be a family of subsets of* X *and let* $\{B_0, B_1\}$ *be a numerable covering of* B *such that* $pX_i \subset B_i$ $(i = 0, 1)$. *Suppose that* X_i *is sectionable over* B_i $(i = 0, 1)$ *and that* $X_0 \cap X_1$ *is contractible over* $B_0 \cap B_1$. *Then there exists a section* $s: B \to X$ *such that* $sB_i \subset X_i$ $(i = 0, 1)$.

In (7.35) we take $(X, Y) = (B, X)$, so that fibre-preserving maps are sections and fibre-preserving homotopies are vertical homotopies. Let $s_i: B_i \to X_i$ be sections $(i = 0, 1)$. Since $X_0 \cap X_1$ is contractible over $B_0 \cap B_1$ the sections $B_0 \cap B_1 \to X_0 \cap X_1$ determined by s_0, s_1 are vertically homotopic. Hence, by (7.35), there exists a section $s: B \to X$ such that $sB_i \subset X_i$ $(i = 0, 1)$, as asserted. This in turn implies

Corollary (7.37). *Let* X, Y *be spaces. Let* $\{X_0, X_1\}$ *be a numerable covering of* X *and let* $\{Y_0, Y_1\}$ *be a numerable covering of* Y. *Let* $\phi: X \to Y$ *be a map such that* $\phi X_i \subset Y_i$ $(i = 0, 1)$. *Suppose that the maps* $X_i \to Y_i$ $(i = 0, 1)$ *and* $X_0 \cap X_1 \to Y_0 \cap Y_1$ *determined by* ϕ *are homotopy equivalences. Then* ϕ *is a homotopy equivalence.*

Let $\phi_i: X_i \to Y_i$ $(i = 0, 1)$ and $\phi': X_0 \cap X_1 \to Y_0 \cap Y_1$ be the maps determined by ϕ. Let W_i $(i = 0, 1)$ and W' be the mapping path-spaces of these maps. Then W_i is contractible over Y_i $(i = 0, 1)$, since ϕ_i is a homotopy equivalence, and W' is contractible over $Y_0 \cap Y_1$. We regard W_i and W' as subspaces of W, the mapping path-space of ϕ. Then W admits a section s over Y such that $sY_i \subset W_i$ $(i = 0, 1)$. Therefore ϕ admits a right inverse $\psi: Y \to X$, up to homotopy, such that $\psi Y_i \subset X_i$ $(i = 0, 1)$.

Now repeat the argument with $\psi: Y \to X$ replacing ϕ. We conclude that ψ admits a homotopy inverse on the right, as well as on the left, and so is a homotopy equivalence. Therefore ϕ is a homotopy equivalence, as asserted.

The fibrewise version of (7.37) is proved similarly, using fibrewise mapping path-spaces. The statement is as follows.

Proposition (7.38). *Let* B *be a space and let* X, Y *be spaces over* B. *Let* $\{X_0, X_1\}$ *be a numerable covering of* X *and let* $\{Y_0, Y_1\}$ *be a numerable covering of* Y. *Let* $\phi: X \to Y$ *be a fibre-preserving map such that* $\phi X_i \subset Y_i$ $(i = 0, 1)$. *Suppose that the maps* $X_i \to Y_i$ $(i = 0, 1)$ *and* $X_0 \cap X_1 \to Y_0 \cap Y_1$ *determined by* ϕ *are fibre homotopy equivalences. Then* ϕ *is a fibre homotopy equivalence.*

Corollary (7.39). *Let B be a space and let X be a space over B. Let $\{X_0, X_1\}$ be a numerable covering of X. If X_0, X_1 and $X_0 \cap X_1$ are contractible over B then X is contractible over B.*

Proposition (7.40). *Let B a space and let X be a space over B. Let $\{X_0, X_1\}$ be a numerable covering of X such that X_0, X_1 and $X_0 \cap X_1$ are fibre spaces over B. Then X is a fibre space over B.*

For let W_i $(i = 0, 1)$ be the mapping path-space of the projection $X_i \to B$, regarded as a subspace of the mapping path-space W of the projection $X \to B$. Since $\{X_0, X_1\}$ is a numerable covering of X, so $\{W_0, W_1\}$ is a numerable covering of W. Also $W_0 \cap W_1$ is the mapping path-space of the projection $X_0 \cap X_1 \to B$. Now the path-spaces $P_i = P(X_i)$ $(i = 0, 1)$ may be regarded as subspaces of the path-space $P = P(X)$. Since X_0, X_1 are fibre spaces over B so, by (6.51), P_0, P_1 are contractible over W_0, W_1 respectively. Moreover $P_0 \cap P_1$ is the path-space of $X_0 \cap X_1$ and so is contractible over $X_0 \cap X_1$. Applying (7.36), therefore, we obtain that P has a section over W and so X is a fibre space over B, by (6.35).

Next I would like to establish two nilpotency results which indicate how, for a fibre space E over B, the Lusternik–Schnirelmann category of B can influence the structure of the monoid $\pi_B(E, E)$ of fibre homotopy classes of fibre-preserving maps of E into itself.

Let B be a space of finite numerable category cat B. Let E be a space over B and let $\Phi(E)$ denote the group of fibre homotopy classes of fibre homotopy equivalences of E with itself. Consider the normal subgroup $\Phi_1(E)$ of $\Phi(E)$ consisting of fibre homotopy classes of fibre homotopy equivalences $\phi: E \to E$ such that ϕ_b is homotopic to the identity on E_b for each point b of B. We prove

Proposition (7.41). *Let E be a fibre space over B. If B is of finite numerable category then $\Phi_1(E)$ is nilpotent of class $<$* cat B.

Let $\{V_i\}$ $(i = 1, \dots, n)$ be a numerically defined covering of B such that V_i is contractible in B for each i. Write $U_k = V_1 \cup \cdots \cup V_k$ $(k = 1, \dots, n)$ and let $\Gamma_k \subset \Phi_1(E)$ denote the normal subgroup consisting of fibre homotopy classes of fibre homotopy equivalences $\phi: E \to E$ such that ϕ is fibre homotopic to the identity over U_k. Then

$$\Phi_1(E) = \Gamma_1 \supset \cdots \supset \Gamma_n = \{1\}$$

is a chain of normal subgroups of $\Phi(E)$. We prove (7.41) by showing that

$$[\Gamma_1, \Gamma_k] \subset \Gamma_{k+1} \qquad (k = 1, \dots, n - 1).$$

For this purpose let $\{\rho_i\}$ $(i = 1, \dots n)$ be a numeration of $\{V_i\}$, and write

$$\beta_k = \rho_1 + \cdots + \rho_k \qquad (k = 1, \dots n).$$

Then $\{\alpha_k, \pi_{k+1}\}$ is a numeration of the covering $\{U_k, V_{k+1}\}$ of U_{k+1}, where

$$\alpha_k = \frac{\beta_k}{\beta_{k+1}}, \qquad \pi_{k+1} = \frac{\rho_{k+1}}{\beta_{k+1}}.$$

Let $\phi: E \to E$ represent an element of Γ_k and let $\psi: E \to E$ represent an element of Γ_1. By definition there exists a fibre homotopy $G_t: E_{U_k} \to E_{U_k}$ of ϕ_{U_k} into the identity. Also, since V_{k+1} is contractible in B, there exists, by (6.58), a fibre homotopy $H_t: E_{V_{k+1}} \to E_{V_{k+1}}$ of $\psi_{V_{k+1}}$ into the identity. Consider the fibre-preserving maps

$$K, L: I \times I \times E_{U_k \cap V_{k+1}} \to E_{U_k \cap V_{k+1}}$$

which are given, for $s, t \in I$ and $e \in E_{U_k \cap V_{k+1}}$ by

$$K(s, t, e) = H(s, G(t, e)), \qquad L(s, t, e) = G(t, H(s, e)).$$

A fibre homotopy $h_t: E_{U_{k+1}} \to E_{U_{k+1}}$ of $(\psi\phi)_{U_{k+1}}$ into $(\phi\psi)_{U_{k+1}}$ is given, for $t \in I$, $b \in U_{k+1}$ and $e \in E_b$, by the formula

$$h(t, e) = \begin{cases} L(2t, 4t\alpha_k(b), e) & (t \leq \frac{1}{2}) \\ K(2 - 2t, 4\alpha_k(b) - 4t\alpha_k(b), e) & (t \geq \frac{1}{2}) \end{cases}$$

when $\alpha_k(b) \leq \frac{1}{2}$ and $b \in U_k \cap V_{k+1}$; by

$$h(t, e) = \begin{cases} L(4t\pi_{k+1}(b), 2t, e) & (t \leq \frac{1}{2}) \\ K(4\pi_{k+1}(b) - 4t\pi_{k+1}(b), 2 - 2t, e) & (t \geq \frac{1}{2}) \end{cases}$$

when $\alpha_k(b) \geq \frac{1}{2}$ and $b \in U_k \cap V_{k+1}$; by

$$h(t, e) = \begin{cases} \phi H(2t, e) & (t \leq \frac{1}{2}) \\ H(2 - 2t, \phi e) & (t \geq \frac{1}{2}) \end{cases}$$

when $b \in V_{k+1} \setminus U_k$; and by

$$h(t, e) = \begin{cases} G(2t, \psi e) & (t \leq \frac{1}{2}) \\ \psi G(2 - 2t, e) & (t \geq \frac{1}{2}) \end{cases}$$

when $b \in U_k \setminus V_{k+1}$.

For the second nilpotency result, which is of rather a different type, we first need to prove

Lemma (7.42). *Let $\{U, V\}$ be a numerable covering of the space B. Let E be a fibre space over B and let $\theta, \phi: E \to E$ be maps over B such that θ is nul-homotopic over U and ϕ is nul-homotopic over V. Then $\phi\theta$ is nul-homotopic over B.*

For let $\{\pi, \rho\}$ be a numeration of $\{U, V\}$. Let $\sigma: U \to E_U$, $\tau: V \to E_V$ be sections. Let $K_t: E_U \to E_U$ be a fibre homotopy of θ_U into σp_U and let $L_t: E_V \to E_V$ be a fibre homotopy of ϕ_V into τp_V. Without real loss of generality we may suppose K_t and L_t to be bordered. Now a section $s: B \to E$ is given for $b \in B$ by

$$s(b) = \begin{cases} L_{\rho(b)}\sigma(b) & (b \in U \cap V) \\ \phi\sigma(b) & (b \in U \setminus V) \\ \tau(b) & (b \in V \setminus U). \end{cases}$$

I assert that $\phi\theta$ is fibre-homotopic to sp.

To prove this consider the convex hull C of $\{(0, 0), (1, 0), (0, 1)\}$ in $\mathbb{R}^2$. Observe that $(I \times I)\setminus(\{1\} \times I)$ is homeomorphic to $C\setminus\{1, 0\}$ under the transformation

$$(\xi, \eta) \mapsto (\xi, (1 - \xi)\eta).$$

Consider the fibre-preserving map

$$F: I \times I \times E_{U \cap V} \to E_{U \cap V}$$

which is given for $\xi, \eta \in I$, $b \in U \cap V$ and $e \in E_b$ by

$$F((\xi, \eta), e) = L_\xi K_\eta(e).$$

Since F is independent of η when $\xi = 1$ there is an induced map

$$G: C \times E_{U \cap V} \to E_{U \cap V}$$

which is given on $\partial C \times E_{U \cap V}$ by

$$G((t, 1 - t), e) = L_t \sigma p(e),$$

$$G((0, t), e) = \theta K_t(e),$$

$$G((t, 0), e) = K_t \phi(e).$$

Hence a fibre homotopy $H_t: E \to E$ of $\phi\theta$ into sp is given, for $b \in B$ and $e \in E_b$, by

$$H(t, e) = \begin{cases} G(t\rho(b), t(1 - \rho(b)), e) & (b \in U \cap V) \\ \phi K_t(e) & (b \in U \setminus V) \\ L_t \theta(e) & (b \in V \setminus U). \end{cases}$$

This establishes the lemma, and now we are ready to prove

Proposition (7.43). *Let E be a fibre space over B, where B has numerable category $\leq n$. Consider the set of fibre-preserving maps $\phi: E \to E$ with the property that $\phi_b: E_b \to E_b$ is nul-homotopic for each point $b \in B$. The composition of any n such maps is nul-homotopic over B.*

For let $\{V_i\}$ $(i = 1, \ldots, n)$ be a numerically defined covering of B such that V_i is contractible in B for each i. As before write $U_k = V_1 \cup \cdots \cup V_k$ $(k = 1, \ldots n)$. Let $\phi_i: E \to E$ $(i = 1, \ldots, n)$ be fibre-preserving maps with $(\phi_i)_b$ nulhomotopic for each b. Suppose that for some $k < n$ there exists a fibre homotopy of $(\phi_k \circ \cdots \circ \phi_1)_{U_k}$ into $\sigma_k p$ for some section σ_k of E over U_k. Since V_{k+1} is contractible in B there exists a fibre homotopy of $(\phi_{k+1})_{V_{k+1}}$ into $\tau_{k+1} p$ for some section τ_{k+1} of E over V_{k+1}. Since $\{U_k, V_{k+1}\}$ is a numerable covering of U_{k+1}, as we have seen, it follows from (7.42) that $(\phi_{k+1} \circ \phi_k \circ \cdots \circ \phi_1)_{U_{k+1}}$ is fibre homotopic to $\sigma_{k+1} p$ for some section σ_{k+1} of E over U_{k+1}. This establishes the inductive step and so proves (7.43).

In Chapter 3 we made some preliminary remarks about fibre bundles, i.e. locally trivial spaces, over a given base space B. Recall that E is a fibre bundle over B if there exists an open covering of B such that E is trivial over each member of the covering. If there exists a numerable covering with this property then the bundle is said to be numerable. Clearly the fibre product of numerable bundles over B is again a numerable bundle over B. Moreover if E is a numerable bundle over B then $\xi^* E$ is a numerable bundle over B' for each space B' and map $\xi: B' \to B$.

If B is paracompact then every bundle over B is numerable, by (7.31).

If E is a numerable bundle over B then (7.27) shows that there exists a countable numerable covering such that E is trivial over each member of the covering. Moreover (7.24) shows that the numeration $\{\pi_j\}$ $(j = 1, 2, \ldots)$ may be chosen so that E is trivial over the support $\mathscr{Cl}\, \pi_j^{-1}(0, 1]$ of π_j for each index j.

We are now going to discuss bundles over the cylinder $I \times A$, where A is any space. If D is a bundle over $I \times A$ then D_t $(0 \le t \le 1)$ will denote the restriction of D to $\{t\} \times A$. We may also regard D_t as a bundle over A by ignoring the $\{t\}$ factor. We need the preliminary

Proposition (7.44). *Let D be a numerable bundle over $I \times A$. Then there exists a numerable covering $\{V_k\}$ $(k \in K)$ of A such that D is trivial over $I \times V_k$ for each index k.*

For let $\{U_j\}$ $(j \in J)$ be a numerable covering of $I \times A$ such that D is trivial over U_j for each index j. By (7.28) there exists a numerable covering $\{V_k\}$ $(k \in K)$ of A and a family $\{\varepsilon_k\}$ of positive real numbers such that D is trivial over $[\alpha, \beta] \times V_k$ whenever $|\alpha - \beta| < \varepsilon_k$. By (3.52) this implies that D is trivial over $I \times V_k$, for each index k, as asserted. Moreover we can go on, using (7.27), and replace $\{V_k\}$ by a countable numerable covering $\{W_n\}$ $(n = 1, 2, \ldots)$, as we shall do in the proof of our next major result.

Proposition (7.45). *Let A be a space and let $\xi: I \times A \to I \times A$ be given by*

$$\xi(t, a) = (1, a) \qquad (t \in I, a \in A).$$

Then for each numerable bundle D over $I \times A$ there exists a pro-equivalence $\phi: D \to D$ over ξ which is the identity on D_1.

Here the term pro-equivalence, as in Chapter 1, means that the map $D \to \xi^*D$ which ϕ determines is an equivalence of spaces over $I \times A$.

To prove (7.45) we choose a countable numerable covering $\{W_n\}$ ($n = 1, 2, \ldots$) of A, as above, such that D is trivial over $I \times W_n$ for each index n. We also choose a trivialization

$$\alpha_n: I \times W_n \times Y \to D_{I \times W_n},$$

where Y is the fibre. Using (7.24) there exists a numeration $\{\pi_n\}$ of $W_n\}$ such that $\mathscr{Cl}\ \pi_n^{-1}(0, 1] \subset W_n$ for each index n. Write

$$\mu(a) = \max_{n=1,2,\ldots} \pi_n(a) \qquad (a \in A)$$

as before, so that an envelope of unity $\{\mu_n\}$ ($n = 1, 2, \ldots$) is given by $\mu_n(a) = \pi_n(a)/\mu(a)$. Consider the sequence of maps $\xi_n: I \times A \to I \times A$ ($n = 1, 2, \ldots$) given by

$$\xi_n(t, a) = (\max(t, \mu_n(a)), a) \qquad (t \in I, a \in A).$$

Note that $\xi_n(t, a) = (t, a)$ when $\mu_n(a) > 0$ or when $t = 1$. Consider also the sequence of maps $\tilde{\xi}_n: D \to D$ ($n = 1, 2, \ldots$) over ξ_n which is given by the identity on $I \times \mu_n^{-1}(0)$ and by

$$\tilde{\xi}_n \alpha_n(t, a, y) = \alpha_n(\xi_n(t, a), y)$$

for $t \in I$, $a \in W_n$ and $y \in Y$. Note that $\tilde{\xi}_n$ is a pro-equivalence over ξ_n, and that $\tilde{\xi}_n$ maps D_1 identically.

We now form the "compositions"

$$\cdots \to I \times A \xrightarrow{\xi_n} I \times A \to \cdots \to I \times A \xrightarrow{\xi_2} I \times A \xrightarrow{\xi_1} I \times A,$$

$$\cdots \to D \xrightarrow{\tilde{\xi}_n} D \to \cdots \to D \xrightarrow{\tilde{\xi}_2} D \xrightarrow{\tilde{\xi}_1} D.$$

These are both well-defined since each point $(t, a) \in I \times A$ is contained in at most a finite number of the sets $I \times W_n$. Moreover they are both continuous since for each such point there exists a neighbourhood which meets at most a finite number of the sets $I \times W_n$. In fact the first composition coincides with the given map ξ, since

$$\max_{n=1,2,\ldots} \mu_n(a) = 1 \qquad (a \in A),$$

and the second defines a map $\tilde{\xi}: D \to D$ over ξ. Furthermore $\tilde{\xi}$ is a pro-equivalence over ξ, since $\tilde{\xi}_n$ is a pro-equivalence over ξ_n for each n, and $\tilde{\xi}$ is the identity on D_1, since $\tilde{\xi}_n$ is the identity on D_1 for each n.

Proposition (7.46). *Let E be a numerable bundle over B. If $\phi_0, \phi_1: A \to B$ are homotopic then ϕ_0^*E and ϕ_1^*E are equivalent as spaces over A.*

To see this, choose a homotopy $\phi: I \times A \to B$ of ϕ_0 into ϕ_1, and take $D = \phi^*E$. Then D is numerable and so D is equivalent to $\xi^*D = I \times D_1$, as a space over $I \times A$. Similarly D is equivalent to $I \times D_0$. Thus $I \times D_0$

is equivalent to $I \times D_1$, as a space over $I \times A$, and so D_0 is equivalent to D_1, as a space over A. Since $D_t = \phi_t^* E$ this proves (7.46).

Corollary (7.47). *Let E be a numerable bundle over the contractible space B. Then E is trivial over B.*

Proposition (7.48). *Let E be a numerable bundle over B. Then the projection $p: E \to B$ is a fibration.*

We have to show that the homotopy lifting property is satisfied. So let X be a space, let $f: X \to E$ be a map, and let $g_t: X \to B$ be a homotopy such that $g_1 = pf$ (the substitution of $t = 1$ for $t = 0$ is of course a formality). Consider the numerable bundle $D = g^*E$, where $g: I \times X \to B$. The map $f: X \to E$ determines a section $s: X \to D_1 = g_1^*E$, and hence a section

$$\mathrm{id} \times s: I \times X \to I \times g_1^*E.$$

Now $I \times g_1^*E = \xi^*D$, where $\xi: I \times X \to I \times X$ is as in (7.45). Let $\tilde{\phi}: D \to \xi^*D$ be the equivalence constructed in (7.45). Then $(\tilde{\phi})^{-1}(\mathrm{id} \times s): I \times X \to D$ is a section which agrees with s on $\{1\} \times X$ since ϕ is the identity on D_1. By composing this section with the canonical map $D \to E$ we obtain a lifting of g as required.

It is clear from the construction that the lifting $h_t: X \to E$ of $g_t: X \to B$ thus constructed has the further property that if g_t is stationary on a subset of X then h_t is also stationary on that subset. Fibrations where the homotopy lifting property has this further property are called regular fibrations.

In case E is a bundle, without the assumption of numerability, the argument used to prove (7.48) shows that p has the homotopy lifting property for domains X which are paracompact.

We now return to the subject of (principal) G-bundles, where G is a topological group. We recall, from Chapter 4, that if X and Y are G-bundles over B then a map $\phi: X \to Y$ over B is an equivalence of G-bundles if ϕ is an equivariant equivalence of spaces over B. Now the argument used to prove (7.46) also proves

Proposition (7.49). *Let E be a numerable G-bundle over B. If $\phi_0, \phi_1: A \to B$ are homotopic then ϕ_0^*E and ϕ_1^*E are equivalent as G-bundles over A.*

With the machinery we have developed it is now possible to give a classification of G-bundles, for any topological group G. This uses an important construction, due to Milnor, which associates to each topological group G a numerable G-bundle with contractible total space, as follows. The total space of the bundle is the coarse join E_G of a countable number of copies of G. Thus points of E_G may be represented in the form

$$\langle t_1 x_1, t_2 x_2, \ldots \rangle,$$

where $t_1, t_2, \ldots \geq 0$, with at most a finite number of the t_i non-zero and with $\Sigma t_i = 1$, and where $x_1, x_2, \ldots \in G$. We make G act on E_G so that

$$\langle t_1 x_1, t_2 x_2, \ldots \rangle . g = \langle t_1 x_1 g, t_2 x_2 g, \ldots \rangle \quad (g \in G).$$

The action is obvious free; I assert that E_G is a G-bundle over the orbit space space B_G.

To see this, first observe that the coordinate function t_i is invariant under the action and so induces a map $u_i\colon B_G \to I$. I say that E_G has a section over the cozero set $U_i = u_i^{-1}(0, 1]$. For consider the fibre-preserving map $\phi_i\colon E_i \to E_i$, where $E_i = E_{U_i}$, given by

$$\phi_i(x) = x.(x_i(x))^{-1} \qquad (x \in E_i).$$

For each element $g \in G$ we have

$$\begin{aligned} \phi_i(xg) &= (xg).(x_i(xg))^{-1} \\ &= (xg).g^{-1}(x_i(x))^{-1} = x.(x_i(x))^{-1} = \phi_i(x). \end{aligned}$$

Thus ϕ_i is invariant and induces a section $U_i \to E_i$. Therefore E_G is a G-bundle over B_G. Moreover since the family $\{U_i\}$ $(i = 1, \ldots, n)$ constitutes a partition of unity on B_G the bundle is numerable, by (7.23).

Proposition (7.50). *Let B be a space and let E be a numerable G-bundle over B. Then there exists a map $\phi\colon B \to B_G$ such that E is equivalent to $\phi^* E_G$, as a G-bundle over B.*

In view of (7.27) we can assume that B admits a countable partition of unity $\{u_n\}$ $(n = 1, 2, \ldots)$ such that E is trivial over each of the cozero sets $U_n = u_n^{-1}(0, 1]$. Choose a trivialization

$$f_n\colon U_n \times G \to E_n = E_{U_n}$$

over each U_n, and let $k_n\colon E_n \to G$ be the second projection of f_n^{-1}. These projections $\{k_n\}$ define a map $k\colon E \to E_G$, where k is given at the point $x \in E_b$ by

$$k(x) = \langle u_1(b) k_1(x), \ldots, u_n(b) k_n(x), \ldots \rangle.$$

Clearly k is a G-map and so induces a map

$$\phi = \frac{k}{G}\colon B \to B_G.$$

Moreover the standard map $\tilde{k}\colon E \to \phi^* E_G$ over B determined by k is a G-map, and hence an isomorphism by (4.70).

Before stating our next result we need to recall from Chapter 5 some further properties of the infinite join. Let E'_G denote the subspace of E_G consisting of points $\langle t, x \rangle$ such that $t_i = 0$ for all odd values of i, and let E''_G denote the subspace of E_G consisting of points $\langle t, x \rangle$ such $t_i = 0$ for all even values of *i*. Write $B'_G = pE'_G$, $B''_G = pE''_G$, where $p: E_G \to B_G$ denotes the natural projection. Now E'_G and E''_G are weak G-deformation retracts of E_G, as we have seen in Chapter 5, and hence B'_G and B''_G are weak deformation retracts of B_G. We use this to prove

Proposition (7.51). *Let $\phi', \phi'': B \to B_G$ be maps such that ϕ'^*E_G and ϕ''^*E_G are equivalent, as G-bundles over B. Then ϕ' and ϕ'' are homotopic.*

In view of what we have just said there exist maps $\psi': B \to B'_G$, $\psi'': B \to B''_G$ such that $u'\psi' \simeq \phi'$, $u''\psi'' \simeq \phi''$, where u', u'' are the inclusions. Then $\psi'^*E'_G \approx \phi'^*E_G$, since $E'_G \approx u'^*E_G$, and similarly $\psi''^*E''_G \approx \phi''^*E_G$. Therefore $\psi'^*E'_G \approx \psi''^*E''_G$. Now let E be any G-bundle over B in the same isomorphism class as these, so that there exist pro-equivalences $\theta': E \to E'_G$, $\theta'': E \to E''_G$ respectively. Consider the composition

$$I \times E \xrightarrow{\theta} E'_G * E''_G \xrightarrow{\xi} E_G,$$

where $\theta(s, e) = (s, \theta'e, \theta''e)$ and where ξ is as in Chapter 5. Since θ and ξ are equivariant, so is $\xi\theta$. Also $\xi\theta_0$ is over ϕ' and $\xi\theta_1$ is over ϕ''. Hence the induced map $I \times B \to B_G$, defined by factoring out the action of G, constitutes a homotopy of ϕ' into ϕ'' as required.

We summarise these conclusions in

Theorem (7.52). *For each space B there is a natural equivalence between the set $\pi(B, B_G)$ of homotopy classes of maps, and the set of isomorphism classes of numerable G-bundles, given by the correspondence $\alpha \mapsto \alpha^*E_G$.*

To complete the discussion let us show, using the classification theorem, that E_G is contractible. First observe that the bundle over E_G induced from E_G by the projection $p_G: E_G \to B_G$ admits a section and is therefore trivial. Consequently p_G is nul-homotopic, by the classification theorem. This implies, by the homotopy lifting theorem, that the identity on E_G is homotopic to a map $f: E_G \to E_G$ of which the image fE_G is contained in some fibre of E_G. In fact f can be chosen so that the image is contained in a specific fibre, since B_G is path-connected. For convenience, choose the fibre

$$G * \{e\} * \{e\} * \cdots * \{e\} * \ldots,$$

which is obviously contractible. Then f is nulhomotopic and so E_G is contractible as asserted.

In fact any numerable G-bundle with contractible total space is universal in the same sense as E_G is universal, so that the base space of the bundle in question will serve as a classifying space for G-bundles in place of B_G. However it follows quite formally from the classification theorem that all such classifying spaces have the same homotopy type.

The argument used to prove (7.50) can also be used to obtain a characterization of the triviality category of G-bundles in the following terms.

Proposition (7.53). *Let E be a numerable G-bundle over B, with classifying map $k\colon B \to B_G$. Then* $\operatorname{trivcat}_B E \le n$ *if and only if k can be deformed into $B_G^{(n)}$.*

Here $B_G^{(n)} = E_G^{(n)}/G$, where $E_G^{(n)}$ denotes the n-fold coarse join of G with itself. Thus $B_G^{(1)} = *$, $B_G^{(2)} = \Sigma G$, and so forth.

To prove (7.53) in the "only if" direction, recall from (4.18) that $\operatorname{trivat}_B E_G^{(n)} \le n$. If k can be deformed into $B_G^{(n)}$, say into $\xi\colon B \to B_G^{(n)}$, then E is equivalent to $\xi^* E_G^{(n)}$ and so

$$\operatorname{trivcat}_B E = \operatorname{trivcat}_B \xi^* E_G^{(n)} \le \operatorname{trivcat}_{B_G} E_G^{(n)} \le n.$$

Conversely suppose that $\operatorname{trivcat}_B E \le n$. Then there exists an open covering $U_1, \ldots, U_n$ of B and trivializations $\xi_1, \ldots, \xi_n$ of $E_{U_1}, \ldots, E_{U_n}$. Applying the same argument as was used in the proof of (7.50) we obtain a map $\xi\colon B \to B_G^{(n)}$ such that E is equivalent to $\xi^* E_G^{(n)}$. It follows at once that the inclusion of $B_G^{(n)}$ in B_G, precomposed with ξ, is a classifying map for E. This completes the proof.

Finally we come to another major result due to Dold. The proof involves the useful notion of halo, which is closely related to the notion of Urysohn function mentioned earlier.

Definition (7.54). Let A, V be subspaces of the space X, with $A \subset V \subset X$. Then V is a halo of A in X if there exists a map $\pi\colon X \to I$ (the haloing function) such that $\pi = 1$ throughout A and $\pi = 0$ away from V.

For example X itself is a halo of each subset A of X with haloing function the constant map at 1. Note that if V is a halo of A in X then so is each subset U of X such that $V \subset U$; the same haloing function may be used. Also note that if V is a halo of A in X then V is also a halo of $\mathscr{Cl}\, A$ in X, since if $A \subset \pi^{-1}(1)$ then $\mathscr{Cl}\, A \subset \pi^{-1}(1)$ because $\pi^{-1}(1)$ is closed.

Example (7.55). If X is normal then each neighbourhood V of a closed set A of X is a halo of A in X.

This follows at once from (7.29), since $X - A$ and V form an open covering of X. The use of haloes enables us to avoid a restriction to normal spaces, as in the following result.

Lemma (7.56). *Let E and F be spaces over the given space B, and let $\phi\colon E \to F$ be a map over B. Regard the fibrewise mapping path-space $W = W_B(\phi)$ as a space over F in the usual way. Let A be a subset of B and let V be a halo of*

A in B. Let s be a section of W over F_V. If ϕ is a homotopy equivalence over a subset U of B then there exists a halo V′ of A in B and a section S of W over $F_{V' \cup U}$ such that $V' \subset V$ and S coincides with s on $F_{V'}$.

We shall show that a section S satisfying the consistency condition exists with $V' = \pi^{-1}(\frac{3}{4}, 1]$, where π is the haloing function for V. Of course we can replace the interval $(\frac{3}{4}, 1]$ by the interval $(\alpha, 1]$, for any α such that $0 < \alpha < 1$, and obtain the same result.

Choose a fibre homotopy inverse $\psi: F_U \to E_U$ of ϕ_U, and let $G_t: E_U \to E_U$, $H_t: F_U \to F_U$ be fibre homotopies such that

$$G_0 = \psi\phi_U, \quad G_1 = \mathrm{id}_{E_U}; \qquad H_0 = \phi_U\psi, \quad H_1 = \mathrm{id}_{F_U}.$$

Regarding H as a map $F_U \to PF_U$ we consider the section σ of W over F_U, where

$$\sigma(f) = (\psi f, Hf) \qquad (f \in F_U).$$

We construct a fibre homotopy $K_t: W_U \to W_U$ of $\sigma\rho_1$ into id_{W_U} as follows. Recall that $I \times \dot{I} \cup \{0\} \times I$ is a retract of $I \times I$; choose a retraction R. Consider the map

$$\theta: (I \times \dot{I} \cup \{0\} \times I) \times W_U \to F_U.$$

which is given on $((s, t), (e, \lambda))$ by

$$\begin{aligned} &H_s\lambda(0) && \text{for } t = 0,\\ &\lambda(s) && \text{for } t = 1,\\ &\phi_U G_t(e) && \text{for } s = 0. \end{aligned}$$

A homotopy $\Theta_t: W_U \to PF_U$ is given by

$$\Theta_t(e, \lambda)(s) = \theta(R(s, t), (e, \lambda))$$

and we define K_t by

$$K_t(e, \lambda) = \begin{cases} (\psi\lambda(1 - 2t), H\lambda(1 - 2t)) & (0 \le t \le \frac{1}{2})\\ (G_{2t-1}(e), \Theta_{2t-1}(e, \lambda)) & (\frac{1}{2} \le t \le 1). \end{cases}$$

Next we transform K_t into a bordered homotopy L_t by reparametrization, so that $L_t = \sigma\rho_1$ for $t \le \frac{1}{4}$ and $L_t = \mathrm{id}_{W_U}$ for $t \ge \frac{3}{4}$. Finally we define S on $F_{V' \cup U}$ by

$$S(f) = \begin{cases} \sigma(f) & (0 \le \pi(b) \le \frac{1}{4})\\ L_{\pi(b)}\sigma(f) & (\frac{1}{4} \le \pi(b) \le \frac{3}{4})\\ s(f) & (\frac{3}{4} \le \pi(b) \le 1), \end{cases}$$

where $f \in F_b$, $b \in V' \cup U$. Clearly S is a section of W over $F_{V' \cup U}$ which agrees with the given section s on $F_{V'}$. This proves the lemma, which we now use to establish

Theorem (7.57). *Let E and F be spaces over a given space B and let $\phi: E \to F$ be a fibre-preserving map. Suppose that there exists a numerable covering of B such that $\phi_U: E_U \to F_U$ is a homotopy equivalence over U for each member U of the covering. Then ϕ is a homotopy equivalence over B.*

By (7.24) and (7.27) there exists a countable numeration $\{\pi_n\}$ $(n = 1, 2, \ldots)$ of B such that ϕ is a homotopy equivalence over a neighbourhood U_n of $\sup \pi_n$ for each n. Write $A_n = \alpha_n^{-1}(1)$, where $\alpha_n = \pi_1 + \cdots + \pi_n$. Then $V_n = \alpha_n^{-1}(1 - 2^{-n}, 1]$ is a halo of A_n in B. We will construct a sequence of sections

$$s_n: F_{V_n} \to W = W_B(\phi) \qquad (n = 1, 2, \ldots)$$

such that s_n coincides with s_{n-1} on $F_{A_{n-1}}$. Assuming this, we define a section s of W over F by $s | F_{A_n} = s_n$. Since W has a section over F we have, by (5.16), that ϕ has a fibre homotopy inverse on the right, say $\psi: F \to E$. Now ψ is a homotopy equivalence over U_n, for each n, since ϕ is a homotopy equivalence over U_n. So we can repeat the argument, with ψ in place of ϕ, and obtain that ψ has a fibre homotopy inverse on the right, as well as on the left, and so is a fibre homotopy equivalence. Therefore ϕ is a fibre homotopy equivalence, as asserted.

It remains for the sequence to be constructed. To start the induction, choose any section s_1 of W over V_1; this exists, by (5.54), since ϕ is a homotopy equivalence over U_1 and so over V_1. Now assume that s_{n-1} has been constructed, for $n \geq 2$. Since ϕ is a homotopy equivalence over U_n we can apply the lemma to obtain a section s_n over V_n, satisfying the consistency condition. This completes the proof.

One of the most useful consequences of the above theorem is

Proposition (7.58). *Let E and F be fibre spaces over the numerably categorical space B. Let $\phi: E \to F$ be a map over B such that $\phi_b: E_b \to F_b$ is a homotopy equivalence for one point b in each path component of B. Then ϕ is a fibre homotopy equivalence.*

For let $\{V_j\}$ $(j \in J)$ be a numerable open covering of B such that the inclusion $V_j \to B$ is nul-homotopic for each index j. If the inclusion is nul-homotopic to the constant at b_j, say, then we may deform the constant at b_j into the constant at the point over which ϕ is a homotopy equivalence. Then ϕ is a homotopy equivalence over V_j and (7.58) follows at once from (7.57).

Proposition (7.59). *Let X be a fibre space over the numerably categorical space B. Suppose that in each path component of B one of the fibres is a retract up to homotopy. Then X is trivial in the sense of fibre-homotopy type.*

It is sufficient to deal with the case when B is path-connected. Let $r: X \to X_b$ be a retraction, up to homotopy, and let $\phi: X \to B \times X_b$ be given by the projection in the first component and by the retraction r in the second. Then ϕ is a map over B which satisfies the hypothesis of (7.58) and so is a fibre homotopy equivalence.

Corollary (7.60). *Let X be path-connected and numerably categorical. If X is a Hopf space then X admits a homotopy inverse.*

To see this we regard $X \times X$ as a space over X using the first projection. Consider the shearing map

$$\phi: X \times X \to X \times X$$

over X, where the first component is the first projection and the second is the multiplication on X. Now ϕ is a homotopy equivalence over $*$, by the Hopf condition, and so is a homotopy equivalence over X, by (7.59). Choose an inverse fibre homotopy equivalence ψ and let $u: X \to X$ be given by the composition

$$X \xrightarrow{(\mathrm{id},\, c)} X \times X \xrightarrow{\psi} X \times X \xrightarrow{\pi_2} X.$$

Then u is a homotopy inverse on the left.

References

E. H. Brown. Twisted tensor products, I. *Ann. of Math.* **69** (1959), 223–246.

J. Derwent. A note on numerable covers. *Proc. Amer. Math. Soc.* **19** (1968), 1130–1132.

T. tom Dieck. Klassifikation numerierbar Bündel. *Arch. Math.* **17** (1966), 395–399.

T. tom Dieck, K. H. Kamps and D. Puppe. *Homotopietheorie*. Springer Lecture Notes, Vol. 157, 1970.

T. tom Dieck. Partitions of unity in homotopy theory. *Comp. Math.* **23** (1971), 159–167.

A. Dold. Partitions of unity in the theory of fibrations. *Ann. of Math.* **78** (1963), 223–255.

A. Dold. Die Homotopieerweiterungseigenschaft ist eine lokale Eigenschaft. *Invent. Math.* **6** (1968), 185–189.

A. Dold. *Lectures on Algebraic Topology*. Springer-Verlag, New York, 1972.

J. Dugundji. *Topology*. Allyn and Bacon, Boston, 1966.

D. Husemoller. *Fibre Bundles*. McGraw-Hill, New York, 1966.

I. M. James. On fibre spaces and nilpotency, I. *Math. Proc. Cambridge Philos. Soc.* **84** (1978), 57–60.

I. M. James. On fibre spaces and nilpotency, II. *Math. Proc. Cambridge Philos. Soc.* **86** (1979), 215–217.

J. P. May. *Classifying Spaces and Fibrations*. Memoir 155. Amer. Math. Soc., Providence, RI, 1975.

H. Meiwes. On fibrations and nilpotency—some remarks upon two articles by I. M. James. *Manuscripta Math.* **39** (1982), 263–270.

J. Milnor. Construction of universal bundles, II. *Ann. of Math.* **63** (1956), 430–436.

CHAPTER 8

Extensors and Neighbourhood Extensors

In Chapter 6 the extension problem for mappings has been considered using the notion of cofibration. There is, however, a different approach to the subject which is also important. In this alternative approach the main restriction is placed on the codomain of the mappings in question. In fact it is necessary to restrict the class of domains to some extent as well. For what we are going to do it will be convenient to work with paracompact domains, although some of the results hold more generally for normal domains.

In the literature absolute retract is the traditional term for what we shall call extensor (apart from differences in the class of domains); however, the traditional terminology is inappropriate for the approach we are going to adopt.

Definition (8.1). The space E is an extensor if E has the following extension property for all paracompact pairs (X, A): every map of A into E can be extended to a map of X into E.

Since $(*, \varnothing)$ is a paracompact pair we see at once that extensors are non-empty. Since $(I, \dot{I})$ is a paracompact pair we also see that extensors are path-connected. More generally we have

Proposition (8.2). *Let E be an extensor. Then all maps of a paracompact space X into E are homotopic. In particular E is contractible when E is paracompact.*

For consider the paracompact pair $(I \times X, \dot{I} \times X)$. A pair of maps $X \to E$ determines, in the obvious way, a map $\dot{I} \times X \to E$, and then the extension $I \times X \to E$ constitutes a homotopy between the given maps.

The Tietze extension theorem shows that the real-line $\mathbb{R}$, and more generally the euclidean space $\mathbb{R}^n$ ($n = 0, 1, \ldots$), is an extensor. In fact $\mathbb{R}^n$ has the extension property for normal pairs.

Definition (8.3). The space E is a neighbourhood extensor if E has the following neighbourhood extension property for all paracompact pairs (X, A): every map of A into E can be extended to a map of some neighbourhood of A into E.

An open subset of a neighbourhood extensor (and hence of an extensor) is a neighbourhood extensor. A retract of an extensor is an extensor. A neighbourhood retract of a neighbourhood extensor is a neighbourhood extensor. Products of extensors are extensors and products of neighbourhood extensors are neighbourhood extensors. These results all follow more or less immediately from the definitions given and will be left to serve as exercises.

In the case of a paracompact pair (X, A) a necessary condition for A to be a neighbourhood extensor is that A be a neighbourhood retract of X. This shows, for example, that the subset

$$\left\{(0) \cup \left(\frac{1}{n}\right) : n = 1, 2, \ldots\right\}$$

of $\mathbb{R}$ is not a neighbourhood extensor.

Proposition (8.4). *Let E be a space and let $\{E_1, E_2\}$ be an open covering of E. If E_1, E_2 and $E_1 \cap E_2$ are extensors then so is E.*

Let (X, A) be a paracompact pair and let $f: A \to E$ be the map which has to be extended over X. The closed subspace A of X is covered by the open sets $f^{-1}(E_1), f^{-1}(E_2)$. Hence X itself is covered by the open sets

$$W_1 = f^{-1}(E_1) \cup (X - A), \qquad W_2 = f^{-1}(E_2) \cup (X - A).$$

Since X is paracompact we can shrink $\{W_1, W_2\}$ to a closed covering $\{X_1, X_2\}$ of X, where $X_1 \subset W_1$, $X_2 \subset W_2$. Write $A_1 = X_1 \cap A$, $A_2 = X_2 \cap A$. Then we have

$$f(A_1) \subset E_1, \qquad f(A_2) \subset E_2, \qquad f(A_1 \cap A_2) \subset (E_1 \cap E_2).$$

Since $(X_1 \cap X_2, A_1 \cap A_2)$ is a paracompact pair and since $E_1 \cap E_2$ is an extensor we can extend $f|(A_1 \cap A_2)$ over $X_1 \cap X_2$. By glueing f to such an extension we obtain a map $g: (X_1 \cap X_2) \cup A \to E$. Now because g maps $(X_1 \cap X_2) \cup A_1$ into the extensor E_1 we can extend $g|(X_1 \cap X_2) \cup A_1$ to a map $h_1: X_1 \to E_1$. Similarly we can extend $g|(X_1 \cap X_2) \cup A_2$ to a map $h_2: X_2 \to E_2$. Since h_1 and h_2 agree on $X_1 \cap X_2$ we can glue them together and obtain an extension

$$h: X = X_1 \cup X_2 \to E_1 \cup E_2 = E$$

of f, as required.

Proposition (8.5). *Let E be a space which admits a numerable open covering, each member of which is a neighbourhood extensor. Then E itself is a neighbourhood extensor.*

For paracompact spaces, according to this result, one may regard the property of being a neighbourhood extensor as a local property. This implies, for example, that paracompact manifolds are neighbourhood extensors.

We begin by proving (8.5) in the case of an open covering $\{E_1, E_2\}$ of E with just two members. The proof in this case begins in the same way as the proof of (8.4). However since $E_1 \cap E_2$ is a neighbourhood extensor, rather than an extensor, the extension of $f|(A_1 \cap A_2)$ is to a neighbourhood M of $A_1 \cap A_2$ in $X_1 \cap X_2$. Since $X_1 \cap X_2$ is paracompact we can shrink M to a closed neighbourhood $\mathscr{Cl}\ N$ of $A_1 \cap A_2$. Then $\mathscr{Cl}\ N \cap A = A_1 \cap A_2$ and a map $g: \mathscr{Cl}\ N \cup A \to E$ is defined by glueing to f an extension of $f|(A_1 \cap A_2)$ over $\mathscr{Cl}\ N$. Since g maps $\mathscr{Cl}\ N \cup A_1$ into the neighbourhood extensor E_1 we can extend $g|(\mathscr{Cl}\ N \cup A_1)$ to a map $h_1: V_1 \to E_1$, where V_1 is a neighbourhood of $\mathscr{Cl}\ N \cup A_1$ in X_1. Similarly we can extend $g|(\mathscr{Cl}\ N \cup A_2)$ to a map $h_2: V_2 \to E_2$, where V_2 is a neighbourhood of $\mathscr{Cl}\ N \cup A_2$ in X_2.

Since X_i $(i = 1, 2)$ is paracompact we can shrink V_i to a closed neighbourhood $\mathscr{Cl}\ U_i$ of $\mathscr{Cl}\ N \cup A_i$ in X_i. Since $(X_1 \cap X_2)\backslash N$ and A are disjoint closed sets of the paracompact space X there exists a neighbourhood W of A in X such that $\mathscr{Cl}\ W$ does not meet $(X_1 \cap X_2)\backslash N$. Consider the closed sets

$$F_1 = \mathscr{Cl}(U_1\backslash X_2) \cap \mathscr{Cl}\ W, \qquad F_2 = \mathscr{Cl}(U_2\backslash X_1) \cap \mathscr{Cl}\ W.$$

Here $F_1 \subset V_1$, $F_2 \subset V_2$ and $F_1 \cap F_2 \subset N$. Therefore $G_1 \subset V_1$, $G_2 \subset V_2$ and $G_1 \cap G_2 = \mathscr{Cl}\ N$, where $G_i = F_i \cup \mathscr{Cl}\ N$ $(i = 1, 2)$. Since h_1 and h_2 agree with g on $\mathscr{Cl}\ N$ we may glue $h_1|G_1$ and $h_2|G_2$ together to define a map $h: G \to E$, where $G = G_1 \cup G_2$. Obviously h is an extension of the original map f. Moreover the set

$$H = (U_1\backslash X_2) \cup (U_2\backslash X_1) \cup N$$

is open in X, hence $H \cap W$ is open in X. However $A \subset H \cap W \subset G$, and so $H \cap W$ is a neighbourhood of A over which f can be extended.

We see, from this special case, that the open sets of E which are neighbourhood extensors satisfy the domino condition of the previous chapter. Hence (8.5) in full generality now follows at once from (7.32).

There are analogous results for closed, as distinct from open, coverings but some further restriction on the class of domains seems to be necessary. Following the terminology used in the case of normal spaces, we say that a space X is completely paracompact if each subspace of X is paracompact. We also say that a pair (X, A) is completely paracompact if X is completely paracompact and A is closed in X. We prove

Proposition (8.6). *Let E be a space and let $\{E_1, E_2\}$ be a closed covering of E. Suppose that E_1, E_2 and $E_1 \cap E_2$ are extensors (resp. neighbourhood extensors)*

for the class of completely paracompact pairs. Then E is itself an extensor (resp. neighbourhood extensor) for the same class of domains.

Let (X, A) be a completely paracompact pair and let $f: A \to E$ be the map which has to be extended. Consider the inverse images $A_1 = f^{-1}(E_1)$, $A_2 = f^{-1}(E_2)$, which are closed in X. Since X is completely paracompact and since $A_1 - A_1 \cap A_2$, $A_2 - A_1 \cap A_2$ are mutually separated there exists, by (7.7), a neighbourhood N of $A_1 - A_1 \cap A_2$ in X such that

$$\mathscr{Cl}\, N \subset X - (A_2 - (A_1 \cap A_2)) = (X - A_2) \cup (A_1 \cap A_2).$$

Write

$$X_1 = \mathscr{Cl}\, N \cup (A_1 \cap A_2), \qquad X_2 = (X - N) \cup (A_1 \cap A_2).$$

Then X_1, X_2 are closed in X, and

$$X = X_1 \cup X_2, \qquad A_1 = X_1 \cap A, \qquad A_2 = X_2 \cap A.$$

Since $(X_1 \cap X_2, A_1 \cap A_2)$ is completely paracompact and since $E_1 \cap E_2$ is an extensor (resp. neighbourhood extensor) it follows that the restriction $f|(A_1 \cap A_2)$ can be extended to $X_1 \cap X_2$ (resp. to a neighbourhood of $A_1 \cap A_2$ in $X_1 \cap X_2$). From this stage the proof of (8.6) is essentially the same as that in the case of open coverings.

Of course (8.6) immediately generalizes to finite closed coverings which are closed under the intersection operation: this implies, for example, that (finite) polyhedra are neighbourhood extensors.

Proposition (8.7). *Let E be a contractible space. If E is a neighbourhood extensor then E is an extensor.*

For let $h: I \times E \to E$ be a homotopy of the identity into the nul-map at some point $*$ of E. Let (X, A) be a paracompact pair and let $f: A \to E$ be the map which has to be extended. Since E is a neighbourhood extensor there exists an extension g of f over some neighbourhood U of A in X. Since X is paracompact there exists a neighbourhood V of A in X such that $\mathscr{Cl}\, V \subset U$. Let $\alpha: X \to I$ be a Urysohn map with value 0 on A, 1 on $X - V$. Then an extension $F: X \to E$ of f is given by

$$\begin{aligned} F(x) &= h(\alpha(x), g(x)) && \text{for } x \in \mathscr{Cl}\, V, \\ &= * && \text{for } x \in X - V. \end{aligned}$$

Our next result deals with the behaviour of extensors and neighbourhood extensors when we pass to mapping spaces. Let $\phi: T_0 \to E$ be a map, where T_0 is a closed subspace of the space T, and let map$(T, E; \phi)$ denote the subspace of map(T, E) consisting of extensions of ϕ to maps $T \to E$. We prove

Proposition (8.8). *If E is an extensor then so is $map(T, E; \phi)$ for all locally compact regular T. If E is a neighbourhood extensor then so is $map(T, E; \phi)$ for all compact regular T.*

This result shows, for example, that if E is an extensor or neighbourhood extensor then so (taking $(T, T_0) = (I, \varnothing)$) is the space PE of paths in E and so (taking $(T, T_0) = (I, \dot{I})$) is the space ΩE of loops based at a given point of E.

So let (X, A) be a paracompact pair and let $f: A \to \mathrm{map}(T, E; \phi)$ be the map which has to be extended to X. Since T is locally compact regular in either case the adjoint $\hat{f}: T \times A \to E$ is defined, and maps $T_0 \times A$ by $\phi\pi_1$. We glue $\hat{f}$ to $\phi\pi_1: T_0 \times X \to E$ so as to obtain a map

$$\hat{g}: T \times A \cup T_0 \times X \to E.$$

Now $(T \times X, T \times A \cup T_0 \times X)$ is a paracompact pair. If E is an extensor then $\hat{g}$ can be extended to a map $\hat{h}: T \times X \to E$ and then the adjoint $h: X \to \mathrm{map}(T, E; \phi)$ of $\hat{h}$ is an extension of the original map f. If E is a neighbourhood extensor then $\hat{g}$ can be extended to a map $\hat{h}: U \to E$ where U is a neighbourhood of $T \times A \cup T_0 \times X$ in $T \times X$. If, further, T is compact, then there exists a neighbourhood V of A in X such that $T \times V \subset U$. The adjoint of $\hat{h}|(T \times V)$ provides an extension of f over V as required.

Our next result shows that neighbourhood extensors have the homotopy extension property for paracompact domains:

Proposition (8.9). *Let E be a neighbourhood extensor. Let (X, A) be a paracompact pair, let $f: X \to E$ be a map and let $h_t: A \to E$ be a homotopy of $f \mid A$. Then there exists an extension of h_t to a homotopy of f itself.*

For consider the paracompact pair $(I \times X, \{0\} \times X \cup I \times A)$. The given map f and homotopy h_t combine to form a map

$$\phi: \{0\} \times X \cup I \times A \to E.$$

Since E is a neighbourhood extensor there exists a neighbourhood U of $\{0\} \times X \cup I \times A$ in $I \times X$ over which ϕ can be extended. Since I is compact there exists a neighbourhood V of A in X such that $I \times V \subset U$. Since $\{0\} \times X \cup I \times V \subset U$ there is an extension ψ of ϕ over $\{0\} \times X \cup I \times V$. Choose a Urysohn map $\alpha: X \to I$ with value 0 on $X - V$ and 1 on A. Then an extension $\psi': I \times X \to E$ of ϕ is given by

$$\psi'(t, x) = \psi(\alpha(x).t, x) \qquad (x \in X, t \in I).$$

Now (8.9) follows at once.

One consequence of (8.9) is that if a given map $A \to E$ can be extended over X then so can any homotopic map. In other words extendability in this sense is a property of the homotopy class rather than the individual map.

Corollary (8.10). *Let E be a neighbourhood extensor. Let (X, A) be a paracompact pair such that A is a retract of X. Let $c: X \to E$ be a nul-map and let $f: X \to E$ be a map such that* (i) *$f \simeq c$ and* (ii) *$f|A = c|A$. Then $f \simeq^A c$.*

For let $h_t: X \to E$ be a homotopy of f into c. Then $h_t r$ is a homotopy of c into itself, where $r: X \to A$ is a retraction. Hence a map $H: I \times X \to E$ is given by

$$H(t, x) = \begin{cases} h_{2t}(x) & (0 \le t \le \frac{1}{2}) \\ h_{2-2t}r(x) & (\frac{1}{2} \le t \le 1). \end{cases}$$

If $x \in A$ then $H(t, x) = H(1 - t, x)$ for all t. Hence $H|(I \times A)$ is homotopic, rel$(\dot{I} \times A)$, to the stationary homotopy. Hence H itself is homotopic, rel$(\dot{I} \times X)$, to a homotopy which is stationary on A. This proves (8.10).

Corollary (8.11). *Let X be a paracompact neighbourhood extensor. Let A be a closed retract of X. Suppose that there exists a retraction $r: X \to A$ such that $ur \simeq \mathrm{id}_X$, where $u: A \subset X$. Then A is a deformation retract of X.*

For let $h_t: X \to X$ be a homotopy of id_X into ur. Then $h_t ur$ is a homotopy ur into itself, since r is a retraction. Consider the map $h': I \times X \to X$ given by

$$h'(t, x) = \begin{cases} h_{2t}(x) & (0 \le t \le \frac{1}{2}) \\ h_{2-2t}ur(x) & (\frac{1}{2} \le t \le 1). \end{cases}$$

If $x \in A$ then $h'(t, x) = h'(1 - t, x)$ for all t. Hence there exists a homotopy rel$(\dot{I} \times A)$ of $h'|(I \times A)$ into $u\pi_2: I \times A \to X$. Extend this to a homotopy rel$(\dot{I} \times X)$ of h' into h'', say. Then h'' constitutes a homotopy rel A of the identity into ur. This proves (8.11).

Proposition (8.12). *Let E be a neighbourhood extensor. Let (X, A) be a paracompact pair, let $f_0, f_1: X \to E$ be maps, and let $h_t: A \to E$ be a homotopy of $f_0|A$ into $f_1|A$. Then there exists a neighbourhood U of A in X and an extension of h_t to a homotopy of $f_0|U$ into $f_1|U$.*

For consider the paracompact pair $(I \times X, \dot{I} \times X \cup I \times A)$. The given maps f_0, f_1 and homotopy h_t combine to form a map

$$\phi: \dot{I} \times X \cup I \times A \to E.$$

Since E is a neighbourhood extensor there exists a neighbourhood V of $\dot{I} \times X \cup I \times A$ in $I \times X$ over which ϕ can be extended. Now $I \times U \subset V$, for some neighbourhood U of A in X, since I is compact. Hence there exists an extension of $\phi|(\dot{I} \times U \cup I \times A)$ over $I \times U$, and this constitutes an extension of h_t to a homotopy of $f_0|U$ into $f_1|U$, as required.

Corollary (8.13). *Let E be a paracompact neighbourhood extensor. Then E is locally categorical.*

For let $*$ be any point of E. Take $(X, A) = (E, *)$, in (8.12), with f_0 the identity and f_1 the nul-map at $*$. The neighbourhood U of $*$ is then contractible in E.

Another application of (8.12) occurs in the proof of

Proposition (8.14). *Consider the triad*

$$E_0 \overset{p_0}{\to} E \overset{p_1}{\leftarrow} E_1,$$

where E, E_0 and E_1 are neighbourhood extensors. The homotopy pull-back $W = W(p_0, p_1)$ is a neighbourhood extensor.

Recall that W is defined as the subspace of $E_0 \times E_1 \times PE$ consisting of triples (e_0, e_1, λ) such that $\lambda(0) = p_0(e_0)$, $\lambda(1) = p_1(e_1)$. To prove (8.14), let (X, A) be a paracompact pair and let $f: A \to W$ be the map which has to be extended over a neighbourhood of A in X. Consider the first and second components

$$E_0 \overset{f_0}{\leftarrow} A \overset{f_1}{\to} E_1$$

of f. Since E_0 and E_1 are neighbourhood extensors there exist extensions of f_0 and f_1 over neighbourhoods U_0 and U_1, respectively, of A in X. Hence, by restriction, there exist extensions

$$E_0 \overset{g_0}{\leftarrow} \mathscr{Cl}\, U \overset{g_1}{\to} E_1$$

of f_0, f_1, respectively, where U is a neighbourhood of A in X such that $\mathscr{Cl}\, U \subset U_0 \cap U_1$. Regard the adjoint $h: I \times A \to E$ of the third component of f as a homotopy of $p_0 f_0$ into $p_1 f_1$. Since E is a neighbourhood extensor there exists, by (8.12), a neighbourhood V of A in X, contained in $\mathscr{Cl}\, U$, such that h can be extended to a homotopy $k: I \times V \to E$, say, of $p_0 g_0 | V$ into $p_1 g_1 | V$. Hence an extension over V of the original map f is given by g_0 in the first component, by g_1 in the second component, and by the adjoint $V \to PE$ of k in the third component. This proves (8.14).

It is natural to ask whether there is a similar result for homotopy push-outs. Is it true, for example, that the suspension ΣE of a neighbourhood extensor E is again a neighbourhood extensor? The answer is yes if E is a compact metric space but I do not know of any really simple proof of this fact.

In the next result I return to the subject of (Lusternik–Schnirelmann) category and, following G. W. Whitehead, prove

Theorem (8.15). *Let X be a paracompact space of finite category* cat X. *Let G be a path-connected topological group which is also a neighbourhood extensor. Then the group $\pi(X, G)$ is nilpotent of class $<$*cat X.

The hypotheses are satisfied, for example, when G is a connected Lie group, since the underlying space is a manifold and therefore a neighbourhood extensor, as we have seen.

It is essential for G to be path-connected since, for example, the group $\pi(S^1, O(2))$ turns out to be the infinite dihedral group which is not nilpotent.

The proof of (8.15) depends on the following observation, which is an immediate consequence of the homotopy extension theorem (8.9). Let A be a closed subspace of the paracompact space X such that A is contractible in X. Then any map of X into G is homotopic to a map which is neutral on A, i.e. which maps A into the neutral element of G.

Suppose, then, that $\{A_1, \ldots, A_n\}$ is a closed covering of X such that each A_j is contractible in X. Write $B_j = A_1 \cup \cdots \cup A_j$, so that

$$A_1 = B_1 \subset B_2 \subset \cdots \subset B_j \subset \cdots \subset B_n = X.$$

Define $\pi_j \subset \pi = \pi(X, G)$ to be the set of homotopy classes of maps of X into G which are neutral on B_j. Clearly π_j is a subgroup of π and

$$\pi_1 \supset \cdots \supset \pi_j \supset \cdots \supset \pi_n = \{1\}.$$

By applying the observation in the previous paragraph to the case $A = A_1$ we see that $\pi_1 = \pi$. I assert that π_j $(j = 2, \ldots, n)$ contains the commutator $[\pi, \pi_{j-1}]$, so that the subgroups form a central chain and π is nilpotent of of class $<n$.

To see this, let $\alpha \in \pi$, $\beta \in \pi_{j-1}$. The above observation, with $A = A_j$, shows that α can be represented by a map $\theta: X \to G$ which is neutral on A_j. Also β, by definition of π_{j-1}, can be represented by a map $\phi: X \to G$ which is neutral on B_{j-1}. Hence the commutator $\xi: X \to G$, given by

$$\xi(x) = \theta(x).\phi(x).(\theta(x))^{-1}.(\phi(x))^{-1} \qquad (x \in X),$$

is neutral on $A_j \cup B_{j-1} = B_j$. Since ξ represents $[\alpha, \beta]$ this shows that $[\alpha, \beta] \in \pi_j$, as asserted.

Finally we note that a covering of X by cat X open sets, each of which is contractible in X, can be shrunk to a covering of X by cat X closed sets, each of which is contractible in X. Hence we can take $n \leq \text{cat } X$, in the above, and the proof of (8.15) is complete.

There are a number of variants of this result, such as

Theorem (8.16). *Let X be a paracompact space of finite category* cat X. *Let G be a group-like space which is a neighbourhood extensor. Then the group $\pi^*(X, G)$ is nilpotent of class* $<$cat X.

In particular this applies when $G = \Omega E$, the space of loops on a neighbourhood extensor E. I leave the proof of (8.16) to serve as an exercise.

Let us turn now to the fibrewise version of extensor theory. Up to a point this is just a routine generalization of what we have already done but in due course we shall encounter results which have no counterpart in the ordinary theory.

In the basic definitions we are presented with the situation displayed in the following diagram.

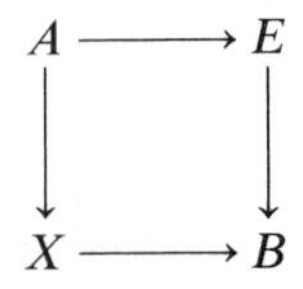

Definition (8.17). Let E be a space over B. Then E is an extensor over B if the following fibrewise extension property holds for all paracompact pairs (X, A) over B: every map of A into E over B can be extended to a map of X into E over B.

Proposition (8.18). *If E is an extensor over the paracompact space B then E admits a section over B, moreover all such sections are vertically homotopic.*

Proposition (8.19). *Let E be an extensor over B. If X is paracompact over B then all maps $X \to E$ over B are homotopic over B. In particular if E is paracompact and admits a section over B then E is contractible over B to that section.*

The proofs of these two results are obvious.

Definition (8.20). Let E be a space over B. Then E is a neighbourhood extensor over B if the following fibrewise neighbourhood extension property holds for all paracompact pairs (X, A) over B: every map of A into E over B can be extended to a map of some neighbourhood of A into E over B.

An open subset of a neighbourhood extensor (and hence of an extensor) over B is a neighbourhood extensor over B. A fibrewise retract of an extensor over B is an extensor over B. A fibrewise neighbourhood retract of a neighbourhood extensor over B is a neighbourhood extensor over B. Fibre products of extensors over B are extensors over B and fibre products of neighbourhood extensors over B are neighbourhood extensors over B. These results all follow more or less immediately from the definitions and will be left to serve as exercises.

If E is an extensor over B then $E|B'$ is an extensor over B' for each subspace B' of B; in particular the fibres of E are extensors. Moreover the pullback ξ^*E is an extensor over B' for each space B' and map $\xi: B' \to B$. Similarly in the case of neighbourhood extensors over B.

Either as a special case of this, or directly, one sees that $B \times T$ is an extensor (resp. neighbourhood extensor) over B if (and only if) T is an extensor (resp. neighbourhood extensor). Since $B \times \mathbb{R}^n$ $(n = 0, 1, \dots)$ is an extensor over B, the fibrewise neighbourhood retracts of $B \times \mathbb{R}^n$ are neighbourhood extensors over B. These are called euclidean neighbourhood retracts over

B and have been recently studied by Dold in connection with fibrewise fixed-point theory.

I now state a series of results each of which is a routine generalization of a result proved earlier in the ordinary case. I leave the proofs to serve as exercises.

Proposition (8.21). *Let E be a space over B and let $\{E_1, E_2\}$ be an open covering of E. If E_1, E_2 and $E_1 \cap E_2$ are extensors over B then so is E.*

Proposition (8.22). *Let E be a space over B which admits a numerable open covering, each member of which is a neighbourhood extensor over B. Then E is a neighbourhood extensor over B.*

Proposition (8.23). *Let E be a space over B and let $\{E_1, E_2\}$ be a closed covering of E. Suppose that E_1, E_2 and $E_1 \cap E_2$ are extensors (resp. neighbourhood extensors) over B, for the class of completely paracompact pairs. Then E is an extensor (resp. neighbourhood extensor) over B for the same class of domains.*

Proposition (8.24). *Let E be contractible over B. If E is a neighbourhood extensor over B then E is an extensor over B.*

These routine generalizations are not as important as the results to be stated next, which have no counterpart in the ordinary theory.

Proposition (8.25). *Let E be a space over B. Suppose that there exists an open covering $\{V_j\}$ $(j \in J)$ of B such that $E_j = E \mid V_j$ is an extensor (resp. neighbourhood extensor) over V_j for each index j. Then E is an extensor (resp. neighbourhood extensor) over B.*

Note the difference between (8.22) and (8.25). As an immediate consequence of (8.25) we have

Corollary (8.26). *Let E be a numerable fibre bundle over B. If the fibres of E are extensors (resp. neighbourhood extensors) then E is an extensor (resp. neighbourhood extensor) over B.*

It follows at once that if E is a numerable bundle over the paracompact space B with extensor fibres then E admits a section over B. As an application, suppose that we have a closed subgroup H of the topological group G. Suppose further that G admits local sections over G/H and that G/H is an extensor. By combining (8.26) with (4.74) we see that every G-bundle over the paracompact space B is reducible to an H-bundle. In fact there is a general result to the effect that every Lie group G contains a compact subgroup H such that G/H is euclidean and hence the above is applicable. One important example is the general linear group $Gl(n, \mathbb{R})$ which contains the orthogonal group $O(n, \mathbb{R})$ as a subgroup with factor space euclidean of dimension $\frac{1}{2}n(n+1)$. So for paracompact spaces B we conclude that the structure group of a $Gl(n, \mathbb{R})$-bundle over B is always reducible to $O(n, \mathbb{R})$.

I give the proof of (8.25) in the extensor case; the modifications required in the neighbourhood extensor case are routine and will be left to serve as an exercise. First of all let us reformulate the definition (8.17) as follows:

Definition (8.27). Let E be a space over B. Then E is an extensor over B if for all paracompact pairs (X, A) over B the induced space ξ^*E over X has the following section extension property: every section of ξ^*E over A can be extended to a section over X.

Here $\xi: X \to B$ is the projection. Of course a similar reformulation can be made of the definition of neighbourhood extensor over B.

Comparing (8.27) with the statement of (8.25) we see that it is sufficient to consider the case when B is paracompact and to show that a given section s of E over a closed subspace A of B can be extended over B. Since B is normal we can shrink the given covering and then, using the Milnor procedure, replace it by a countable covering. Specifically we may begin with a sequence $A_0, \ldots, A_n, \ldots$ of closed sets of B such that each point of B lies in the interior of one of the A_n and such that $E_n = E|A_n$ is an extensor over A_n for each n.

So consider the closed filtration

$$A = B_0 \subset B_1 \subset \cdots \subset B_n \subset \cdots \subset \mathrm{B},$$

where $B_{n+1} = B_n \cup A_n$ $(n = 0, 1, \ldots)$. We already have the section s defined on B_0; write $s = s_0$ and make the inductive hypothesis that s_{r-1} can be extended to a section s_r over B_r, for $r = 1, \ldots, n$. I assert that s_n can be extended to a section s_{n+1} over B_{n+1}. For since E_{n+1} is an extensor over A_{n+1} we can extend $s_n|(A_{n+1} \cap B_n)$ to a section over A_{n+1}, and then define s_{n+1} by glueing this to s_n. This establishes the inductive step, and then an extension s' of s over the whole of B is defined by $s'|B_n = s_n$, for all n. This proves (8.25) in the extensor case, and the neighbourhood extensor case is similar.

Proposition (8.28). *Let E be a neighbourhood extensor over B. Let B' be a closed subspace of B and write $E' = E|B'$, $E'' = E|B''$, where $B'' = B - B'$. Suppose that E' is an extensor over B' and that E'' is an extensor over B''. Then E is an extensor over B.*

Again it is sufficient, in view of (8.27), to consider the case when B is paracompact and to show that a section s of E over a closed subspace A of B can be extended over B. We proceed as follows. Since E' is an extensor over B' and since $(B', A \cap B')$ is a paracompact pair we can extend $s|(A \cap B')$ to a section over B'. Glueing this to s produces a section t, say, over $A \cup B'$. Since E is a neighbourhood extensor over B and since $(B, A \cup B')$ is a paracompact pair there exists a neighbourhood V of $A \cup B'$ in B and an extension

τ of t to a section over V. Since B is paracompact we can find a closed subset $B_0 \subset B$ such that $B_0 \subset V$ and $A \cup B' \subset \mathcal{I}nt\, B_0$. Then $B'' - \mathcal{I}nt\, B_0 = B - \mathcal{I}nt\, B_0$ is closed and so paracompact. Since E'' is an extensor over B'' and since $(B'' - \mathcal{I}nt\, B_0, B_0 - \mathcal{I}nt\, B_0)$ is a paracompact pair we can extend $\tau|(B_0 - \mathcal{I}nt\, B_0)$ to a section over $B'' - \mathcal{I}nt\, B_0$. Glueing this to $\tau|B_0$ produces the required extension of s to a section over B. This proves (8.28).

Our next result deals with the behaviour of fibrewise extensors and fibrewise neighbourhood extensors when we pass to fibrewise mapping spaces. Let $\phi: T_0 \to E$ be a map over B, where T_0 is a closed subspace of the space T over B, and let $\mathrm{map}_B(T, E; \phi)$ denote the subspace of $\mathrm{map}_B(T, E)$ consisting, for each point $b \in B$, of extensions of ϕ_b to maps $T_b \to E_b$. By a routine generalization of the argument used to prove (8.8) we obtain

Proposition (8.29). *If E is an extensor over B then so is $\mathrm{map}_B(T, E; \phi)$ for all locally compact regular T over B. If E is a neighbourhood extensor over B then so is $\mathrm{map}_B(T, E; \phi)$ for all compact regular T over B.*

This result shows, for example, that if E is an extensor or neighbourhood extensor over B then so is the space $P_B E$ of fibrewise paths in E and so is the space $\Omega_B E$ of fibrewise loops based at a given section of E.

Let E be a space over B and let σ be a section of E over a closed subspace B_0 of B. Consider the subspace $\mathrm{sec}_B(E; \sigma)$ of $\mathrm{sec}_B E$ consisting of sections of E over B which extend σ. The argument used to prove (8.8) also proves

Proposition (8.30). *If E is an extensor (resp. neighbourhood extensor) over the compact regular space B then $\mathrm{sec}_B(E; \sigma)$ is an extensor (resp. neighbourhood extensor) for all sections σ of E over the closed subspace B_0 of B.*

Proposition (8.31). *Let $\phi: E \to F$ be a map, where E and F are neighbourhood extensors. Then the mapping path-space $W = W(\phi)$ of ϕ is a neighbourhood extensor over F.*

Of course we already know from (8.14) that W is a neighbourhood extensor in the ordinary sense; however the result stated above is much stronger.

In the proof of (8.31) we describe maps into W in the usual way by giving their components in E and in PF; the latter we at once rewrite in the adjoint form. So let (X, A) be a paracompact pair and let

$$\alpha: \{0\} \times A \to E, \qquad \beta: I \times A \to F, \qquad \gamma: \{1\} \times X \to F$$

be maps such that

$$\beta|(\{0\} \times A) = \phi\alpha, \qquad \beta|(\{1\} \times A) = \gamma|(\{1\} \times A).$$

Since E is a neighbourhood extensor there exists a closed neighbourhood $\mathcal{C}\ell\, V$ of A in X and an extension $\alpha': \{0\} \times \mathcal{C}\ell\, V \to E$ of α. Now β can be regarded as a homotopy of $\phi\alpha'|(\{0\} \times A)$ into $\gamma|(\{1\} \times A)$. Since F is a neighbourhood extensor there exists, by (8.12), a neighbourhood U of A in X, contained in V, and a homotopy $\beta': I \times U \to F$ of $\phi\alpha'|(\{0\} \times U)$ into $\gamma|(\{1\} \times U)$ which extends β. Using these components we construct an extension over U of the original map $A \to W$, as required.

The next set of results are again routine generalizations of the corresponding results in the ordinary theory and so proofs are omitted.

Proposition (8.32). *Let E be a neighbourhood extensor over B. Let (X, A) be a paracompact pair over B. let $f: X \to E$ be a map over B, and let $h_t: A \to E$ be a homotopy of $f|A$ over B. Then there exists an extension of h_t to a homotopy over B of f itself.*

We may refer to (8.32) as the fibre homotopy extension theorem.

Corollary (8.33). *Let E be a neighbourhood extensor over B. Let (X, A) be a paracompact pair over B such that A is a fibrewise retract of X. Let $c: X \to E$ be a nul-map over B and let $f: X \to E$ be a map over B such that* (i) *$f \simeq_B c$ and* (ii) *$f|A = c|A$. Then $f \simeq_B^A c$.*

Corollary (8.34). *Let X be a paracompact neighbourhood extensor over B. Let A be a closed fibrewise retract of X. Suppose that there exists a fibrewise retraction $r: X \to A$ such that $ur \simeq_B \mathrm{id}_X$ where $u: A \subset X$. Then A is a fibrewise deformation retract of X.*

Proposition (8.35). *Let E be a neighbourhood extensor over B. Let (X, A) be a paracompact pair over B, let $f_0, f_1: X \to E$ be maps over B, and let $h_t: A \to E$ be a homotopy of $f_0|A$ into $f_1|A$. Then there exists a neighbourhood U of A in X and an extension of h_t to a homotopy over B of $f_0|U$ into $f_1|U$.*

Corollary (8.36). *Let E be a paracompact neighbourhood extensor over B. Then for each closed section $s: B \to E$ there exists a neighbourhood U of sB in E such that the inclusion $U \to E$ is nul-homotopic over and under B.*

In other words E is fibrewise locally categorical.

Proposition (8.37). *Consider the triad*

$$E_0 \overset{p_0}{\rightarrow} E \overset{p_1}{\leftarrow} E_1,$$

where E, E_0, E_1 are neighbourhood extensors over B. The fibre homotopy pull-back $W_B(p_0, p_1)$ is a neighbourhood extensor over B.

Proposition (8.38). *Let $\phi: E \to F$ be a map over B, where E and F are neighbourhood extensors over B. Then the fibrewise mapping path-space $W_B(\phi)$ of ϕ is a neighbourhood extensor over F.*

This last result is used in the proof of

Theorem (8.39). *Let E, F be neighbourhood extensors over the paracompact space B. Let $\phi: E \to F$ be a map over B. Suppose that there exists an open covering of B such that $\phi_U: E_U \to F_U$ is a homotopy equivalence over U for each member U of the covering. Then ϕ is a homotopy equivalence over B.*

For consider the fibrewise mapping path-space $W_B(\phi)$ of ϕ, as a space over F. The fibrewise mapping path-space $W_U(\phi_U)$ of ϕ_U, as a space over U, is just the restriction of $W_B(\phi)$ to U. Now $W_B(\phi)$ is a neighbourhood extensor over F, by (8.38), while its restriction $W_U(\phi_U)$ to U is an extensor over F_U, by (8.24), for each member U of the given covering. Also the given covering is numerable, since B is paracompact, and so its pull-back to F is also numerable. Therefore $W_B(\phi)$ is an extensor over F, by (8.25). In particular $W_B(\phi)$ admits a section over F. Hence ϕ has a fibre homotopy right inverse $\phi': F \to E$, say. Now the restriction $\phi'_U: F_U \to E_U$, where U is as before, is a fibre homotopy right inverse of the fibre homotopy equivalence ϕ_U, and so is itself a fibre homotopy equivalence. Therefore we can repeat the argument we have just given, with $\phi': F \to E$ in place of $\phi: E \to F$, and show that ϕ' has a fibre homotopy right inverse. Since ϕ' also has the fibre homotopy left inverse ϕ we conclude that ϕ', and hence also ϕ, is a fibre homotopy equivalence. This proves (8.39), and by a similar argument based on (8.28) instead of (8.25) we obtain

Theorem (8.40). *Let E, F be neighbourhood extensors over the paracompact space B. Let $\phi: E \to F$ be a map over B. Suppose that ϕ is a fibre homotopy equivalence over a closed subspace B' of B and also over the complement $B'' = B - B'$. Then ϕ is a fibre homotopy equivalence over B.*

We now combine the ideas of the present chapter with the idea of fibration as discussed in Chapter 6, as follows:

Proposition (8.41). *Let E be a neighbourhood extensor over B such that the projection $p: E \to B$ has the weak homotopy lifting property for paracompact domains. Then p has the homotopy lifting property for paracompact domains.*

For let X be a paracompact space over B and let

$$\theta:\{0\} \times X \to E, \qquad \phi: I \times X \to B$$

be maps over B such that $p\theta = \phi|(\{0\} \times X)$. By hypothesis there exists a map

$$\psi: \{0\} \times I \times X \cup I \times \{0\} \times X \to E$$

such that $\psi(0, 1, x) = \theta(0, x)$ and such that

$$p\psi(0, t, x) = \phi(0, x), \qquad p\psi(s, 0, x) = \phi(s, x)$$

for all $s, t \in I$ and $x \in X$. Since $(I \times X, \{0\} \times X)$ is a paracompact pair we can extend ψ to a map $\lambda: I \times I \times X \to E$ such that

$$p\lambda(s, t, x) = \phi(s, x) \qquad (s, t \in I, x \in X).$$

Then $\mu: I \times X \to E$, given by

$$\mu(s, x) = \lambda(s, 1, x),$$

is a lifting of ϕ such that $\mu(0, x) = \theta(0, x)$. Thus E is a fibre space over B, rather than just a weak fibre space over B.

Proposition (8.42). *Let E be a neighbourhood extensor over B such that the projection $p: E \to B$ has the homotopy lifting property for paracompact domains. Then p has the relative homotopy lifting property for paracompact pairs.*

For if (X, A) is a paracompact pair over B then so is $(I \times X, \{0\} \times X \cup I \times A)$. Hence there exists a neighbourhood filler for the following diagram.

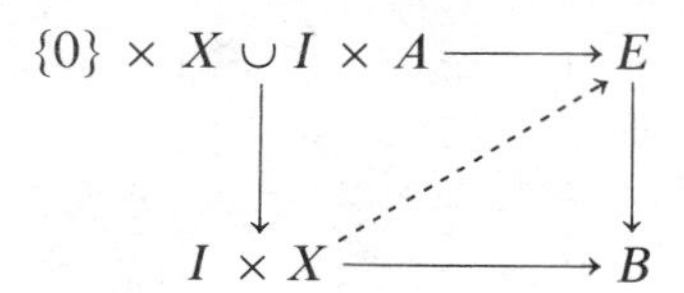

Using a Urysohn map exactly as in the proof of the homotopy extension theorem (8.9) we obtain an absolute filler as indicated above.

Finally we turn to the equivariant theory of extensors. Everything we have done in the ordinary case can be extended to the equivariant case in a routine fashion. Unless otherwise stated G is a topological group, without any restriction.

Definition (8.43). The G-space E is a G-extensor if E has the following equivariant extension property for all paracompact G-pairs (X, A): every G-map of A into E can be extended to a G-map of X into E.

Clearly the fixed-point set E^G of a G-extensor is path-connected.

Definition (8.44). The G-space E is a neighbourhood G-extensor if E has the following neighbourhood equivariant extension property for all paracompact G-pairs (X, A): every G-map of A into E can be extended to a G-map of an invariant neighbourhood of A into E.

In the representation theory of topological groups it is shown that if $\mathbb{R}^n$ is a G-space, where G is compact Hausdorff, then $\mathbb{R}^n$ is a G-extensor. The argument depends on the averaging effect of Haar integration. One first uses the ordinary Tietze theorem to extend the given G-map $A \to E$ to a map $X \to E$ and then averages the extension to obtain an equivariant extension.

An open (invariant) subset of a neighbourhood G-extensor (and hence of a G-extensor) is a neighbourhood G-extensor. An equivariant retract of a G-extensor is a G-extensor. A neighbourhood equivariant retract of a neighbourhood G-extensor is a neighbourhood G-extensor. Products of G-extensors are G-extensors, and products of neighbourhood G-extensors are neighbourhood G-extensors. These results all follow more or less immediately from the definitions we have given and will be left to serve as exercises.

It should be quite clear, after a little thought, that all the results about extensors and neighbourhood extensors given in the first part of this chapter can be generalized to the equivariant case in a routine fashion. For example we have the G-homotopy extension theorem, which asserts that if E is a neighbourhood G-extensor then for each paracompact G-pair (X, A), G-map $f: X \to E$ and G-homotopy $h_t: A \to E$ of $f \mid A$, there exists a G-homotopy $X \to E$ of f which extends h_t. To avoid much tedious repetition we we shall not write out all these results but simply refer to each of them, on occasion, as "the equivariant version" of whatever the result may be in the ordinary case.

What is much more interesting, however, are results which have no counterpart in the ordinary case, such as

Proposition (8.45). *Suppose that E is a G-extensor (resp. neighbourhood G-extensor) where G is compact. Then the fixed point set E^H is an extensor (resp. neighbourhood extensor) for each closed subgroup H of G.*

This follows easily from the definitions and from (4.24) above. Thus (to take the extensor case) let (X, A) be a paracompact pair. Each map $A \to E^H$ determines a G-map $G/H \times A \to E$, where G acts trivially on A and by left translation on G/H. Now G/H is compact Hausdorff, hence

$$G/H \times (X, A) = (G/H \times X, G/H \times A)$$

is a paracompact pair, by (7.17) above. Therefore the G-map $G/H \times A \to E$ can be extended to a G-map $G/H \times X \to E$, since E is a G-extensor. The restriction of the extension to $H/H \times X$ determines an extension $X \to E^H$ of the map of A originally given.

Again if B is a G-space we can work in the category of G-spaces over B and make the appropriate definitions as follows.

Definition (8.46). The G-space E over B is a G-extensor over B if E has the following fibrewise equivariant extension property for all paracompact G-pairs (X, A) over B: every G-map of A into E over B can be extended to a G-map of X into E over B.

Definition (8.47). The G-space E over B is a neighbourhood G-extensor over B if E has the following fibrewise neighbourhood equivariant extension property for all paracompact G-pairs (X, A) over B: every G-map of A into E over B can be extended to a G-map of a neighbourhood of A into E over B.

An open (invariant) subset of a neighbourhood G-extensor over B (and hence of a G-extensor over B) is a neighbourhood G-extensor over B. A fibrewise equivariant retract of a G-extensor over B is a G-extensor over B. A fibrewise neighbourhood equivariant retract of a neighbourhood G-extensor over B is a neighbourhood G-extensor over B. Fibre products of G-extensors over B are G-extensors over B, and fibre products of neighbourhood G-extensors over B are neighbourhood G-extensors over B. These results all follow more or less immediately from the definitions we have given and will be left to serve as exercises.

Note that if E is a G-extensor over B then ξ^*E is a G-extensor over B' for each G-space B' and G-map $\xi: B' \to B$. Similarly for neighbourhood G-extensors over B.

It should be obvious enough that all the results about extensors and neighbourhood extensors over B given earlier in this chapter can be generalized in a routine fashion to the equivariant case. Of greater interest are results which have no counterpart in the ordinary theory such as the following. Recall from Chapter 4 the colon construction (:) which provides such a useful bridge between the equivariant theory and the ordinary theory. We refer to this in

Proposition (8.48). *Let B be a G-space and let E be a G-space over B, with G compact. Then E is a G-extensor (resp. neighbourhood G-extensor) over B if and only if $(Y : E)$ is an extensor (resp. neighbourhood extensor) over $(Y : B)$ for all paracompact G-spaces Y.*

I give the proof in the extensor case; the neighbourhood extensor case is very similar and may serve as an exercise.

To prove (8.48) in the "if" direction, let (X, A) be a paracompact G-pair, and let θ, ϕ be G-maps such that the diagram on the left is commutative.

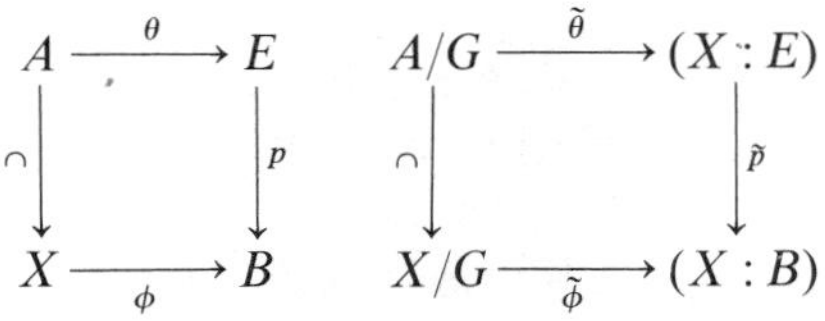

$$\begin{array}{ccc} A & \xrightarrow{\theta} & E \\ \cap\downarrow & & \downarrow p \\ X & \xrightarrow[\phi]{} & B \end{array} \qquad \begin{array}{ccc} A/G & \xrightarrow{\tilde{\theta}} & (X : E) \\ \cap\downarrow & & \downarrow \tilde{p} \\ X/G & \xrightarrow[\tilde{\phi}]{} & (X : B) \end{array}$$

Then the diagram on the right also commutes, where $\tilde{p} = (\mathrm{id}_X : p)$ and where $\tilde{\theta}$, $\tilde{\phi}$ are obtained from the graph functions of θ, ϕ in the obvious way. If $(X : E)$ is an extensor over $(X : B)$ there exists a filler $\tilde{\psi}: X/G \to (X : E)$ of the diagram on the right. The corresponding G-map $\psi: X \to E$ is an extension of θ over ϕ, as required.

To prove (8.48) in the "only if" direction let (X, A) be a paracompact pair (no longer a G-pair) and let θ', ϕ' be maps such that the following diagram commutes.

$$\begin{array}{ccc} A & \xrightarrow{\theta'} & (Y : E) \\ \cap\downarrow & & \downarrow\tilde{p} \\ X & \xrightarrow[\phi']{} & (Y : B) \end{array}$$

Define $A^* \subset A \times Y$ to be the pull-back of the G-space Y with respect to the composition

$$A \to (Y : E) \to Y/G,$$

and define $X^* \subset X \times Y$ similarly with respect to the composition

$$X \to (Y : B) \to Y/G.$$

As in (4.32) above we see that $(X^*/G, A^*/G)$ can be identified with (X, A). Now θ', ϕ' determine sections

$$\tilde{\theta}: A \to (A^* : E), \qquad \tilde{\phi}: X \to (X^* : B)$$

and hence determine G-maps θ, ϕ such that the diagram shown below is commutative.

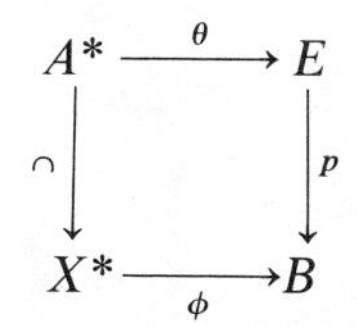

When E is a G-extensor over B there exists an extension $\psi: X^* \to E$ of θ over ϕ. Combining this with the canonical map $X^* \to Y$ determines an extension $X \to (Y : E)$ of θ' over ϕ' in the original diagram. This completes the proof of (8.48).

We now turn our attention to the case when G is a compact Lie group. Our main purpose is to prove

Theorem (8.49). *Let G be a compact Lie group. Let B be a paracompact G-space and let E be a neighbourhood G-extensor over B. Suppose that E^H is an extensor over B^H for all closed subgroups H of G. Then E is a G-extensor over B.*

Before giving the proof of (8.49), which involves a number of steps, we will show that (8.49) implies the useful

Proposition (8.50). *Let G be a compact Lie group. Let A, B be paracompact neighbourhood G-extensors. Let $\phi: A \to B$ be a G-map which determines a homotopy equivalence $\phi^H: A^H \to B^H$ for all closed subgroups H of G. Then ϕ is a G-homotopy equivalence.*

Corollary (8.51). *Let G be a compact Lie group. Let A be a paracompact neighbourhood G-extensor such that A^H is contractible for all closed subgroups H of G. Then A is G-contractible.*

To deduce (8.50) from (8.49) we apply the latter to the mapping path-space $W = W(\phi)$, with the obvious action of G. Then W is a neighbourhood G-extensor over B, by the equivariant version of (8.31) above, and so W^H is a neighbourhood extensor over B^H for all closed subgroups H of G, by (8.45). Now W^H is the mapping path-space of $f^H: A^H \to B^H$, which is a homotopy equivalence by hypothesis. Therefore W^H is contractible over B^H, and so is an extensor over B^H, by (8.24). Hence W is a G-extensor over B, by (8.49), and so W admits an equivariant section over B. Hence ϕ admits a G-homotopy right inverse $\phi': B \to A$, say. Now $\phi'^H: B^H \to A^H$ is a homotopy right inverse of ϕ^H and so is a homotopy equivalence. We may therefore repeat the argument, with ϕ' in place of ϕ, and obtain that ϕ' admits a G-homotopy right inverse. But ϕ' also admits ϕ as a G-homotopy left inverse and so is a G-homotopy equivalence. Therefore ϕ is a G-homotopy equivalence as asserted.

In view of (8.48) we may reformulate the conclusion of (8.49) in the following form:

Proposition (8.52). *Under the same hypotheses as* (8.49) *the space* $(Y: E)$ *is an extensor over the space* $(Y: B)$ *for all paracompact G-spaces Y.*

The proof of (8.52) proceeds in four stages of increasing generality, according to the complexity of the orbit structure of the G-space Y.

First suppose (case I) that only one orbit type $[H]$ occurs in Y. Let N be the normalizer of H in G with Weyl group $K = N/H$. Recall from (4.43) that the obvious maps η shown below are equivalences.

$$\begin{array}{ccc} Y^H \times_K E^H & \xrightarrow{\ \eta\ } & (Y: E) \\ \Big\downarrow{\scriptstyle \mathrm{id} \times p^H} & & \Big\downarrow{\scriptstyle \tilde{p}} \\ Y^H \times_K B^H & \xrightarrow[\ \eta\ ]{} & (Y: B) \end{array}$$

It is therefore sufficient for our purpose to show that $Y^H \times_K E^H$ is an extensor over $Y^H \times_K B^H$ with projection $\mathrm{id} \times p^H$. Now observe that K acts freely (and of course properly) on Y^H; moreover the projection $Y^H \to Y^H/K$ is that of a principal K-bundle. Thus $Y^H \times_K E^H$, $Y^H \times_K B^H$ are the associated bundles with fibres E^H, B^H, respectively. If U is a member of a triviality covering for the principal bundle then $U \times E^H$ is an extensor over $U \times B^H$, with projection $\mathrm{id} \times p^H$, since E^H is an extensor over B^H with projection p^H. Hence it follows at once, using (8.25), that $Y^H \times_K E^H$ is an extensor over $Y^H \times_K B^H$ with projection $\mathrm{id} \times p^H$. This proves (8.52) in case I.

Next suppose (case II) that only a finite number of orbit types occur in Y. We partially order these in the usual way and list them in order as $[H_1], \dots, [H_n]$. Let $Y_0 = \emptyset$ and let Y_j $(j = 1, \dots, n)$ denote the subset of Y where the stabilizer is conjugate to H_i for some $i \leq j$. Then Y_j is closed and

$$\emptyset = Y_0 \subset Y_1 \subset \cdots \subset Y_n = Y.$$

Also only one orbit type occurs in each of $Y_1 - Y_0, Y_2 - Y_1, \dots, Y_n - Y_{n-1}$. Hence $(Y_j - Y_{j-1} : E)$ is an extensor over $(Y_j - Y_{j-1} : B)$ for $j = 1, \dots, n$, by case I. Now (8.52) in case II follows by applying (8.28) successively with respect to the closed pairs

$$((Y_1 : B), (Y_0 : B)), \dots, ((Y_n : B), (Y_{n-1} : B)).$$

Next suppose (case III) that the orbit structure of Y is locally finite. This means that there exists an open covering of Y such that only a finite number of conjugacy classes of stabilizers occur in each member U of the covering. By case II, therefore, $(U : E)$ is an extensor over $(U : B)$ for all such U, and so (8.52) in case III follows at once from (8.25).

Finally we turn to the general case (case IV) where no assumption is made about the orbit structure of Y. Write $Y = Y_0 \cup Y_1 \cup \cdots \cup Y_d$, where $d = \dim G$ and where

$$Y_i = \{y \in Y \mid \dim G_y = i\} \qquad (i = 1, \dots, d).$$

Then $Y_i \cup \cdots \cup Y_d$ is closed for each i. Also the orbit structure of each Y_i is locally finite, since there exists a neighbourhood of each point y of Y such that the stabilizers of points in that neighbourhood are all subconjugate to G_y, while no compact Lie group can contain more than a finite number of closed subgroups having the same dimension as itself. Hence $(Y_i : E)$ is an extensor over $(Y_i : B)$ by case III. Hence and from (8.28) it follows by descending induction on i that $(Y_i \cup \cdots \cup Y_d : E)$ is an extensor over $(Y_i \cup \cdots \cup Y_d : B)$ for $i = d, \dots, 1$ and so that $(Y : E)$ is an extensor over $(Y : B)$ as required. This completes the proof of (8.52) and hence of (8.49).

References

K. Borsuk. *Theory of Retracts*. Polish Scientific Publishers, Warsaw, 1967.
S-T. Hu. *Theory of Retracts*. Wayne State Univ. Press, Detroit, 1965.
I. M. James and G. B. Segal. *On Equivariant Homotopy Theory*. Springer Lecture Notes, Vol. 788, 1980.

Index